D1521041

Cambridge Studies in Ecology presents balanced, comprehensive, up-to-date, and critical reviews of selected topics within ecology, both botanical and zoological. The Series is aimed at advanced final-year undergraduates, graduate students, researchers, and university teachers, as well as ecologists in industry and government research.

It encompasses a wide range of approaches and spatial, temporal, and taxonomic scales in ecology, including quantitative, theoretical, population, community, ecosystem, historical, experimental, behavioural and evolutionary studies. The emphasis throughout is on ecology related to the real world of plants and animals in the field rather than on purely theoretical abstractions and mathematical models. Some books in the Series attempt to challenge existing ecological paradigms and present new concepts, empirical or theoretical models, and testable hypotheses. Others attempt to explore new approaches and present syntheses on topics of considerable importance ecologically which cut across the conventional but artificial boundaries within the science of ecology.

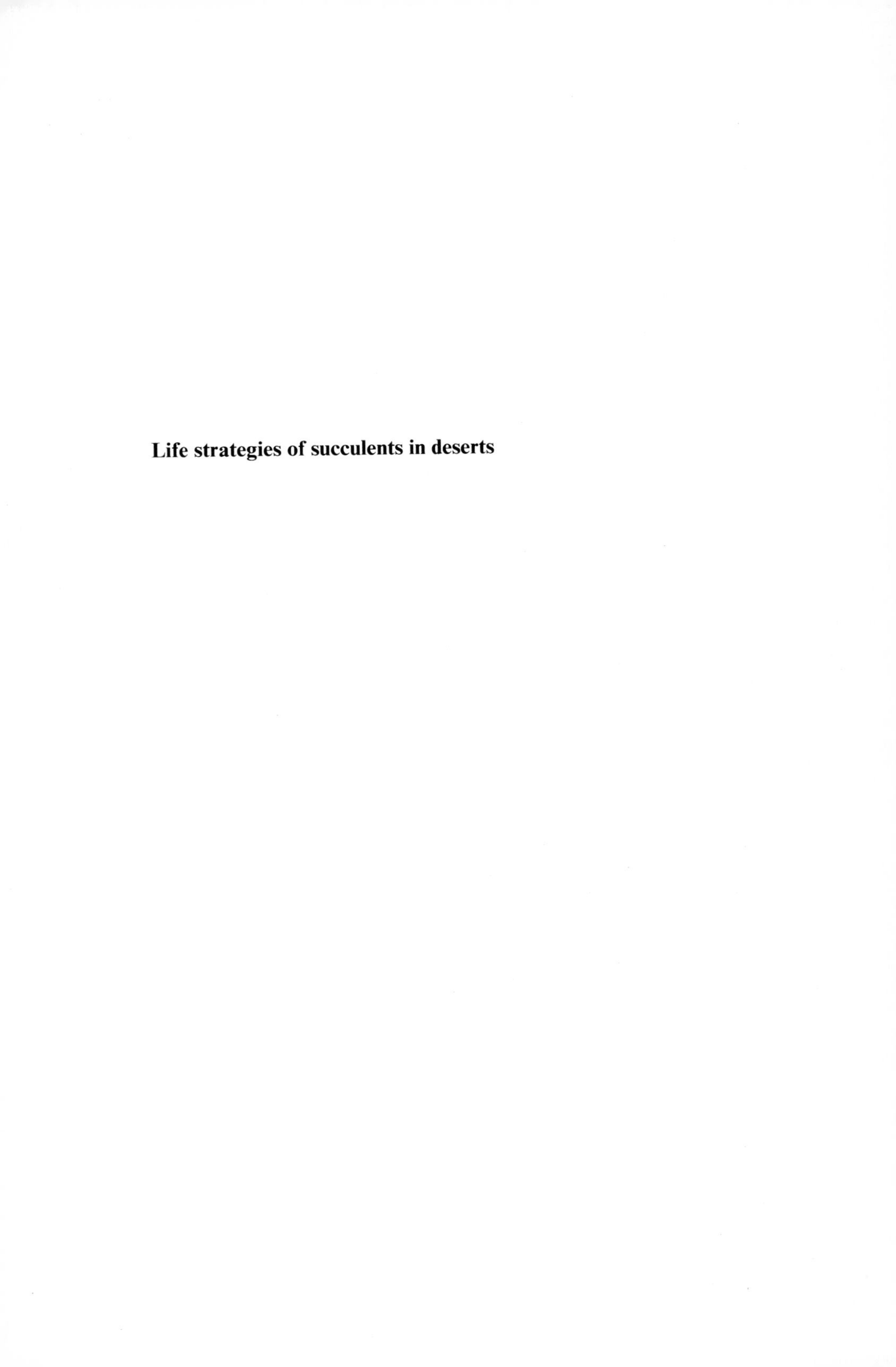

Life strategies of succulents in deserts

CAMBRIDGE STUDIES IN ECOLOGY

ALSO IN THE SERIES

H. G. Gauch, Jr *Multivariate Analysis in Community Ecology*

R. H. Peters *The Ecological Implications of Body Size*

C. S. Reynolds *The Ecology of Freshwater Phytoplankton*

K. A. Kershaw *Physiological Ecology of Lichens*

R. P. McIntosh *The Background of Ecology: Concept and Theory*

A. J. Beattie *The Evolutionary Ecology of Ant–Plant Mutualisms*

F. I. Woodward *Climate and Plant Distribution*

J. J. Burdon *Diseases and Plant Population Biology*

J. I. Sprent *The Ecology of the Nitrogen Cycle*

N. G. Hairston, Sr *Community Ecology and Salamander Guilds*

H. Stolp *Microbial Ecology: Organisms, Habitats and Activities*

R. N. Owen-Smith *Megaherbivores: the Influence of Large Body Size on Ecology*

J. A. Wiens *The Ecology of Bird Communities:*
Volume 1 Foundations and Patterns
Volume 2 Processes and Variations

N. G. Hairston, Sr *Ecological Experiments*

C. Little *The Terrestrial Invasion: an Ecophysiological Approach to the Origins of Land Animals*

P. Adam *Saltmarsh Ecology*

R. Hengeveld *Dynamic Biogeography*

M. F. Allen *The Ecology of Mycorrhizae*

Life strategies of succulents in deserts

with special reference to the Namib desert

DIETER J. VON WILLERT
University of Münster, Germany
BENNO M. ELLER
University of Zürich, Switzerland
MARINUS J. A. WERGER
University of Utrecht, The Netherlands
ENNO BRINCKMANN
University of Bayreuth, Germany
HANS-DIETER IHLENFELDT
University of Hamburg, Germany

CAMBRIDGE UNIVERSITY PRESS
Cambridge
New York Port Chester
Melbourne Sydney

Published by the Press Syndicate of the University of Cambridge
The Pitt Building, Trumpington Street, Cambridge CB2 1RP
40 West 20th Street, New York, NY 10011-4211, USA
10 Stamford Road, Oakleigh, Victoria 3166, Australia

First published 1992

Printed in Great Britain at the University Press, Cambridge

A catalogue record of this book is available from the British Library

Library of Congress cataloguing in publication data

Life strategies of succulents in deserts : with special reference to the Namib desert / Dieter J. von Willert . . . [et al.].
p. cm. – (Cambridge studies in ecology)
Includes bibliographical references and index.
ISBN 0-521-24468-4
1. Succulent plants. 2. Succulent plants–Physiological ecology. 3. Succulent plants–Namibia–Namib Desert. 4. Desert flora–Physiological ecology. 5. Desert flora–Namibia–Namib Desert. I. Willert, Dieter J. von. II. Series.
QK922.L54 1991
582′.05′2652–dc20 91-18832 CIP

ISBN 0 521 24468 4 hardback

SE

Contents

	Preface	xi
	Abbreviations, symbols and units	xiv
1	**The succulent**	1
1.1	Towards a definition of a succulent	2
1.2	Morphological and anatomical considerations	6
	1.2.1 Amount of utilizable water	7
	1.2.2 Static problems	13
	1.2.3 Protective tissues of leaves	17
	1.2.4 Protective tissues of stems	20
	1.2.5 Remarks on the evolution of succulence	23
1.3	The distribution of succulents	24
2	**Climate and vegetation of deserts**	28
2.1	Environmental features of deserts	28
	2.1.1 Water	30
	2.1.2 Radiation	37
	2.1.3 Temperature	40
	2.1.4 Air humidity	44
	2.1.5 Soil	47
2.2	Plant life in deserts	49
	2.2.1 Life cycles	51
	2.2.1.1 Seedbanks and seed germination	52
	2.2.1.2 Seedling establishment	58
	2.2.1.3 Phenological rhythms and reproduction	60
	2.2.1.4 Dissemination and dispersal	63
	2.2.2 Vegetation	65
3	**The Namib desert**	71
3.1	Geomorphology	73
3.2	Climate	75
	3.2.1 The southern Namib	83
3.3	Plant geography and vegetation	89

4 Physiological implications 99
4.1 Energy fluxes 100
4.1.1 The physical energy fluxes around a plant 100
4.1.2 Optical properties of plants and the soil 102
4.1.3 Solar and thermal radiation 113
4.1.4 The radiation budgets 116
4.1.4.1 The solar radiation budget 117
4.1.4.2 The thermal radiation budget 127
4.1.4.3 The total radiation budget 131
4.1.5 Latent and sensible heat 132
4.1.6 Convection and wind 136
4.1.7 The plant's physical energy balance 141
4.1.8 The significance of succulence for the plant's energy balance 151
4.2 Water fluxes 155
4.2.1 The driving forces for water fluxes 156
4.2.1.1 Transpiration 160
4.2.1.2 Water uptake from soil 163
4.2.1.3 Water translocation 163
4.2.2 Water status of succulents 164
4.2.2.1 Osmotic potential 164
4.2.2.2 Turgor potential 172
4.2.2.3 Leaf water potential 177
4.2.3 Transpirational water loss 178
4.2.4 Water uptake 188
4.2.4.1 Uptake from the soil 189
4.2.4.2 The abundance of CAM succulents in the Richtersveld 197
4.2.4.3 Uptake from the atmosphere 201
4.2.4.4 The hydathodes of crassulas 203
4.2.5 The significance of succulence for the water balance 209
4.2.5.1 Utilizable water 210
4.2.5.2 Water translocation 215
4.3 Carbon fluxes 224
4.3.1 The CO_2 budget 225
4.3.1.1 Photorespiration 225
4.3.1.2 Mitochondrial respiration in the light 226
4.3.1.3 Mitochondrial respiration in the dark 227
4.3.1.4 Gross photoynthesis 228
4.3.2 Modes of CO_2 fixation 230
4.3.2.1 C_3 plants 230
4.3.2.2 C_4 plants 231
4.3.2.3 CAM plants 234

4.3.2.4 Intermediates 237
4.3.2.5 Comparative aspects 242
4.3.3 Carbon investments for succulence 245
4.3.4 The significance of succulence for the carbon budget 252
4.3.4.1 The short-term drought 254
4.3.4.2 The long-term drought 264
4.3.4.3 The role of CAM 275

5 Life strategies of succulents 298
5.1 The water and carbon balance within the life cycle 300
5.1.1 Ephemerals and annuals 304
5.1.2 Hapaxanthous plants 306
5.1.3 Geophytes 307
5.1.4 Deciduous perennials 308
5.2 Life strategies 312

References 319
Index 332

Preface

We have been working on various aspects of the biology of succulents for many years. We have studied aspects of their taxonomy and anatomy, their biogeography and local distribution patterns, their flowering biology and dispersal, their germination ecology, their energy, water and carbon balances, their growth and their stress tolerances. Some of these aspects received considerably more of our attention than others. As well as performing numerous experiments and making many measurements in the laboratory, the greenhouse and under simulated climatic conditions in plant growth chambers, we have experimented extensively with succulents in their natural habitats. During our many field expeditions which repeatedly covered the same area we had the opportunity to make investigations on the same individual plants subject to different climatic conditions as a consequence of changing seasons and/or different constraints. In this book we try to outline our experiences. We deal mainly with two questions: firstly, what is a succulent and, secondly, what features, if any at all, justify the separation of this group of plants from others?

Many people, not necessarily just botanists, love succulents and cultivate them. Yet, we were surprised to find that, except for cacti and agaves, little is known about the physiology and ecology of succulents. In particular, knowledge about the performance of succulents in their natural habitat seems very limited.

Geographically, the main area of our field studies is located in the arid parts of southwestern and southern Africa. Therefore, in this book we focus on the succulents of that area. For succulents it is one of the world's richest areas. Over wide stretches the landscape is dominated by succulents. Moreover, the species number and form diversity of the succulents there is astonishing. It was easy to become fascinated by this wealth of extraordinary plants living under such harsh climatic conditions. Our field studies in the area cover more than 12 years, spanning series of exceptionally dry years as well as sequences of years with more precipitation. Many of the data presented in this book are original and have not been published before. We did not publish them in separate papers because we wanted to present them in an integrated text. We believe that this increases the significance of our results for an understanding of the physiology and ecology of succulents.

This book aims at several different goals. First of all it should present for colleagues an overview of all interesting facts and results from our studies that help explain the biology of succulents in relation to their habitat conditions. This requires a scientific presentation and the publication of research data. Secondly, it should be a textbook for students, introducing the main aspects of ecophysiology and applying this knowledge to the group of succulents. Therefore it was necessary to present some basic features of related subjects. We tried to reduce this to the necessary facts and to write these parts as simply as possible. Finally, this book addresses a broader circle of interested people, not necessarily botanists. We hope to get a broad spectrum of people interested in the amazing world of succulents and their biology. But we are aware of the fact that some chapters require a basic knowledge of physics and biochemistry.

One of our main concerns in writing this book was to reveal the functional aspects of the life strategies of succulents and not merely to classify their life strategies. Therefore we present a detailed classification of the life strategies for only one particular, though representative, group of succulents.

Intensive, lively, and sometimes heated discussions preceded and accompanied the writing of this book. While writing we were confronted with many unsolved problems. We try to discuss these or draw attention to them in order to initiate further investigations. Many parts of the book have a hypothetical character and we hope that we make this clear. In this way we intend to stimulate discussion rather than to make final statements that are not sufficiently proven by sound experimental work.

We have to thank Cambridge University Press for its patience and constructive co-operation. Our special thanks go to Mettler Waagen AG for subsidizing the colour plates. The generous financial support of the Deutsche Forschungsgemeinschaft (Bonn) and the Swiss National Science Foundation over the last decade enabled us to do the necessary field work. In addition we were generously sponsored by Boehringer (Mannheim), Eppendorf (Hamburg), South African Airways (Frankfurt) and by Mr A. P. Lötter of the former Ochta Diamond Mine (Sendlingsdrif). We gratefully acknowledge the assistance of the Botany Departments of the

Universities of Stellenbosch and Pretoria and the permissions of the Nature Conservation Divisions of the Provincial Administrations in the Republic of South Africa and Namibia to collect plant material and to perform research work in areas under their administration. Most of all we thank all our assistants in the field and in the laboratory, without naming them individually, for their enthusiastic help.

We gained much satisfaction in carrying out the research for this book. We had a good time writing it. Now we hope that many readers enjoy its contents.

Abbreviations, symbols and units

a	absorptivity, a_u upper, a_l lower surface, $\bar{a}$ mean
a_G	absorptance for global radiation, a_{Gu} upper, a_{Gl} lower surface
a_R	absorptance for thermal radiation
AAT	aspartate amino transferase
AMP, ADP, ATP	adenosine mono-, di-, tri-phosphate
C	capacitance, C' per unit water volume
C	convection, convective energy flux
c_i	concentration of substance i
C_v	volumetric specific heat
C_3	photosynthetic pathway with RuBP as CO_2 acceptor
C_4	photosynthetic pathway with PEP as CO_2 acceptor
CAM	Crassulacean acid metabolism
cm	centimetre (10 mm)
CN	cyanide
CO_2	carbon dioxide
D	distance
d	diameter
dm	decimetre (100 mm)
dw	dry weight
e_a	water vapour pressure of unsaturated air
e_R	emissivity for thermal radiation
e_s	saturation water vapour pressure
F_c	total biomass of whole inflorescence, F_{cs} non-utilizable, F_{cu} utilizable biomass
F_{ws}	structural water of the whole inflorescence
F_{wu}	utilizable water of the whole inflorescence
fw	fresh weight
G	global radiation

g	gram (1 US lb = 453.6 g)
G_a	absorbed global radiation
g_{wv}	transpirational conductivity
h	height
h	hour
h_c	convective heat transfer coefficient
h_o	reference level for height
H_2O	water
ha	hectare
J	joule (1 J = 0.2388 calories)
J	electric current
J_c	convective energy flux
J_c	flux of carbon
J_c(net)	CO_2 gas exchange rate, rate of photosynthesis
J_G	irradiance, flux of global radiation, J_{Gu} upper, J_{Gl} lower surface
J_H	heat energy flux
J_{mi}	flux of carbon connected with mitochondrial respiration
J_L	energy flux by transpiration
J_P	energy flux into photosynthesis
J_{ph}	flux of carbon in photosynthesis
J_{phr}	flux of carbon in photorespiration
J_R	flux of thermal radiation, J_{Ru} upper, J_{Rl} lower surface
J_R(net)	net thermal radiation flux
J_{sd}	flux of direct sunlight, $J_{sd}(0)$ at angle 0, $J_{sd}(\alpha)$ at angle $\propto$
J_w	water flux
J_w(net)	transpirational water loss
J_{wg}	water fluxes expressing water gain
J_{wl}	water fluxes expressing water loss
J_{wr}	water flux entering the root
J_{wv}	flow of water vapour in transpiration
K	kelvin (1 K is equivalent to 1 °C, 0 °C = 273.16 K)
k	thermal conductivity
k_c	convective heat transfer coefficient
k_δ	constant for the determination of boundary layer thickness
kJ	kilojoule
kg	kilogram (1000 g)
Km	Michaelis constant

km	kilometre (1000 m, 1 km = 0.6214 US miles)
kPa	kilopascal (1000 Pa)
kW	kilowatt (1000 W)
l	litre (1 l = 10^{-3} m^3)
L_c	loss of biomass, L_{cs} loss of non-utilizable, L_{cu} loss of utilizable biomass
ln	natural logarithm
log	logarithm to the base 10
L_w	loss of water, L_{ws} loss of structural, L_{wu} loss of utilizable water
m	meter (1 m = 3.2818 ft)
M_c	total biomass, M_{cu} utilizable, M_{cs} non-utilizable biomass
M_c(net)	carbon turnover, M_{cu}(net) turnover to/from utilizable, M_{cs}(net) turnover to non-utilizable biomass
M_{cd}	shift from utilizable to non-utilizable biomass
M_{wd}	shift from utilizable to structural water
M_{ws}	total structural water, M_{ws}(net) shift to structural water
M_{wu}	total utilizable water of a plant, M_{we}(net) shift to/from utilizable water
MDH	malate dehydrogenase
MJ	megajoule (10^6 J)
ml	millilitre (10^{-3} l)
mm	millimetre (10^{-3} m)
mol	mole, mass equal to molecular weight in grams
mRNA	messenger ribonucleic acid
N_2	nitrogen
n_{ch}	number of moles in the cell sap of the chlorenchyma
n_i	number of moles of substance i
n_{ws}	number of moles in the cell sap of the water storage tissue
NAD^+	nicotine adenine dinucleotides, NADH reduced form of NAD^+, NADP = NAD-phosphate
nm	nanometre (10^{-9} m)
NW	northwest
O_2	oxygen
O_3	ozone
OAA	oxaloacetic acid

p	partial pressure
P-	phospho-
P–V	pressure–volume
Pa	pascal (1 atm = 1.01325 bar = 101.25 kPa)
PAR	photosynthetically active radiation
PEP	phosphoenolpyruvate
PEPC	phosphoenolpyruvate carboxylase
PEPCK	phosphoenolpyruvate carboxykinase
PGA	3-phosphoglycerate
pH	− log of proton concentration
P_i	inorganic phosphate
ppm	parts per million
pyr	pyruvate
Q_H	heat energy flux by conduction
Q_M	metabolic heat energy
Q_P	energy consumed by photosynthesis
Q_W	energy flux by phase transition of water
R	gas constant
R	resistance, R_s stomatal, R_{ss} substomatal, R_{es} epistomatal, R_a boundary layer, R_c cuticular resistance
R	reflectance for leaf without wax bloom
R*	reflectance for leaf with wax bloom
r_R	reflectance of thermal radiation
R	thermal radiation, R_a absorbed thermal radiation, R_e emitted thermal radiation
r	reflectivity
R_{wr}	water uptake resistance by roots
R_{wv}	transpirational resistance
R_{wvl}	leaf resistance to diffusion of water vapour
rh	relative air humidity
Rubisco	ribulose-bisphosphate carboxylase/oxygenase
RuBP	ribulose 1,5-bisphosphate
RWC	relative water content
S	leaf expansion coefficient
s	second
S_m	mesophyll succulence
SCoA	co-enzyme A-SH
SE	southeast
SEM	scanning electron micrograph

SH	sun's height
t	transmissivity
t	time
T	absolute temperature (in K), T_a of ambient air, T_c of black body, T_u upper surface, T_l lower surface or of leaf, T_{ss} of the air in the substomatal cavity
t_R	transmittance
TCA	tricarbonic acid cycle
U	electric potential
UV	ultraviolet radiation
V	volume
v	velocity
V_{ch}	volume of the chlorenchyma
V^o	water volume at saturation
V_w	partial molal volume of water
VPD	water vapour partial pressure deficit
W	Watt
W	west
WUE	water use efficiency
x	length
δ	boundary layer thickness
Δ	difference, Δh in height, ΔM_c in total biomass ΔM_{cu} in utilizable biomass, ΔM_{cs} in non-utilizable biomass, ΔT in temperature, ΔV in volume, Δx in length, $\Delta \Psi$ in water potentials
Δ_{CO_2}	CO_2 budget
ΔG_{CO_2}	Gibbs free energy change of incorporated CO_2
Δt	timespan
ΔW	difference in molar fraction of water vapour between substomatal cavity and ambient air
ϵ	emissivity
λ	wavelength
μm	micrometre (10^{-6} m)
μ mol	micromole (10^{-6} mol)
σ	Stefan–Boltzmann constant
Σ	sum of
Ψ	water potential
Ψ_a	water potential of the air

Ψ_c	water potential of the cell, Ψ_{cw} in the cell wall, Ψ_{ca} of the air adjacent to the cell wall
Ψ_h	gravitational water potential
Ψ_M	matric water potential
Ψ_P	pressure potential of the solution
Ψ_r	water potential of the root
Ψ_s	water potential of the soil, Ψ_{ss} water potential of the air in the substomatal cavity
Ψ_t	total water potential of the plant
Ψ_π	osmotic water potential
°C	degree Celsius
°N	degree north
°S	degree south
$^0/_{00}$	parts per thousand

1

The succulent

Although the term succulent is widely and frequently used to characterize plants that can store water in their organs, there appears to be no clear definition that is generally accepted. Perhaps this situation persists because succulence is a quality which can be possessed to a higher or lower degree whereas the term succulent is intended as a typification or classification: a plant is or is not a succulent.

It has become common usage that plants whose leaves have xeromorphic features (small in stature, small cells, thickened outer cell walls of the epidermal cells, thick cuticle, 'submerged' stomata, sclerotic cells and fibres, etc.) are called xerophytes. In analogy one might call plants exhibiting succulence and which are succomorphous, succophytes. However, this term has never been proposed even though it might be a useful one to describe plants with succulence. On the other hand, xerophytes are defined as plants of arid habitats just as halophytes are understood as plants growing in habitats with a high salt content of the soil. A similar definition deriving from the habitat cannot be applied to succulents and 'succophyte' will not make any sense in this context. We will use the term succulent throughout this book.

The foremost sites to look for succulents are arid places in otherwise humid habitats and the arid habitats in various parts of the world. Salty habitats are also often settled by succulents. So far no complete explanation has been given for the special features of the succulent's physiology and ecology that allow understanding of the growth and survival of this type of plant in its characteristic habitats.

We are aware of the limited value of allocating plants to different boxes of which the content is then defined precisely and distinctly in such a way that it is different from that of another box. Nevertheless, we are convinced that one should make an attempt to give the term succulent a meaning that goes beyond a pure morphological or anatomical description. In the following chapters we will outline which physiological or ecophysiological features might be used to provide a meaningful characterization. We want to start with a preliminary definition of a succulent and its morphological and anatomical background, while, wherever possible, including reference to relevant aspects of the growth and life cycle of these plants.

1.1 Towards a definition of a succulent

Commonly succulents are placed within the xerophytes as a special group of water storing plants, but the number of succulents that do not meet the criteria applied to xerophytes is large. This is a strong argument against xeromorphism being a necessary feature of a succulent.

Often the term succulence is used to define a succulent by simply saying that a succulent has at least one tissue which exhibits succulence. In these cases succulence is used to describe a quality of a tissue that appears to have a high water content and that therefore is called succulent, fleshy or juicy. The reason for the apparently high water content of the tissue lies in the huge vacuole of each of its cells which fills 99 per cent or more of the whole cell leaving only a small plasma layer along the cell wall. Although succulence appears to be a quality of a single cell it is normally used to describe a cluster of more or less uniform cells forming a tissue. Succulent tissues can occur in different organs of a plant though only rarely are all cells and tissues of an organ really succulent. Most often succulence is restricted to parts of a plant organ, such as the epidermis of a leaf, the cortical tissue of a stem or a root, the exocarp of a fruit, etc. Nevertheless, the term succulent commonly and unjustifiably refers to the whole organ regardless of its actual anatomy and morphology.

We believe this usage of the term to be wrong from a morphological and anatomical point of view as well as regarding the physiological consequences resulting from the fact that in an organ only a distinct part is succulent. In particular, the evaluation of the impact of succulence on the physiology will be more difficult if the fact that only some of the organ is succulent is not considered properly. Moreover, it is obvious that despite similar appearance the succulence of different organs serves different functions, e.g. compare the berry of a redcurrant with the leaf of a *Lithops*. The succulence of a halophyte like *Salicornia* has a different physiological meaning to that of the succulence of a barrel cactus.

Some species possess leaves in which, with the exception of the vascular tissue, the stomata and parts of the epidermis, all parenchyma cells store water and are succulent. Such leaves are properly called all-cell succulent leaves. They should be clearly distinguished from those leaves which

Fig. 1.1 (a) Cross-section through the leaf of a succulent (*Prenia sladeniana*) belonging to the all-cell succulent type and (b) through a leaf of a succulent (*Gasteria pillansii*) belonging to the partially succulent type.

contain only one (rarely two) succulent tissues (e.g. epidermis, central or peripherous parenchyma). These leaves should be termed partially succulent leaves. Fig. 1.1 gives examples of both types of leaves.

This distinction is further strengthened by physiological and biochemical features. If we consider water storage a main function of succulence, then leaves of the all-cell succulent type show no internal separation of tasks. All cells store water and all cells photosynthesize since they have chloroplasts. Partially succulent leaves, on the contrary, show a separation of tasks between the different tissues. The chlorenchyma which is generally not succulent takes care of photosynthesis while a weakly differentiated but succulent parenchyma stores water. In the course of this book we will come back to this important distinction as it seems to be the basis of different life strategies of succulents.

Very often the terms weak, medium or highly succulent are used in order to quantify the subjective impression of the succulence of a plant organ and describe it more objectively. For the same purpose the degree of succulence, as first introduced by Delf (1912), is used. This term is defined by

$$\text{degree of succulence} = \frac{\text{water content at saturation (g)}}{\text{surface area of the organ (dm}^2\text{)}} \tag{1.1}$$

With respect to anatomical and morphological considerations as well as to physiological aspects, the usefulness of this quotient seems questionable as has been shown by Starnecker (1984) for the genus *Peperomia*. In a survey of 46 *Peperomia* species he could not find a clear correlation between the mode of CO_2 fixation and the degree of succulence. Two further aspects supporting our objections should be mentioned:

(1) The degree of succulence of a plant organ is always determined on the whole organ even if only part of it is succulent. As long as the different tissues of a plant organ are not separated from each other the fraction of non-succulent cells disturbs the result and the calculated degree of succulence does not reflect the succulence of the succulent tissue. Only leaves of the all-cell succulent type would allow a proper comparison with this method. Unfortunately there is no information available about the degree of succulence of various succulent tissues of partially succulent organs. Consequently we can only speculate about similarities and dissimilarities. However, it seems justified to assume that, in agreement with their different functions, the degree of succulence characteristically varies among the different succulent tissues, and that it might be possible to distinguish between succulent tissues that only store water and those that store both salt and water.

(2) The degree of succulence is determined from the water content at saturation of a plant or a plant organ. It must be emphasized here that it is one characteristic feature of succulent tissues that they have no saturating water content. Even in the natural habitat it can be easily observed that succulent or partially succulent plant organs burst after abundant rainfall, especially if the rain was preceded by a prolonged period of drought (Fig. 1.2). It is impossible to determine experimentally the saturating water content, and even if this could be done it would make no sense because it occurs rarely in nature.

Taking into account these objections it appears that a proper quantitative determination of the degree of succulence in the sense of Delf (1912) is neither possible nor desirable. Therefore, as this quotient has no ecophysiological meaning, it should not be used to describe succulence.

Similar objections emerge with other attempts to quantify succulence, e.g. as the surface expansion defined by

$$\text{surface expansion} = \frac{\text{surface (cm}^2\text{)}}{\text{fresh weight (g)}} \qquad (1.2)$$

We will return to other descriptive quotients like mesophyll succulence and their meaning in Section 4.3 where we will discuss the 'cost–benefit' balance of evergreen and deciduous succulent leaves.

Fig. 1.2 Burst leaves of *Stoeberia beetzii* after abundant rainfall in the southern Namib. It is noticeable that the crevices sealed quickly and that the burst leaves did not dry out.

Before we deal in more detail with the anatomical and morphological features of succulents we have to outline the term succulent and give a meaningful definition. We hope to prove the applicability of our definition in the final chapter. Before we can do this we have to demonstrate the significance of succulence for the plant.

We repeatedly point out that one should not look upon succulence solely as a morphological and anatomical description of a succulent but that one should also take into account the function of succulence. This, unequivocally, is the storage of water. However, we have to emphasize that one cannot judge a succulent on the total amount of water that is stored in its succulent tissues. One can only consider that fraction of stored water which is available to the rest of the plant by withdrawal from the storage tissue either reversibly or irreversibly. With respect to this fraction, which we call the utilizable water, succulents differ significantly from each other.

The growth form of a succulent will mainly determine the functioning of the water storage tissue. In long-lived evergreen leaves withdrawal of the utilizable water from and refilling of the storage tissue will happen repeatedly. This requires distinct anatomical features which will be dealt with in the following section. Short-lived annuals or deciduous plants, mostly of the all-cell succulent type, usually do not show cyclic refilling of the water storage tissue. Their utilizable water is mostly used irreversibly and withdrawal leads to the death of the leaf. Therefore the definition of a succulent takes into consideration the criterion of utilizable water but regardless of the possibility of refilling. Furthermore, it is important to state that only living water storage tissues are relevant for defining succulence. Taking all this into account we define a succulent as follows:

A succulent (or succophyte) is a plant possessing at least one succulent tissue. A succulent tissue is a living tissue that, besides possible other tasks, serves and guarantees an at least temporary storage of utilizable water, which makes the plant temporarily independent from external water supply when soil water conditions have deteriorated such that the root is no longer able to provide the necessary water from the soil.

The special ability of the succulents to store utilizable water enables these plants to live in habitats that endure periodic droughts. It is the amount of utilizable water at the onset of the drought that determines the length of the drought period that can be survived. An overview of the distribution of succulents within the different climatic zones reveals that succulents are quite common in several areas and that even in humid climates succulents often occur most abundantly on dry sites. It is presumably this fact that led to the false conclusion that succulents are well adapted to arid zones and drought. This statement is pertinently in need of correction. Succulents are not generally adapted to aridity but only to periodically occurring droughts. That means the conditions of drought should not pertain for too long to allow succulents to survive. At more or less regular intervals (the length of which depends on the amount of utilizable water within the plant at the beginning of the drought and the climatic conditions during the drought period) the succulent should meet conditions which allow the refilling of its succulent tissues and/or growth. If those intervals are too long, the succulent cannot survive. This is the reason why extremely arid areas lack succulents or support them only in some special sites.

1.2 Morphological and anatomical considerations

In the previous section we emphasized that succulent tissues may develop in different plant organs. It is quite common to characterize succulents as leaf succulents, stem succulents or root succulents, but strictly speaking this characterization is applicable only to plants in which the succulent tissue is limited to a single type of organ. In some genera this is undoubtedly the case, e.g. in *Lithops* (leaf succulent), *Ferocactus* (stem succulent) or *Brachystelma* (root succulent). There are numerous examples of plants possessing a combination of succulent tissues in several types of organs. This usually is not clearly indicated in anatomical and morphological texts on succulents. In some species all organs can contain succulent tissues. This is shown in Fig. 1.3 for *Ceraria namaquensis*. In such cases the relative contribution of the various succulent tissues to the total utilizable amount of water in the plant can be quite different.

A further problem with the simple identification of a plant as a leaf, stem or root succulent is provided by those species where only part of an organ remains and is succulent, e.g. the petiole in *Oxalis succulenta*, or where succulence is restricted to a morphologically hardly identifiable tissue complex, e.g. composed of parts of the root, the hypocotyl, the epicotyl and the shoot as in *Pachypodium brevicaule*, *Ceraria pygmaea*, *Fockea crispa* and *Anacampseros comptonii*. Plants with such tissue complexes are called caudiciforms (Rowley 1987).

1.2.1 Amount of utilizable water

We believe that the basic feature of a succulent is its possession of a storage tissue for utilizable water. The proportion of the succulent tissue in an organ, and the proportion of the water in a succulent tissue that is available as utilizable water determine the plant's water content. The problem of utilizable water will be discussed in more detail in Section 4.2. In Table 1.1 we present the water content values as percentages of the weights at the time of harvesting of a selection of succulents and non-succulent plant species from the southern Namib.

These data do not allow conclusions about the proportion of water that is utilizable. Even so, it is tempting to assume that those organs of a succulent with similar water content values (below 60%) to those of non-succulents will probably not be able to contribute relevantly to the amount of utilizable water. This assumption can be wrong, however. We have already outlined in Section 1.1. that in the tissues of an organ the actual water content can differ greatly. For instance, the stem of *Cotyledon orbiculata* (water content 58.5% see Table 1.1) consists of a woody central supporting tissue which contributes substantially to the dry matter of the stem, and of a succulent peripheral tissue which may contain an appreciable amount of utilizable water.

However, the absolute water content of a succulent should not be expected to be more or less constant. On the contrary, it is highly variable and at any one moment related to the physiological condition of the plant. This depends on the timespan since the last water supply (compare

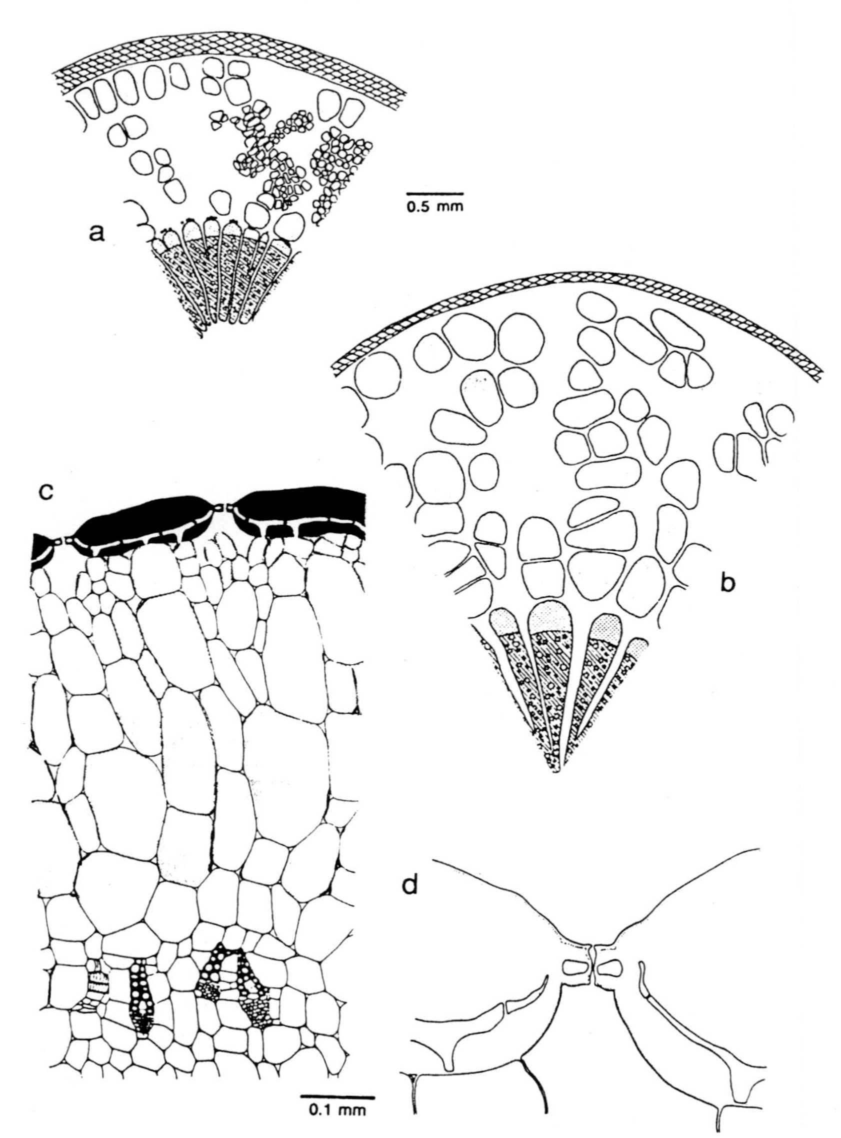

Fig. 1.3 Cross-sections of a stem (a), root (b) and leaf (c) of *Ceraria namaquensis*. A detail of a stoma is given in (d).

Table 1.1. *Water content as a percentage of fresh weight (field situation) in the organs of a number of Namaqualand plants*

Species	Water content in % fresh weight		
	Leaf	Stem	Root
Acanthopsis disperma	83.9		63.2
Aloe pearsonii	89.2	72.1	
Aloe ramosissima	81.6	80.5	—
Sphalmanthus trichotomus	94.0	42.5	—
Psilocaulon subnodosum	91.8	52.5	—
Ceraria fruticulosa	80.3	61.7	79.6
Ceraria namaquensis	80.6	52.0	—
Ceraria pygmaea	82.1	71.6*	—
Cheiridopsis robusta	83.9	42.2	29.9
Conophytum aequale	85.6	55.3	30.8
Conophytum minutum	81.8	28.0	40.5
Cotyledon orbiculata	84.2	58.5	—
Tylecodon paniculatus	89.9	88.3	58.9
Tylecodon pearsonii	86.9	82.3	50.8
Tylecodon reticulatus	91.3	83.3	63.2
Tylecodon wallichii	80.3	69.3	63.9
Crassula deceptor	80.7	37.2	—
Crassula elegans	80.5	40.3	46.2
Delosperma pergamentaceum	77.9	46.0	—
Forsskaolea candida	67.5	68.0	62.1
Hermbstaedtia glauca	70.4	69.9	—
Mesembryanthemum barklyi	93.6	86.0	—
Mesembryanthemum pellitum	91.7	85.7	72.7
Mitrophyllum clivorum	93.0	53.8	—
Opophytum aquosum	90.5	89.1	57.2
Osteospermum microcarpum	87.7	77.8	54.7
Ozoroa dispar	72.3	57.9	—
Pachypodium namaquanum	77.6	87.2	—
Prenia sladeniana	89.1	72.3	—
Ruschia sp.	89.8	39.0	—
Senecio corymbiferus	92.0	82.2	59.2
Stoeberia beetzii	80.1	36.1	—
Trianthema triquetra	93.3	67.1	49.8
Trichodesma africanum	75.3	—	77.7

Note:
* transition of stembase to root.

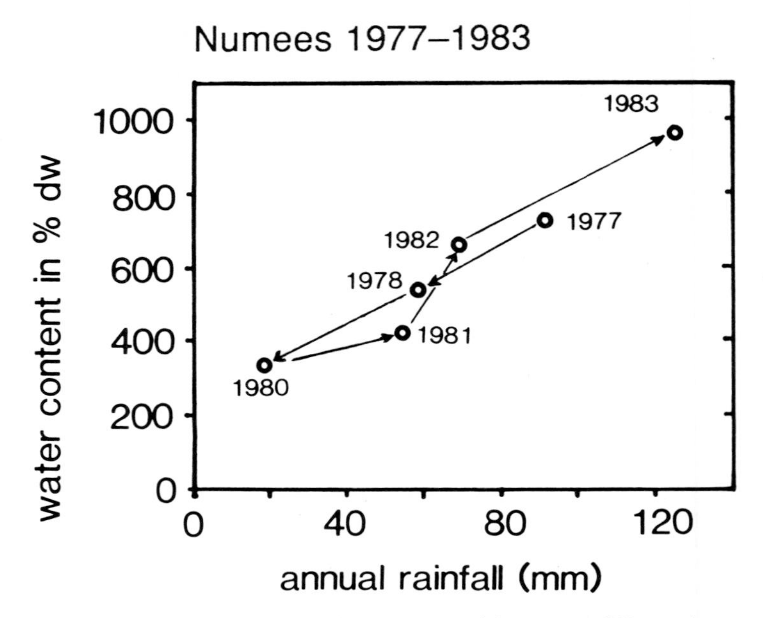

Fig. 1.4 Actual leaf water content in percentage of dry matter of 32 succulents with evergreen leaves of the partially succulent type (Asteraceae, Crassulaceae, Liliaceae, Mesembryanthemaceae, Portulacaceae) plotted over the annual rainfall. Figures at the dots indicate the respective year of the determination. The direction of shrinkage and swelling is indicated by arrows.

Chapter 4). Unfortunately there are virtually no previous studies on the temporal variation in water content of the succulent organs and tissues. Although the variation in absolute water content of an organ does not precisely follow the variation in the water content of the succulent tissue, such data present an approximation, however limited, of the range in values of utilizable water. Fig. 1.4 shows the changes in the actual water content values of a selection of evergreen succulents from the southern Namib during the years 1977 to 1983. It is important to note that this direct correlation between the actual plant water content and the annual rainfall was obtained for succulents of the partially succulent type.

Repetitive filling and emptying of any succulent storage tissue with utilizable water requires special anatomical features that allow changes in volume of the tissue and the organ. In succulents with flat leaves the processes of filling and emptying can proceed as in an inflatable pillow, but round and triangular leaves can also reversibly shrink and swell. Fig. 1.5 illustrates the extent of such changes for a variety of succulent leaves.

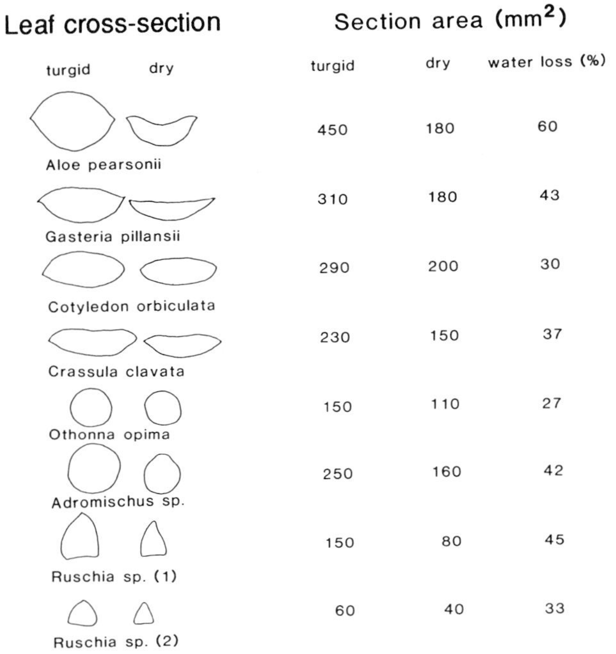

Leaf cross-section	Section area (mm^2) turgid	dry	water loss (%)
Aloe pearsonii	450	180	60
Gasteria pillansii	310	180	43
Cotyledon orbiculata	290	200	30
Crassula clavata	230	150	37
Othonna opima	150	110	27
Adromischus sp.	250	160	42
Ruschia sp. (1)	150	80	45
Ruschia sp. (2)	60	40	33

Fig. 1.5 Cross-sections of leaves of the partially succulent type during a drought period (dry) and after abundant rainfall (turgid) in the southern Namib desert. Figures give the area of the cross-section in mm^2.

Accordingly, *Senecio medley-woodii*, a succulent possessing all-cell succulent leaves, has a strongly reduced ability to vary its leaf thickness as compared to *Aloe pearsonii* or *Gasteria pillansii*.

In leaves with triangular or circular cross-sections the repeated shrinkage and swelling is made possible by special folding structures, illustrated in Fig. 1.6. Here the direction of the shrinkage is anatomically predetermined by the structure of the inner tissues. But there are also good examples where the structure of the epidermis is responsible for the direction of shrinkage and folding (Jürgens 1986).

The anatomical provision allows a shrinkage without damage to the tissue. It can also have other advantages. It allows, for instance, that the leaf after shrinkage reaches a certain form which ensures that the leaf maintains a certain position in relation to the direction of the stem. This can be to the benefit of the light climate inside the leaf, or can contribute to prevent an excessive absorption of global radiation which would unfavourably affect the temperature of the leaf.

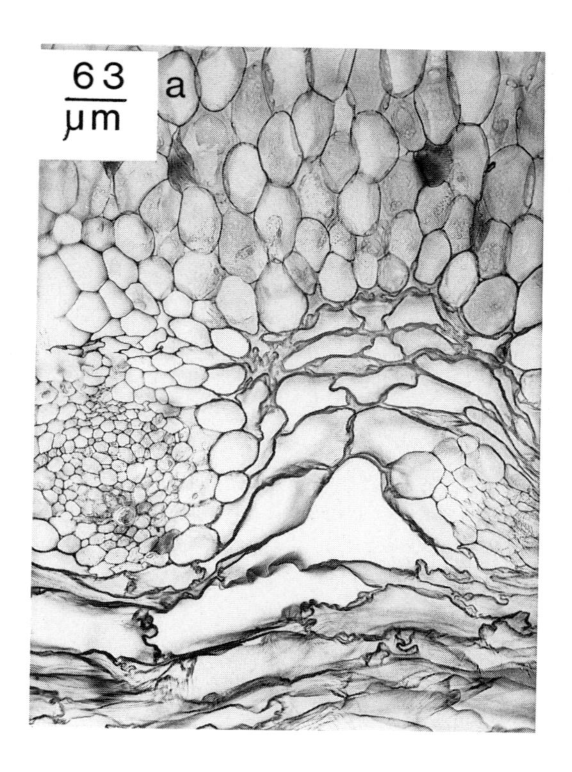

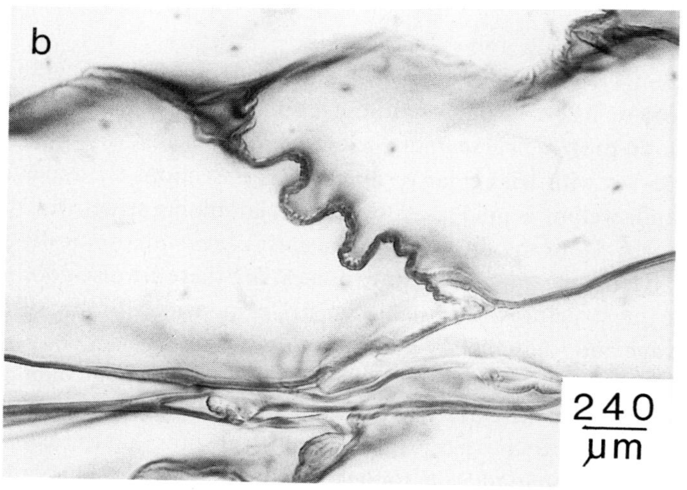

Fig. 1.6 Cross-sections through a drought-stressed leaf of *Aloe pearsonii.* (a) The central water storage tissue (lower part and centre) is more or less dry while the cells around the bundle are still turgid. (b) The larger magnification shows the shrunk cell walls in detail.

1.2.2 Static problems

Obviously succulent organs, particularly if they are filled, have quite a weight. As a consequence such organs constitute a considerable burden for the plant, requiring the development of an adequately strong support system or relevant modifications in the growth form to avoid heavy supportive constructions, or both. Such modifications in growth form certainly contribute to the peculiar habitus shared by many succulents. It is unfortunate that anatomical and morphological studies have hardly addressed this question. Voluminous succulent leaves show a clear tendency to increase the relation of thickness to width of the leaf, resulting in a cuneate cross-section.

Just as in all woody plants, secondary growth of the stem is instrumental in providing stable plant support. We can see beautiful examples of this in the strong support offered by the woody xylem skeleton of the huge candelabra euphorbias (*Euphorbia ingens*, *Euphorbia candelabrum*, *Euphorbia conspicua*, etc.) and columnar cacti (Gibson and Nobel 1986). Much less impressive, but nevertheless comparable, are the woody stems of the many shrubby species in the Mesembryanthemaceae, e.g. *Ruschia utilis*, *Stoeberia beetzii*, *Ruschianthemum gigas*, *Mitrophyllum sp.*, etc. (Fig. 1.7).

Succulents belonging to the monocotyledons (monocots) cannot develop a compact woody stem since they do not exhibit normal secondary growth. As in other woody monocots additional interconnected vascular bundles are added to the original stem by a secondary peripheral cambium. These bundles are accompanied by strands of fibres which form a sponge-like skeleton. Examples are offered by the giant aloes (*Aloe dichotoma*, *Aloe pillansii*, *Aloe ramosissima*) and cross-sections through a stem of dead *Aloe dichotoma* are presented in Fig. 1.8. In agaves the succulent leaves also contain a well-developed kind of skeleton composed of long fibres, but in aloes these are not so obvious.

The mechanical problem of supporting the comparatively large weight of succulent organs can be solved in different ways. In the leaf succulent group there are plants with a dense packing of the succulent leaves forming 'compact' life forms. Three types can be distinguished. They are presented in Fig. 1.9.

(1) The internodes are extremely reduced resulting in a ground rosette with the outer leaves touching the soil (e.g. *Crassula pseudohemisphaerica*, *Aeonium tabuliforme*, *Haworthia tesselata*).

(2) The internodes are short with the subsequent leaves touching and supporting one another at their bases resulting in compact heads

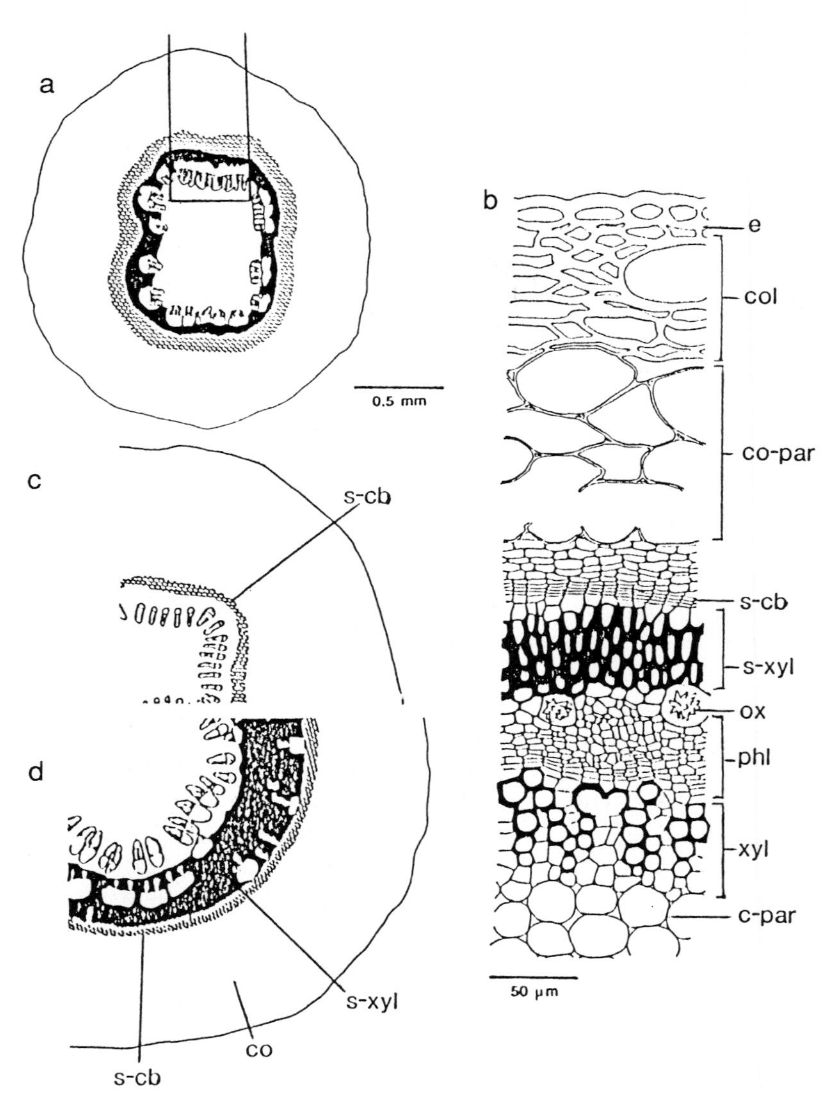

Fig. 1.7 (a) Cross-section through a woody shoot of *Ruschia* sp. and (b) a detail from the cortical tissue to the centre. Cross-sections through a primary (c) and a secondary (d) shoot of *Stoeberia beetzii*. e = epidermis, col = collenchyma, co-par = cortex parenchyma, s-cb = secondary cambium, s-xyl = secondary xylem, ox = oxalate crystals, phl = phloem, xyl = xylem, c-par = central parenchyma, co = cortex.

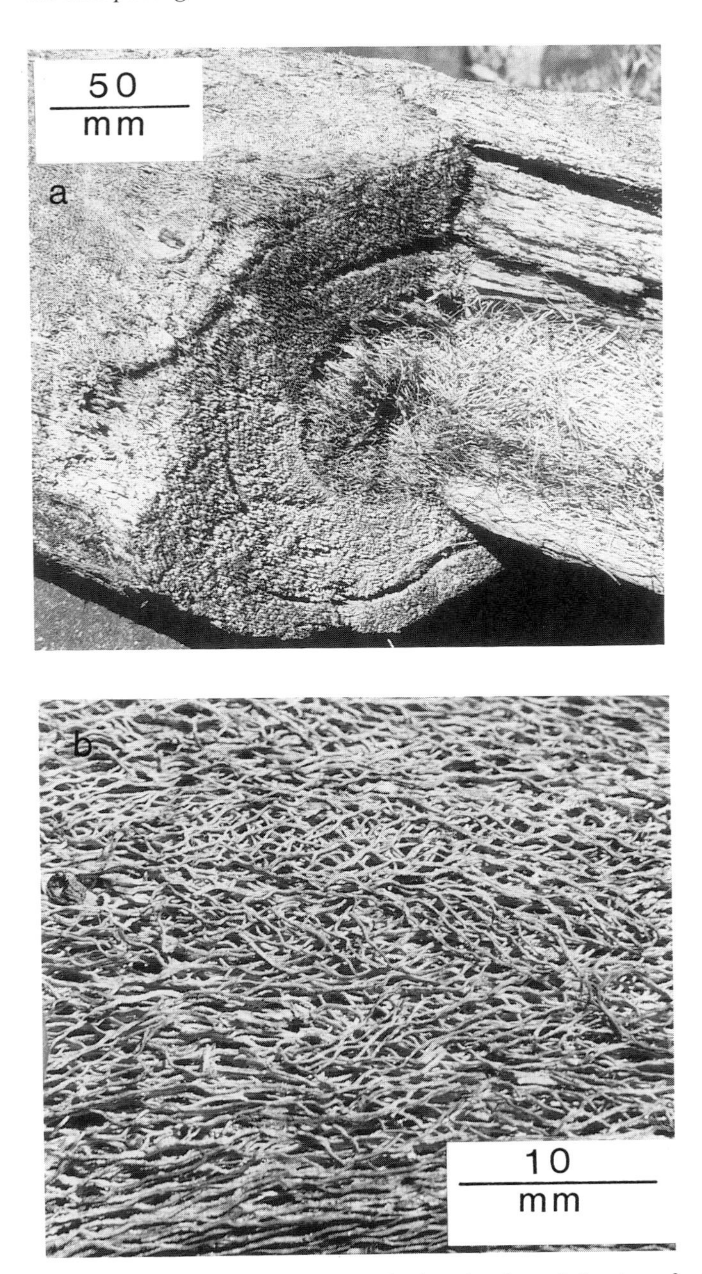

Fig. 1.8 (a) Cross-section and longitudinal cutting through the stem of a dead *Aloe dichotoma* showing the spongy and fibrous structure of the interior. (b) Same as (a) but with a higher magnification.

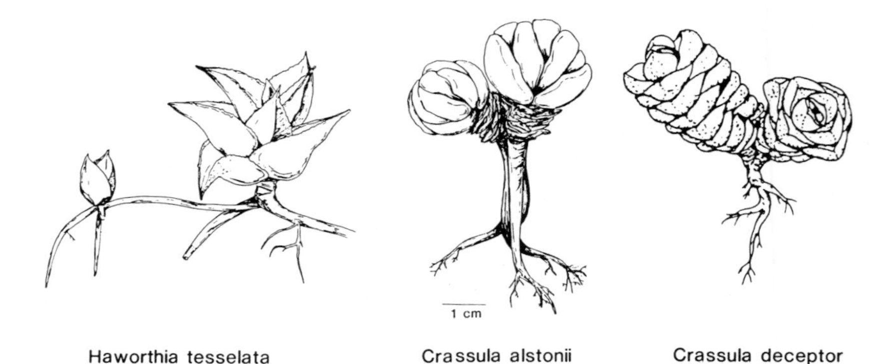

Fig. 1.9 Three types of leaf packing in leaf succulents. (1) Ground rosette (*Haworthia tesselata*), (2) compact heads or cushions (*Crassula alstonii*) and (3) compact columns (*Crassula deceptor*). (Drawings by Diana Rembges.)

or cushions (e.g. *Crassula alstonii*, *Stomatium*, *Chasmatophyllum*, *Faucaria*, *Lapidaria margaretae*).

(3) The internodes are short with the subsequent leaves showing extremely dense packing resulting in compact columns (e.g. *Crassula deceptor*, *Crassula columnaris*).

Another way to circumvent the support problem is shown by prostrate succulents. For example *Malephora uitenhagensis*, because of its creeping habit, is able to possess a considerable number of leaf pairs with a great weight without forming a thick or woody stem. Also stem succulents have developed prostrate growth, e.g. the 'creeping devil' *Stenocereus (Machaerocereus) eruca* in North America or *Haageocereus pacalaensis* in South America. Some of the creeping species are able to climb on appropriate structures (*Sarcostemma viminale*, *Cissus quadrangularis*, *Selenicereus*).

The reduction or the absence of petioles is widespread among leaf succulent species and certainly also contributes to supporting heavy succulent leaves. Similar considerations apply to the erect implantation (in the stem) of the bases of the large spherical leaves of, for example, *Othonna opima*, *Cotyledon orbiculata* and *Augea capensis* as well as the joints of several species of *Opuntia*. Undoubtedly this positioning reduces the momentum of forces working on the supporting stem (Givnish and Vermeij 1976, Givnish 1986). On the other hand, such a positioning can also be adaptive in optimizing the received amount of radiation. Finally we want to point out that species with very large succulent leaves, such as occur in the genus *Aloe*, usually possess well-developed leaf sheaths which most probably also primarily serve the purpose of support, e.g. in *Aloe marlothii*.

1.2.3 Protective tissues of leaves

Since the basic feature of succulents consists of the storage of utilizable water it is clear that prevention of waste of this water can be expected to be an important issue. In arid areas, where the evaporative demand is strong, highly efficient epidermal or peridermal tissues for minimizing water loss are an absolute necessity for all evergreen leaves and stems. Plants in arid areas possess an epidermis which reduces water loss by means of thickened outer cell walls of the epidermal cells and/or a thick cuticle, often covered by epicuticular waxes. This type of epidermis is called the xeromorphic type. It is characteristic for organs of high longevity (leaves or stems) and all giant succulents (*Carnegiea gigantea*, *Euphorbia candelabrum*) seem to possess such an epidermis. Certain features, such as the thickness of the outer cell wall and the amount of wax, can be considerably modified by the ecological conditions under which the plant grows (see Section 4.1).

A case of special interest in this respect are plants in which parts of individual leaves are exposed to highly different microclimatic conditions. Examples are provided by the leaves of geophyllous species in the genus *Lithops*. Here the laterally connate pair of leaves is obcone in outline, with virtually all of the leaves subterraneous except for the flat tops. While the flat leaf top is fully exposed to the usually harsh aerial climatic conditions, the rest of the leaf is protected by the soil from convection, direct radiation and rapid changes of air humidity. The thick-walled and protruding epidermal cells are densely packed and resemble a cobble-stone pavement (Fig. 1.10). The visible stomata are located in the narrow fissures between the epidermal cells. On the subterraneous part of the leaf pair the epidermal cells exhibit only a central papilla and the stomata are visible, i.e. they are not concealed by the adjacent epidermal cells (Fig. 1.10).

Even a superficial survey among succulents will reveal, however, that there are many succulents which lack such a type of xeromorphic and homogeneous epidermis. In *Drosanthemum* species (Fig. 1.11), for instance, the epidermis consists of two types of cells. Most of the cells are small (normal epidermal cells) but a number of others are extremely enlarged and swollen like balloons (bladdercell idioblasts). In this type of epidermis normally the outer wall of all epidermal cells is thin, as is the cuticle, and wax covers in many cases are not detectable. This type of epidermis is called the heterogeneous or idioblast type. The idioblasts are capable of storing

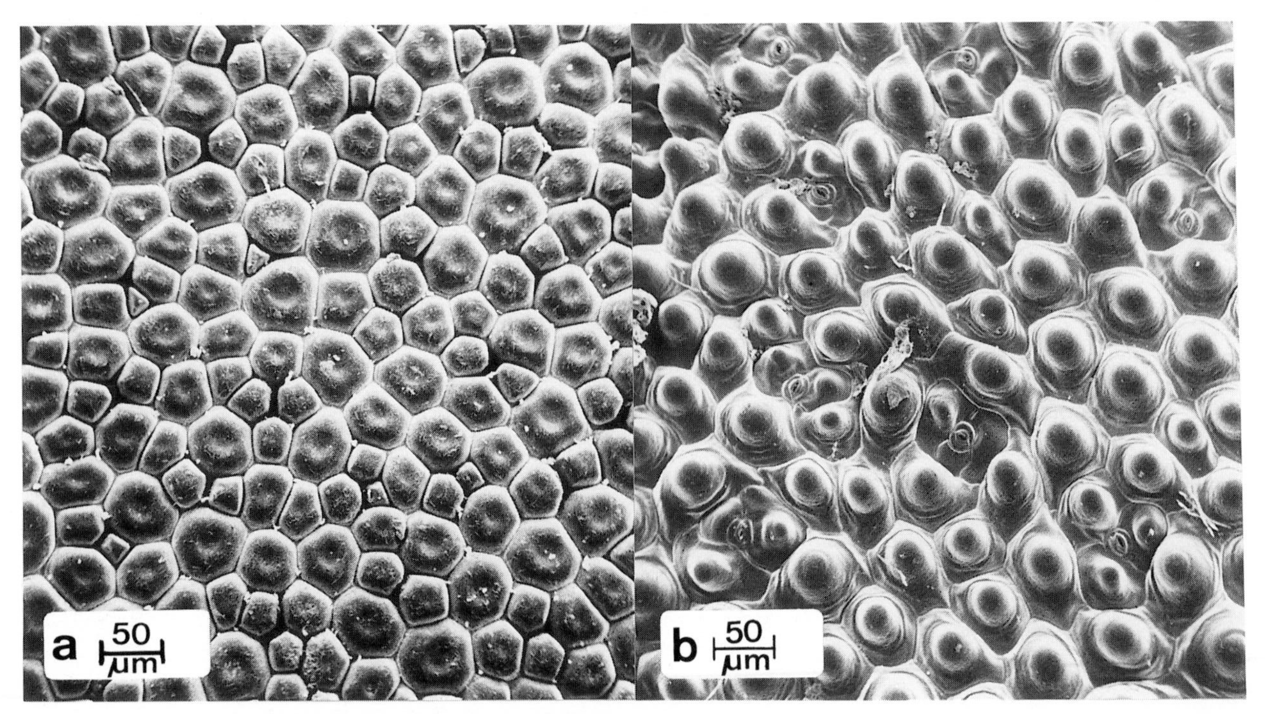

Fig. 1.10 SEM picture of the epidermis of *Lithops lesliei.* (a) The top of the leaf facing air and (b) the lateral surface of the leaf facing soil. Wax bloom and debris which usually cover the epidermis in the habitat have been removed.

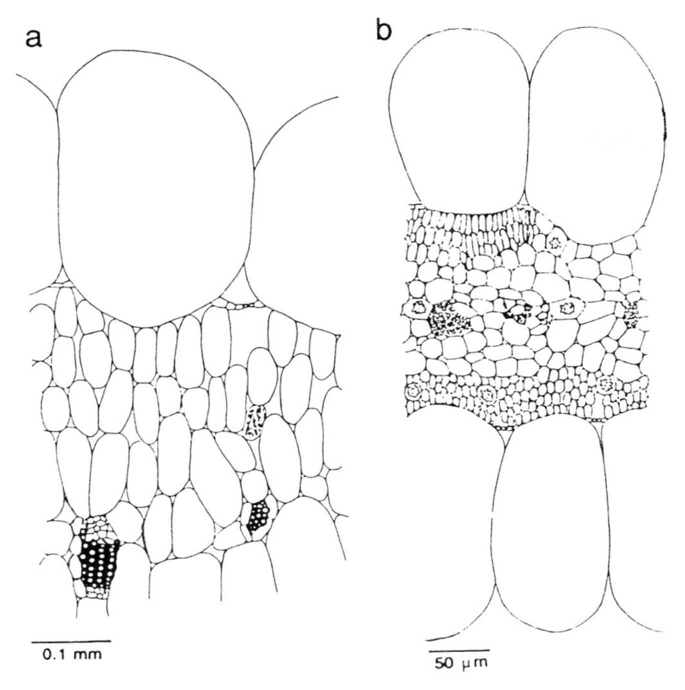

Fig. 1.11 Parts of cross-sections through a leaf of *Drosanthemum* sp. (a) and *Galenia dregeana* (b) showing the epidermal water idioblasts.

water and represent a peripheral water store which can contribute up to more than 50% to the total water storing capacity of the leaf.

This type of epidermis apparently reduces water losses sufficiently. It is characteristic for short-lived organs and restricted to families with predominant leaf succulents. A good example is represented by the Mesembryanthemaceae in which the subfamily Mesembryanthemoideae and parts of the subfamily Ruschioideae exhibit this type. All annual members of this family belong to these two subfamilies. But this type of epidermis also occurs in long-lived perennials of considerable size (e.g. *Sphalmanthus decurvatus*, *Jacobsenia*, *Aridaria*). Similar epidermis types are found in certain members of the Crassulaceae (*Monanthes*, *Tylecodon*) and Oxalidaceae (*Oxalis*).

In succulents, as in all other phanerogams, the gas exchange (CO_2 and H_2O) occurs primarily through the stomata. To reduce water loss these often are anatomically modified. As with other characters which are generally considered adaptations to arid environments (whether or not this is correct we shall discuss in the subsequent chapters on the physiology of succulents), stomatal features also vary with the degree of aridity of the habitat (Cutler *et al.* 1977, Ihlenfeldt and Hartmann 1982) and these variations may have adaptive significance. For instance, epidermal cells which protrude above the level of the stomata as in *Lithops* (Fig. 1.10) or where the epidermis is densely covered with idioblasts (Fig. 1.11) will certainly reduce transpiration because they affect the distance over which gases must diffuse. The same effect can also be achieved in a different way. In *Aloe*, for instance, stomata are deeply sunken (Fig. 1.12a). The occurrence of stomatal chimneys built by cuticular rims, as typical for species of *Haworthia* (Fig. 1.12b), can be interpreted likewise. An ecophysiological evaluation of such anatomical particulars, taking dimensions, frequencies and distribution patterns of stomata into account, will be carried out in our discussion on transpiration (Section 4.2).

1.2.4 Protective tissues of stems

Stomata are supposed to function throughout the life span of the respective organ in order to allow an effective gas exchange. This may become a problem in very long-lived organs, a problem which is obvious in long-living stem succulents incapable of forming leaves that can be replaced after a certain time. Ontogenetical and morphological development of the stem should be such that stomatal functioning over extended periods of time is ensured. To our knowledge it is not known how many years stem stomata can function. This might be an interesting question especially when the mature parts of, for instance, *Cereus* spp. carry many epiphytic Bromeliaceae like *Tillandsia recurvata*. In succulents with both succulent leaves and stems such special requirements for the functioning of stomata in the stem are not necessary. In *Aloe dichotoma*, *Tylecodon paniculatus* and *Othonna herrei*, for example, the epidermis on the young shoots is soon replaced by a thick periderm (Fig. 1.13). It is hard to imagine any efficient gas exchange through such structures. In Chapter 4 we shall try to answer the question as to why even under such a thick periderm the first few layers of the cortical tissue nevertheless contain a substantial amount of chloroplasts.

In this context it is interesting to speculate about the possible functions of the ribs and warts which are so commonly found in cacti and other stem succulents in the Euphorbiaceae and Asclepiadaceae. Such structures certainly are useful in ‘dosing’ the radiation load received by the plant, facilitate shrinkage and swelling of the water storage tissue, and contribute to the mechanical stability of the tall columnar plants. But the same structures also allow for secondary growth without the necessary replace-

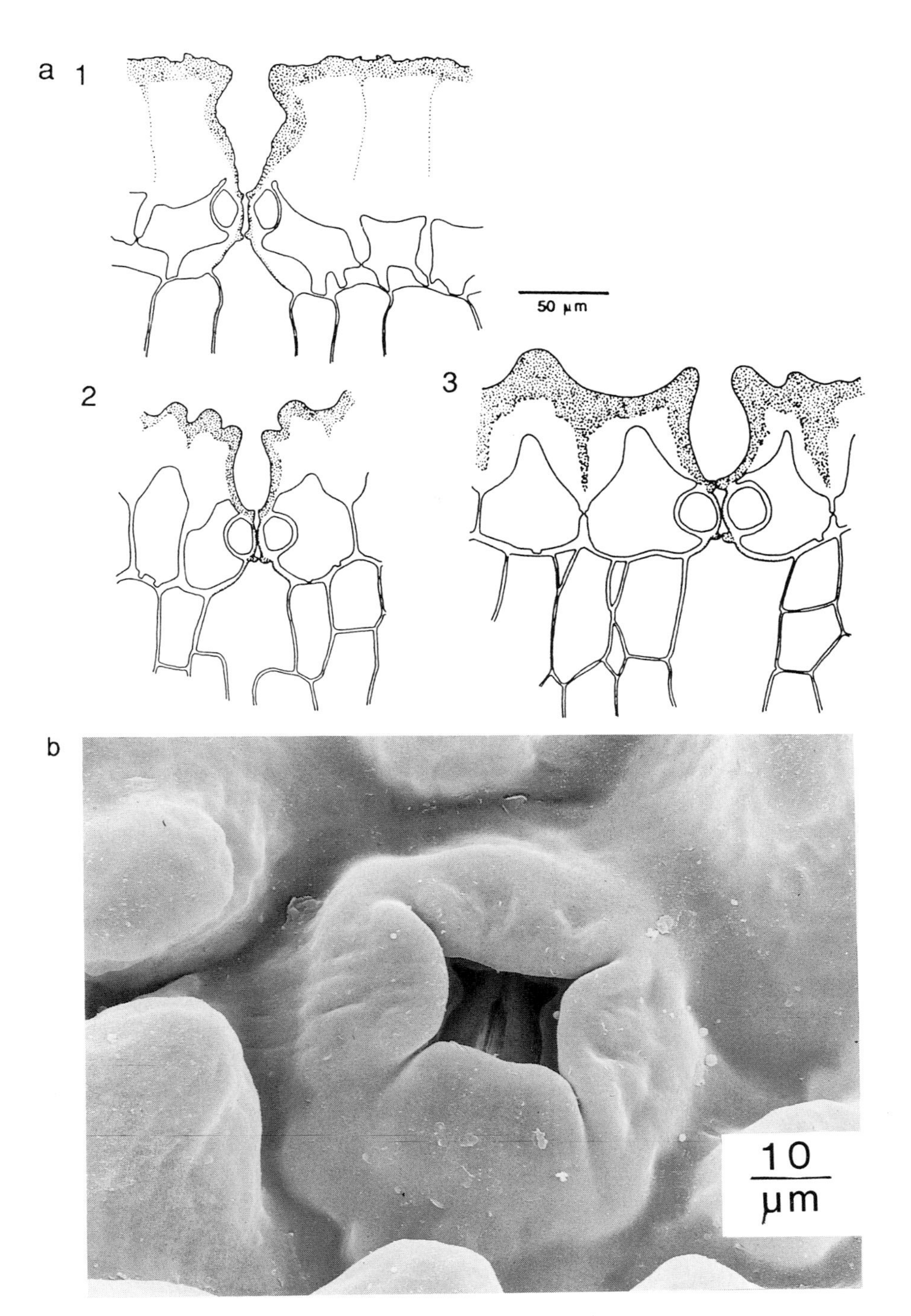

Fig. 1.12(a) Cross-sections through the stomata of three different *Aloe* species (1) *Aloe pillansii*, (2) *Aloe ramosissima*, (3) *Aloe pearsonii*. (b) Stomata of *Haworthia attenuata* with the cuticular rims forming a chimney around the stomatal pore.

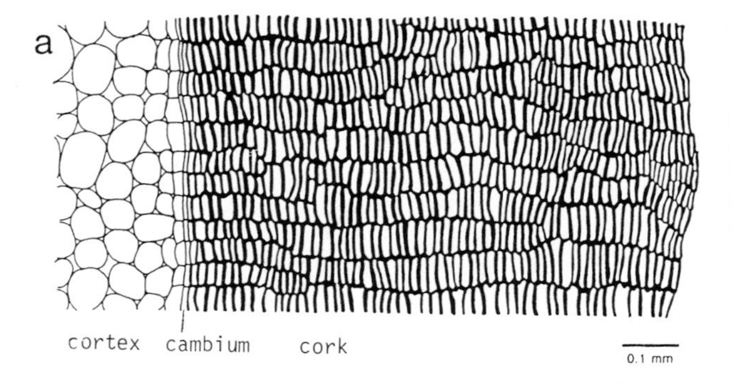

Fig. 1.13 (a) Cross-section through the periderm of *Othonna herrei*. (b) A stem of *Aloe dichotoma* showing the bark of the shoot. Underneath the bark the cortex of the stem is green.

ment of the epidermis with its crucially important stomata by an impervious periderm.

1.2.5 Remarks on the evolution of succulence

When the morphology of succulents is discussed, the morphological convergence in succulent organs between phylogenetically different families is nearly always emphasized. Though the plant kingdom shows convergence all over, it is really eye-catching among the succulents. So much so, that convergence is often presented as a special feature of succulents. If we accept the principle that special features are more apparent the more extreme the environment of the organism is, it seems obvious that morphological convergence is more apparent in succulents than in many other plant types. It suggests that the number of solutions for adaptation to distinct environmental conditions is rather limited as is shown by Rowley (1987) for caudiciform succulents which have evolved in over 100 genera of 37 plant families.

Convergence results from gradual evolution of phylogenetically remote groups under similar selection pressures. Therefore, a comparison of different plant families possessing succulent tissues can be ideal for the study of the evolution of succulent organs. A study on the convergent structures in succulent leaves by Jürgens (1985) pointed out that leaves with a peripheral storage tissue are more frequent than hitherto believed. These peripheral water reserves are located outside the chlorenchyma. They can be formed by single swollen epidermal idioblasts or by a peripheral tissue, often in a uni- or multilayered swollen epidermis.

In a number of important leaf succulent groups of, for instance, Mesembryanthemaceae, Asteraceae and Piperaceae, the peripheral water storage tissues are linked to the central water storage tissue of related members of the same group by a morphological sequence of intermediate forms, raising the question in which direction this morphological sequence might have been developed during evolution.

A central water storage tissue can evolve as a novel structure out of the underlying chlorenchyma via an all-cell-succulent stage by a progressive loss of pigmentation in the central part of the leaf, or it can develop from a peripheral storage tissue by folding up a leaf. During this procedure the originally peripheral water storage tissue becomes enclosed inside the chlorenchyma forming a pseudo-central water tissue. Transitional stages between peripheral and central position of the water storing tissue are best preserved in *Peperomia* (Piperaceae). In this case we observe a translocation of the water storing tissue from the peripheral to a central position which in some cases can still be followed up in the ontogeny of the leaf (Kaul 1977).

For our present purpose it is rather irrelevant, however, to pursue in detail how succulent tissues evolved and from precisely which original tissue. Here we are primarily interested in the ecophysiological aspects of succulence. From that point of view, the fact that a succulent tissue makes utilizable water available to the plant, not the anatomical details, is what matters.

1.3 The distribution of succulents

In a global view succulents are primarily associated with two major types of habitat: dry areas which lack severe frosts during the season of active growth, and salty habitats. We will disregard bulbous geophytes with short-lived, water-rich leaves in this section.

In dry areas succulents can be very common, occurring in high species densities and sometimes reaching amazingly high abundances. Aridity of the habitat does not automatically imply succulence, however. In fact very extensive stretches of dry land lack any succulents. Ellenberg (1981) examined the distribution pattern of tall stem succulents in relation to climate. He found them to occur under a wide range of temperature conditions, even including areas which experience some night frosts. Most characteristic, according to Ellenberg, is that tall stem succulents occur in areas where rainfall is low but some is regularly received. As an index of regularity of rainfall he took the Hellmann quotient, being the ratio of maximum annual precipitation over minimum annual precipitation over a sufficiently long series of years. As such Ellenberg's conclusion substantiates Grisebach's (1872) early hypothesis.

We partly support Ellenberg's conclusions. In our opinion the distribution of succulents in dry areas is limited by two factors: low temperatures during the season of active growth and the length of the dry period. The limitations imposed by low temperatures are apparent from a comparison of the global distribution patterns of temperatures, dry areas and succulents. In the dry areas of the world, where severe night frosts regularly occur till deep into the growing season, e.g. in Central Asia, eastern Siberia, the northern part of the American Great Basin and most of Patagonia, succulents are absent or very scarce. Low temperature limitation becomes even more apparent when one compares the distribution patterns of succulents and non-succulents in mountainous dry areas. On Tenerife, in the Canary Islands, for example, succulents are extremely abundant in the dry, low-lying northwestern part of the island at Teno; they decrease in abundance with altitude, and are virtually lacking on the mountain slopes above the mean cloud line where rather severe night frosts regularly occur throughout the year (Kaemmer 1974). Similarly in Namaqualand the high mountains of the western escarpment support islands of narrow-leaved, sclerophyllous dwarf shrubs and virtually no succulents above the mean cloud (mean night frost) line, while all the remaining, low-lying area is covered by succulents. Again this distribution pattern seems governed by temperatures and Werger (1983a, 1985) stated that succulents become scarce in areas where night frosts of lower than −4 or −5 °C occur during the growing season. Lösch and Kappen (1981, 1983) showed experimentally that freezing point temperatures for leaves of most Canarian succulents, all of the all-cell succulent type, are above −5 °C and only a few species froze at lower temperatures. For all species cold resistance (defined as the temperature causing 10% necrotic leaf area) ranged between −4 and −10 °C. Cold resistance was achieved either through accumulation of osmotics causing freezing point depression, or through tolerance of extracellular ice formation, with species possessing the former mechanism showing the lowest temperature tolerances. Some cacti (partially succulents) have been shown to tolerate temperatures down to −20 °C and the ability to cold harden (Nobel 1982). Thus, very low temperatures outside the season of active growth do not seem to prevent completely the occurrence of succulents, as is shown by the distribution areas of *Sempervivum* spp. in the Alps, *Delosperma nubigenum* in the mountains of Lesotho, some species of cacti reaching Canada or the high Andes, and various crassulacean species in Turkey, Iran and the Himalayas. However, succulents never predominate in these areas.

The length of the dry season certainly is also limiting. Succulents are very common in dry areas where drought conditions are regularly interrupted by weather conditions which allow the uptake of water, either through the roots, or directly from the atmosphere. (We discuss this further in Section 4.2.) Thus, along a climatic gradient at low elevation along which moist atmospheric conditions get scarcer and further apart, many succulents drop out. These are foremost the leaf succulents, while the stem and root succulents continue to occur further along the gradient. However, some leaf succulents, such as species of *Agave* and *Aloe*, are also remarkably persistent along such a gradient.

The frequent interruption of the drought conditions which favours succulents is not necessarily limited to precipitation; it might also be established by dewfall or high air humidities at night. It is somewhat surprising that Ellenberg (1981) found a close correspondence between the predominance of tall stem succulents in an area and the value of the

Hellmann quotient for that area, because that quotient does not express the degree of regularity with which the drought is broken and it is determined by exceptionally rainy as well as extremely dry years.

Also, salty habitats are well known for their abundance of succulents. Such habitats can be coastal as well as inland and can have fine-textured as well as sandy soils or rocky outcrops. We will not deal further with these halophytic succulents. Their physiology and ecology is quite different from non-halophytic succulents discussed in this book.

But succulents are not restricted to the dry and salty habitats discussed so far. There are at least four more habitat types which commonly support succulents, though generally mixed with other types of plants. These include the high, well-lit parts of the canopies of tropical rain forests. Here the exposed branches and stems experience a highly regular and very frequent change from wet to dry conditions, and consequently seem an ideal habitat for succulents. There are many succulent epiphytic species here, belonging to the Orchidaceae, Asclepiadaceae, Rubiaceae, and several other families.

In the temperate zones, sites where ice-cold water seeps to the surface often contain some succulents, e.g. *Saxifraga* spp. in the Alps. Without appropriate research it is risky to explain succulence in this environment. Perhaps these plants have high osmotic values to make up for poor root functioning in this cold environment. Also with high osmotic values much water will be taken into the leaves so that they appear succulent.

Well-drained habitats in the temperate zone, which frequently dry out, such as deep soils of coarse sand, or very thin layers of soil on rocks, also frequently support some succulents, e.g. *Spergula* spp., *Sedum acre*, *Sedum album*, *Sempervivum* spp. In their ecological features these habitats resemble the major dry succulent-supporting habitats of the subtropics discussed above.

Finally, quite a number of species with somewhat succulent leaves occur in the highest vegetation zones near the snow line on high mountains (e.g. *Ranunculus glacialis* and *Primula glutinosa* in the Alps). These are always very short hemicryptophytic or geophytic plants and their leaves, which are only present during summer, are regularly exposed to rather low night temperatures, although they are nearly always covered by thin nightly snowfalls when night temperatures drop considerably below zero. This prevents damage to their leaves. Leaves of *Ranunculus glacialis*, for example, die when exposed to -6 °C. However, they can withstand temperatures as low as -11 °C if their resistance is built up slowly over a period of days. But such low summer night temperatures without snowfall occur only about once in a century at the habitat of *Ranunculus glacialis* at 3000 m in the Alps (Koerner, pers. comm.). Nevertheless, as the leaves meet sub-zero temperatures, they must contain a substantial amount of freezing point lowering substances. *Ranunculus glacialis*, for example, contains 5% of nitrogen (on a dry weight basis) in its leaves and a high amount of soluble carbohydrates. Such substances also have high osmotic values, so that the plant accumulates water in its leaves. Here the plants appear succulent, but succulence has no adaptive value in the sense of storage of utilizable water; succulence here merely occurs as a trade-off of an adaptation against frequently occurring very low ambient temperatures and can be compared with succulence of halophytes.

2

Climate and vegetation of deserts

There is a remarkable variation in the concept of desert adhered to by different people. This seems to be a consequence of the involvement people have with deserts. Desert nomads emphasize somewhat different criteria to oil-drilling engineers, soldiers or ecologists. Nevertheless, there is a substantial overlap between the various concepts of desert, and there clearly are common elements. The most central elements in the various characterizations of deserts concern the harsh, dry climatic conditions and the sparse plant cover. But the limits employed by various people differ. Dry climates as defined by Meigs (1953) and McGinnies *et al.* (1968) cover 26% of the land surface of the world. But all this area is not unanimously conceived as desert. Quite a large area called desert by Americans would, because of its fair cover of vegetation, not be considered desert by South Africans or Australians, who are familiar with similar landscapes in their countries. Instead they would prefer the terms semi-desert or dwarf shrub steppe. Several more examples could be given.

We deal with the characteristic features of climate, soil and vegetation of the whole range of dry areas, be those desert or semi-desert, in this chapter.

2.1 Environmental features of deserts

A lot of argument and scientific effort have been devoted to attempts to present a clear and comprehensive definition of a desert, and to get that definition accepted by the scientific community at large. Much of this energy has been spent to no avail, however, as suggestions did not meet general acceptance, and new definitions are still being proposed quite regularly. The problem with deserts is (and this is rather common wherever attempts are made to define complex concepts), that the essence of a desert is felt to be somewhat different as the type of interest of the researcher in deserts differs, and consequently there is no general agreement on the

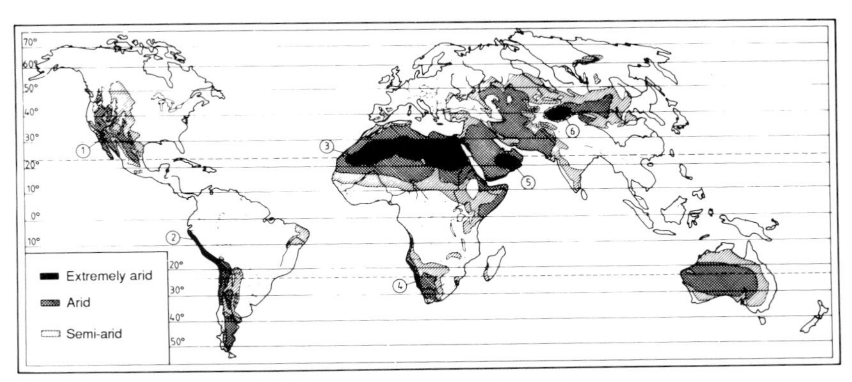

Fig. 2.1 Deserts of the world (after Meigs 1953). Extremely arid areas are: (1) Baja California, (2) Peruvian desert and Atacama, (3) Sahara, (4) Namib, (5) Rub'al Khali and (6) Takla Makan.

critical factors to be included in the definition. For example, to many people deserts are harsh, hot environments that are largely devoid of vegetation, whereas others do not take temperature considerations into account and call the cold, bare areas at high latitudes 'polar deserts', while again others consider the absence of vegetation as sufficient a criterion and even characterize the unvegetated centres of our bigger towns and cities as 'asphalt deserts'. Like many ecologists before us, we shall not attempt to formulate an air-tight definition; we prefer to take the practical approach and regard deserts as those parts of the globe that are mapped by Meigs (1953) (Fig. 2.1) as semi-arid, arid and extremely arid, and that can be loosely though satisfactorily defined solely by mean annual precipitation (Noy-Meir 1973, McGinnies 1979, Shmida 1985):

> Semi-arid areas receive from 150–250 to 250–500 mm precipitation per annum,
> arid areas receive from 60–100 to 150–250 mm,
> extremely arid areas receive less than 60–100 mm.

The higher values in each limit refer to areas of subtropical summer rainfall where evaporation during the growing season is high.

This classification has proved reasonably practical as a tool to correlate precipitation patterns with vegetation patterns as well as with patterns of agricultural practices. If we disregard polar deserts, we may say that water availability is the pre-eminent controlling factor in deserts. Consequently, Noy-Meir (1973) defines a desert ecosystem as a water-controlled ecosystem with infrequent, discrete and largely unpredictable water inputs. For a finer characterization Cloudsley-Thompson (1977), based on the work of

McGinnies *et al.* (1968), used the causes of the prevailing aridity as criteria. Accordingly, he distinguished between

(1) subtropical deserts which largely result from persistent high pressure cells and lie within the 30° N and S parallels of latitude (e.g. parts of the Sahara and the Near East, Thar, part of Australia);
(2) cool, coastal deserts which are due to the combined effects of descending high-pressure air and the proximity of cool ocean currents (e.g. Namib, Atacama);
(3) rain-shadow deserts which result on the leeward side of high mountain ranges that cause orographic precipitation on their windward sides (e.g. Mojave, North American Great Basin, Monte, Patagonia, southwest Madagascar);
(4) interior continental deserts which are arid because of their great distance away from the maritime, moist air and their position in the global wind system (e.g. Takla Makan, Gobi, inner Australia, part of North America, Karoo and Kalahari).

The availability of water can conveniently be characterized as the pre-eminent force regulating desert ecosystems but, of course, several environmental factors contribute to it. These include climatic factors such as rainfall, fog and dew, radiation, air temperature, and air humidity, as well as soil and hydrological factors (Table 2.1). In this chapter we discuss these factors and show how they interact and make deserts a harsh environment for plant life.

2.1.1 Water

Water for plant life in deserts can be provided in the form of soil and subsoil moisture or through rain, fog, dew and high air humidity. We have seen that the annual amount of rain is used to classify desert areas. Of course, the total annual amount of rainfall has little meaning in terms of water availability. A hundred millimetres of rain falling within two months with at least 10 mm at each event will give rise to a comparatively luxuriant vegetation whereas the same amount, but falling in 2 mm batches once a week throughout the year, will support only sparse vegetation cover. The amount of rain falling per occasion and the distribution over the year are much more important than the total amount per year. According to Meigs (1966), parts of the southern Namib desert receiving less than 70 mm annual precipitation are extreme deserts. However, despite the limited amount of rainfall, part of this area carries a rather luxuriant vegetation, mainly made up of succulents. To the casual visitor it never meets the criterion of an extreme desert. Apart from the temporal distribution of the rain, water availability at any one site is also determined by edaphic and topographic conditions. They can lead to a highly non-uniform distribution of water.

Rain can fall predominantly in a definite season (or seasons) or completely erratically. But even in deserts with seasonal rainfall the event of rain is highly unpredictable in time and in amount. In fact, areas with less than 700 mm annual rainfall show a linear correlation between year-to-year variation and the total amount of rainfall (Noy-Meir 1973). The term variability coefficient is often used to describe the year-to-year variation in rainfall. This coefficient, defined as the maximum amount of annual rainfall divided by its minimum over a sufficiently long period of years, is called the Hellmann quotient. It normally ranges from about 6 to 20 but can exceed 100 in extreme deserts. In general, the variability in annual rainfall is lower in deserts with winter rainfall than in deserts with summer rainfall. It is obvious that 100 mm of rain in the hot summer season will not have the same ecological relevance as the same amount in the cooler winter season when, due to the lower air temperatures, the evaporative demand is much lower. This implies that the time of rainfall affects the resulting vegetation. Generally, winter rainfall produces a denser vegetation than the same amount of summer rainfall. We see this well illustrated in the Namib desert, where the southern parts, receiving winter rainfall, support a far more luxuriant vegetation, mainly of succulents, than the corresponding northern parts with equal amounts of summer rainfall. The variability in rainfall over time is largest in desert areas without a distinct rainy season. This is especially clear in deserts where a transition area separates a winter rainfall area from a summer rainfall area. Off-seasonal rains can also occur in pronouncedly seasonal deserts, however, though they seem to be more common in summer rainfall deserts than in winter rainfall deserts. Truly seasonal deciduous species will not grow new leaves, and seeds of annuals will not germinate if the rain falls in the wrong season. One impressive example for this is the Sonora desert which receives both summer and winter rainfall and carries a distinctive set of annuals solely depending upon the season the rainfall occurs (Mulroy and Rundel 1977, MacMahon and Wagner 1985). Winter rain mostly falls in soft drizzles or light showers whereas summer rain falls in heavy showers, thunderstorms or downpours. Both have in common that they are metereologically caused by local cells resulting in a high variability in precipitation over short distances.

Of great importance for the vegetation are the length of the rainy season as well as the length of the time gap between two effective rain falls. A

Table 2.1. *Summary of the environmental characteristics of the world desert regions and the different factors contributing to their existence. (After Meigs 1953 and Shmida 1985.)*

Continent	Altitudinal range (m)	Rainfall: winter	Rainfall: summer	Rainfall: erratic	Cold offshore current (fog)	Degree of aridity: semi-desert	Degree of aridity: true desert	Degree of aridity: extreme desert	Mean temperature °C: coldest month	Mean temperature °C: warmest month	Main vegetation types
North America											
Great Basin	800–2000	+	+			++	+		<0	10–20	Artemisia–Atriplex dwarf shrub steppe and pygmy desert woodland
Sonora	0–1200	+	+			++	+		10–20	20–30	thorny succulent shrubland and savanna
Chihuahua	700–1600		+			++	+		10–20	20–30	open shrubland and thicket
Baja Calif.	0–500	+	+		+	+	+	(+)	10–20	>30	thorny succulent shrubland and dwarf shrub steppe
South America											
Patagonia	0–700		+			++			0–10	10–20	perennial grass steppe
Peru–Chile	0–900			+	+	+	++	+	10–20	20–30	succulent desert dwarf shrubland
Monte	200–1500		+			++			10–20	20–30	thorny succulent shrubland and savanna
Asia											
Gobi	500–1800		+			++	++	(+)	<0	20–30	Chenopod–Tamarix desert shrubland
Takla Makan	800–4300		+			+	++	+	<0	20–30	Chenopod dwarf shrub desert

Turkestan	0–800	(+)	+			++	+		< 0	20–30	Artemisia–Stipa steppe and pygmy desert woodland
Iran	500–3300	+				++	+		0–10	20–30	Chenopod–Artemisia dwarf shrub steppe and pygmy desert woodland
Thar	0–800	(+)	+			+	(+)		10–20	> 30	thorny-rattanoid savanna
Syria	200–800	+				+	++		10–20	> 30	Artemisia–Chenopod dwarf shrub steppe and desert
Arab	0–1200			+		(+)	+	+++	20–30	> 30	Chenopod–Zygophyllum dwarf shrub desert, Artemisia and grass on sands
Africa											
Sahara	0–800	(+)		+		(+)	+	+++	10–20	> 30	Chenopod–Zygophyllum dwarf shrub desert, Artemisia and grass on sands
Sahel	0–1300		+			++	(+)		20–30	> 30	thorny-rattanoid savanna
Somalia	0–900		+	(+)		+	+	+	20–30	> 30	thorny savanna and dwarf shrubland
Namib	0–1100	+	+	+	+	+	+	+	10–20	10–> 30	succulent dwarf shrubland and desert
Kalahari	500–1500		+			++	(+)		10–20	20–> 30	thorny-succulent savanna
Summer rainfall Karoo	600–2000		+			++	(+)		0–10	20–30	asteraceous dwarf shrubland and open thorny shrubland
Western Karoo	0–1400	+	(+)		+	++	(+)		5–15	15–25	succulent dwarf shrubland
Australia	0–700	+	+			+	+		10–20	20–30	Chenopod shrubland and sclerophyll evergreen low woodland

botanically effective rainfall is an amount that is sufficient for a water input into the root zone such that the perennial vegetation benefits. Such an effective rain will also allow seed germination and initial seedling survival. The significance of the length of the gap between two effective rainfall events, particularly for succulents, will be outlined in Section 4.2.

The precipitation of soft drizzles will easily penetrate into a dry soil. Heavy rains are relatively less effective. The sudden large amounts of water cannot be soaked up by the soil quickly enough and much water disappears as run-off. Especially silty and loamy soils tend to seal by generating a crust with the first droplets of rain which then prevents penetration of rain and prohibits establishment of seedlings. On gentle slopes and with limited amounts of precipitation, run-off in thin sheets or very shallow runnels leads to run-on in nearby sites, locally increasing the available water. Such patterns can commonly be seen in greatly different plant density patterns in desert vegetation. With more rain or steeper slopes, run-off water gathers in small creeks that can fuse to form complete draining systems or in depressions forming shallow ponds and lakes. After complete evaporation of the water in the ponds, a crust of loam or salt is left behind.

As the creeks fuse their capacity increases. Fusing creeks (or wadis) are ranked: the initial, shallow creeks are called wadis of the first order, they fuse to wadis of the second order, and so on. There is an inverse relationship between the frequency with which the wadis carry water and the amount they carry. This results from the increasing capacity of wadis as they get bigger and from the frequency distribution in the size of desert showers: as we have already pointed out, small showers are far more frequent than downpours. This gradient in water frequency and water volume carried by the wadis is reflected in the vegetation they support along their edges. In extreme deserts, generally, first-order wadis support only opportunistic annuals or ephemerals, second-order wadis support larger perennial hemicryptophytes and an occasional dwarf shrub, third-order wadis support larger dwarf shrubs, fourth-order wadis support dwarf shrubs and shrubs, and trees are usually found only along wadis of the fifth and higher order, (Shmida 1985).

In deserts the amount of rainfall per occasion is rarely high enough to generate sufficient penetrating water to reach the water table. In topographic depressions where greater amounts of run-off water gather or in river beds that flood after a rain storm this is more common. Nevertheless, the water table in deserts often is not very deep. In several cases subsurface water currents occur. They can easily be spotted from deep rooting trees which exactly follow the current. An example of such rows of trees in a desert plain is presented in Fig. 2.2.

Fig. 2.2 Rows of *Acacia* trees indicating underground watercourses, on the plains of the Central Namib north of Gobabeb.

Fog is a very special phenomenon in deserts. It occurs regularly in the Namib, the Peruvian and Chilean desert and in Baja California. The water input from fog varies considerably depending on the duration and the intensity of fog. In the Central Namib we often experienced an advective fog with definite droplets which precipitate like a soft drizzle. Nevertheless, maximum observed precipitation on a horizontal plate was 0.7 mm within 6 hours of heavy fog. Much higher amounts of water can be obtained if the surface is vertically exposed in the advective fog. Then droplets condense at the surface and drop down. The effectiveness of this interception can be markedly increased if the surface is split, like the branches of a shrub or tree. Such obstacles comb the fog out of the air and substantial amounts of water can be gained. An impressive example of this is the Loma-vegetation in Peru which combs out such quantities of water that the soil underneath the shrubs and trees is effectively wetted down to the rooting horizon (Rauh 1985). It is very difficult to measure properly the possible water gain from fog. The reported values vary, depending on the method used, from 40 to 300 mm per year, with much higher values definitely occuring naturally for the Loma.

Generally, fog occurs more frequently during the wet season. Although precipitation from fog can exceed the amount of annual rainfall at a site, its effectiveness for higher plants is questionable. Only lichens and some

Fig. 2.3 Lichens in the northern Namib desert. (a) 'Strauchflechten' on the bare soil, (b) 'Krustenflechten' on stones, (c) detail of a stone which is completely covered by these lichens.

phanerogamous epiphytes have proved to be able to use fog efficiently as a source of water (Kappen *et al.* 1980, Benzing 1984). In parts of the Namib with frequent fog and dew, rocks and stones are completely covered with lichens (Fig. 2.3).

Higher plants are assumed to make only limited profits from fog. The advantage presumably mainly lies in a temporary relief of the harsh environmental conditions. Only in a few cases will the dripping water intercepted from the air be sufficient to reach the root zone.

Another phenomenon in many deserts is dew. With the exception of the Negev (Israel) there is little knowledge about the frequency and the amount of dew. At low altitude the probability of dew is inversely correlated with the distance from the coast (see Fig. 3.4). In contrast to rainfall, the reliability of dewfall is much higher. Over a period of 22 years the number of days with dewfall in the Negev ranged from 157 to 216 per year, which means a variability coefficient of as low as 1.4. The average annual amount of water gained by dew equalled 33 mm precipitation which corresponds to an average amount of 0.18 mm per dew night (Evenari 1985a). Such a value corresponds with reports on the amount of dewfall in other deserts and with measurements in the Namib (Eller *et al.* 1983).

There is still a controversial discussion going on as to whether or not higher plants can use dew in a direct way by taking up the water through the surfaces of above-ground organs. Only a quantitative uptake would yield a substantial improvement of the water balance and this is in fact very unlikely. We will come back to this problem in Section 4.2 when we discuss the water absorption by above-ground plant surfaces.

2.1.2 Radiation

In the deserts under the Tropics of Cancer and Capricorn (between 20° and 30° latitude, Fig. 2.1) a dry and for most of the year cloudless atmosphere (Fig. 2.4) leads to an extremely high radiation climate.*

The sun radiates with a power of about 1370 W m^{-2} towards the earth (Neckel and Labs 1984). On the way through the atmosphere this so-called extraterrestrial solar radiation, which is a nearly parallel beam, is modified. A fraction of this beam penetrates the atmosphere in a straight way and reaches the earth's surface directly. Therefore it is termed direct solar radiation. Another fraction is reflected or scattered by the molecules and the aerosols of the atmosphere. It is either lost by reflection to outer space or it is incident on the earth's surface in the form of the so-called diffuse solar

* SI terminology and units are used throughout this book. For photon fluxes the recommendations of Bell and Rose (1981) are accepted.

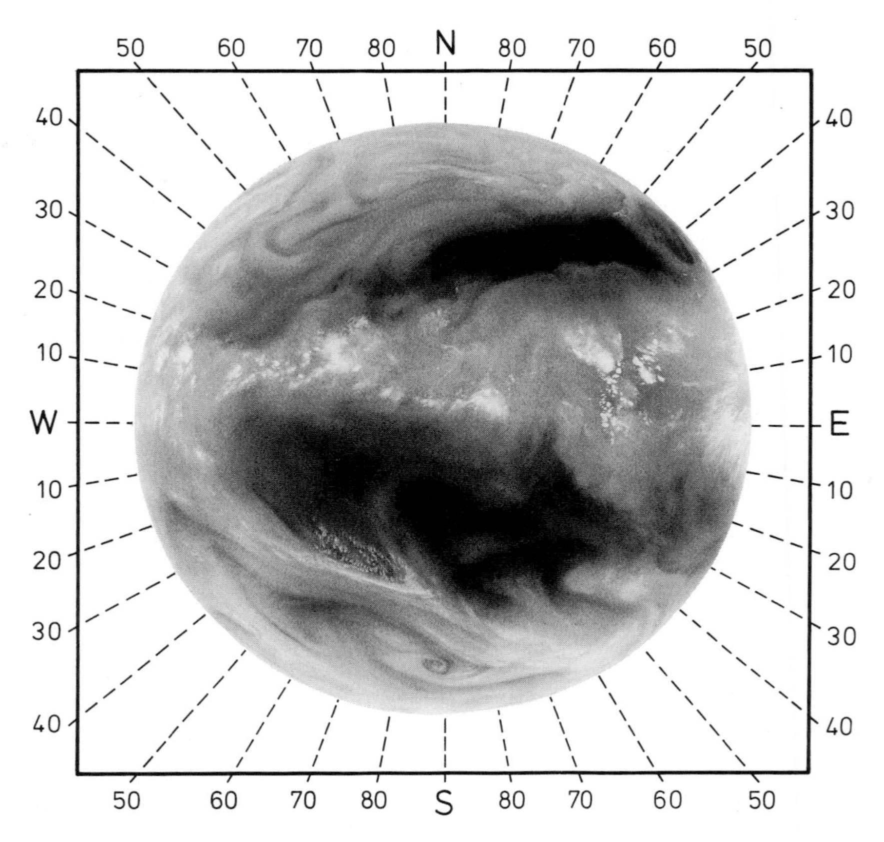

Fig. 2.4 Water vapour content of the upper troposphere (5–10 km). The dry air masses of the subtropics (black areas) and the tropical thunderstorms along the equator (white spots) are clearly visible. The scale indicates the geographical latitude. The picture was taken from the METEOSAT weather satellite on 14 August 1984 at 11:55 Greenwich Mean Time in the water vapour channel (5.7–7.1 μm). The position of the satellite is at 0° longitude above the equator. (With kind permission of the European Space Agency.)

radiation, frequently also termed sky radiation. On the earth's surface a locality of a defined surface dimension can receive additional solar radiation by the reflection of direct or diffuse radiation by clouds or structures of its environment, e.g. nearby hills, rocks or vegetation. The sum of direct, diffuse and reflected solar radiation is the real amount of solar radiation energy impinging on a given surface. It is termed global radiation. Its energy consists of the wavelength range of about 300 to 2500 nm. However, the total amount of energy and the spectral distribution of the global

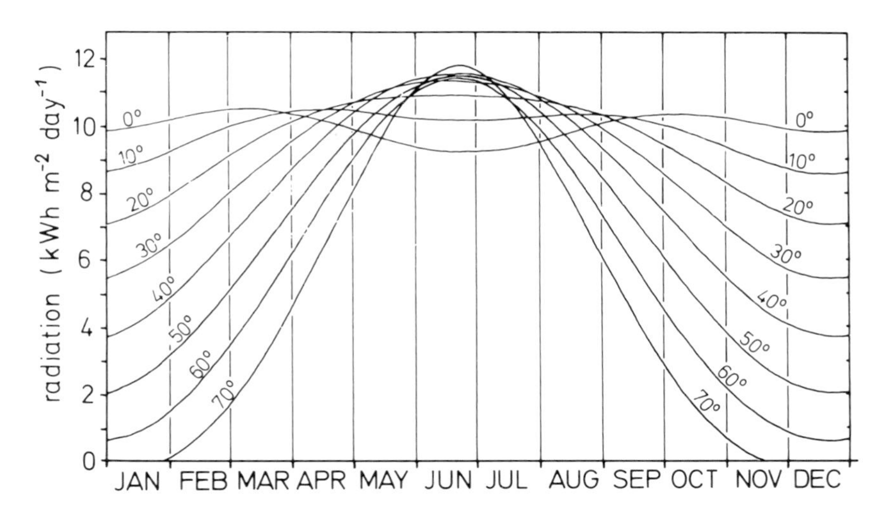

Fig. 2.5 Annual variation of the daily sums of global radiation for different geographical latitudes. (After Schulze 1970.)

radiation varies considerably according to the atmospheric conditions with the daily course of the sun along the sky. At clear sky conditions the maximal values of global irradiation around noon can reach values of 950 W m^{-2} in deserts; under certain conditions, e.g. a sky with white cumuli, even values as high as 1200 W m^{-2} can occur (for details see Section 4.1.3).

On its way through the atmosphere the solar radiation is attenuated and only a fraction of the extraterrestrial solar radiation energy reaches the earth's surface. Schulze (1970) calculated that in extreme deserts only 60 to 70% of the extraterrestrial radiation arrives at the earth's surface (yearly average). This value is higher than in the tropics (50 to 60%) where the annual insolation is reduced by the frequent presence of clouds. Of the global radiation energy impinging on the earth's surface a part is reflected to the atmosphere or to outer space; the rest is absorbed by the earth's surface structures and eventually converted to heat. The reflective power of the earth's surface is called albedo, a term also used to characterize the reflective properties of smaller fractions of the earth's surface, e.g. a cultivated area, a sand dune or stand of vegetation.

Global radiation varies with the seasons and also with geographical latitude. Fig. 2.5 gives an overview of these variations. The solar energy input to deserts situated under the tropics (between about 20 and 30° latitude) varies from 7 to 11 kWh m^{-2} day^{-1}, or ±22% around a mean value of 9 kWh m^{-2} day^{-1}.

We must emphasize that only the part of global radiation that is absorbed by a structure itself, whether or not this structure is a living plant or a dead rock, is the actual energy input from solar radiation to that structure. Consequently, apart from the amount of insolation, the degree of the absorption of a plant is important for this energy input. Thus, determination of the plant's absorption properties and their relevance with respect to the global radiation load is important. This will be discussed in detail in Section 4.1.

Apart from the solar radiation, other radiant energy fluxes occur. This fact is very often overlooked, even by plant ecologists. Basic physics teaches that each substance with a temperature above 0 K (−273.16 °C) radiates according to the Stefan–Boltzmann law (for details see Section 4.1.3). In the biologically relevant range of temperatures of life this radiation is thermal (or infrared) radiation with radiant energy of wavelengths from 2 μm to about 40 μm. A plant surface with a temperature of, for example, 20 °C emits approximately 400 W m^{-2}. The atmosphere also emits thermal radiation, often termed atmospheric reradiation, but under clear sky conditions this radiant energy is substantially lower than that of the earth's surface at the same temperature. Therefore the radiation emitted from the soil or the plants is higher than the thermal radiation received from the sky, and radiation budget considerations reveal that during clear nights a net loss of energy (net radiation loss) from the earth's surface to the atmosphere occurs. This radiant energy loss causes a considerable lowering of the temperature of the earth's surface and the air layer adjacent to it. The prevailing clear sky conditions of deserts together with the high solar radiation input during day and the net thermal radiation loss during night account for the great daily air and surface temperature oscillations. This daily change from hot to cool (or even cold) temperature conditions is supposed to have had an important influence on the evolution of the particular adaptations of plants and animals for their existence in deserts and needs our attention in the subsequent chapters.

2.1.3 Temperature

The monthly mean temperatures correlate with the pattern of the development of plant life in deserts as well as in other habitats. But of greater importance are the absolute maxima and minima. They can determine survival or death. The annual variation in monthly mean temperatures is characteristic for the different deserts. In hot deserts the value of the hottest month generally is 20 to 30 °C or more, but the annual variation can be different from one hot desert to another. Frequently the coldest month has mean temperatures of 10 to 20 °C, e.g. in the Sonora or the Sahara (Table 2.1). The annual amplitude of the monthly mean temperatures is about 20 K but in some hot deserts the variation is smaller. The Arab desert has high values throughout the year and the annual amplitude may be less than 10 K (Table 2.1). This variation is similar to that observed in the Namib desert which, however, has a value of 20 °C for the warmest month instead of more than 30 °C in the Arab desert.

Even more pronounced than the annual variations in air temperature are the daily fluctuations which result from the actual weather conditions and geographic features. The distance from the ocean, the presence or absence of cold currents in a nearby ocean, the altitude and the orographic situation of a habitat determine to a large extent the amplitude and the frequency at which daily variations occur in a particular habitat. Diurnal changes in the direction of winds or sudden changes in the prevailing wind direction (e.g. from a cool southwesterly wind to a hot wind blowing from the east, as is often observed in the Namib desert), can also markedly influence the diurnal variation of air temperature. The pattern in the diurnal variation in air temperature in a desert environment is similar to that of a sunny day in other regions on earth. Yet, in deserts the temperature amplitude is greater and the maxima can be very high (Fig. 2.6), because in deserts weather conditions with a clear sky can persist for weeks or months. The energy input through solar radiation varies with the orographic situation and the exposition of a plant's habitat, so that only one generalizing statement can be made, namely that under clear sky conditions a daily amplitude in air temperature of 20 to 30 K is common in deserts.

The heating and cooling of the lowest part of the atmosphere is caused almost entirely by the flow of sensible heat to and from soil and the vegetation. Heat transfer from surfaces to the air or vice versa is by convection, i.e. by local winds, eddies or local slow air movements (see Section 4.1.5). The nearly complete absence of an energy consuming evapotranspiration means that the solar energy absorbed by the surface of a desert is entirely converted into sensible heat. Moreover, in a dense and well-structured vegetation cover the solar energy is absorbed at different levels above soil surface, resulting in a spaced distribution of the heat energy released. This energy input gives way to an increase of temperature of the vegetation, the soil and their surrounding air, or in other words heat energy is used to warm up a large structured and spaced mass. In deserts which are almost bare of vegetation the same amount of solar energy is converted to heat either on a soil surface or a sparse vegetation with little vertical extension. This results in a 'compression' of the heat energy to the

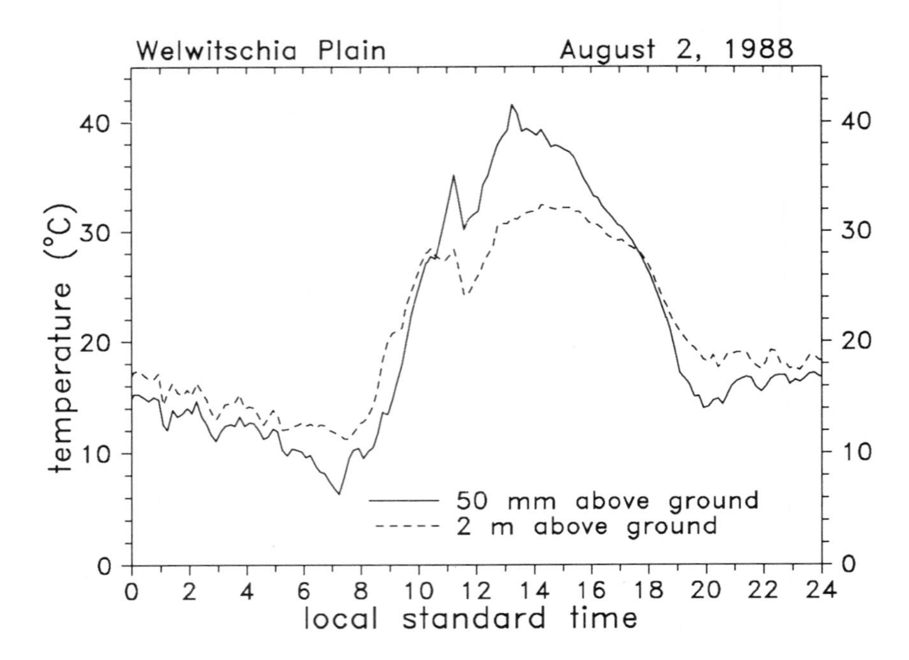

Fig. 2.6 Diurnal variations of the air temperature in the Central Namib desert on a day with clear sky. Amplitude of air temperature is 21 K at 2 m height and 38.5 K 50 mm above soil surface.

small volume of the soil surface and its boundary layer of air and a contingent small vegetation cover. Thus the same amount of heat energy will, in deserts, cause higher surface and air temperatures than in the more structured vegetation cover of humid regions.

The variation in soil surface temperature depends on the varying radiation energy fluxes to or from the soil. The pattern of variation is similar to that of bare soils in other regions but differs in magnitude and maximum and minimum values reached in a 24-hour period. A certain amount of the heat energy released at the soil surface is transferred to the air by convection. From the remainder a substantial amount flows to deeper soil layers by heat conduction. Therefore during daytime the temperature of the upper soil layers (to about 300 mm depth) follows the soil surface temperature. The pattern in the diurnal course of the soil temperature at different depths is very uniform but the maximal and minimal temperatures are reached later in the day at greater depths. The length of this time lag depends on the thermal conductivity of the soil and the actual temperature gradient within the soil. It can be very important for plants. The highest

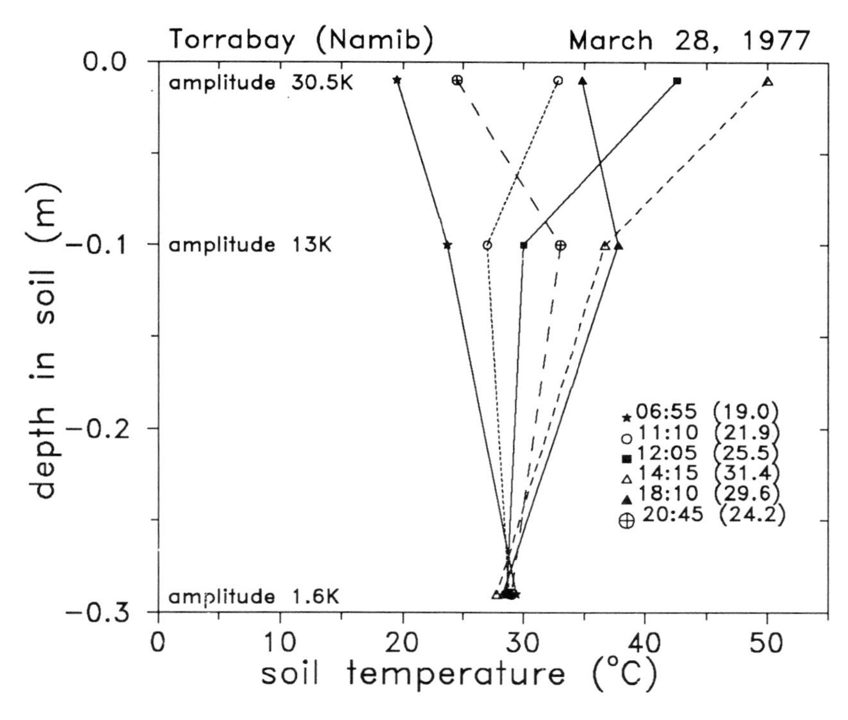

Fig 2.7 Daily variation of the temperatures of bare soil in a habitat of *Welwitschia mirabilis* in the Namib desert on a day with clear sky. Symbols indicate the temperature profiles measured at different times of the day as indicated; air temperatures at 2 m above the ground at those times are given in parentheses. The daily amplitudes of soil temperatures are 30.5 K at 10 mm, 13 K at 0.1 m and 1.6 K at 0.3 m depth.

temperatures in the rooting zone can be reached hours later than at the soil surface. In Fig. 2.7 an example of such a daily variation in soil temperatures is given. During the late afternoon and the night the soil temperature at 0.1 m depth can be considerably higher than close to the soil surface (e.g. 8.5 K at 20:45). Fig. 2.7 also confirms that even in desert soils bare of vegetation the temperature at depths greater than 0.3 m is almost constant over the time period of a day.

For plant organs, especially leaves, the origin and the effect of the heat energy are the same as for the soil. Consequently the pattern in diurnal variations is the same, but firstly, the absorption of solar radiation of a plant's surface can be lower or higher than that of the inorganic environment, secondly, the plant can receive less radiation due to a different exposition angle with respect to the radiation incident on its surface, and thirdly, the convective energy dispersion can be much better. A detailed presentation of the energy fluxes which determine the actual temperature of a leaf will be given in Section 4.1.

2.1.4 Air humidity

For most people deserts are always hot and dry. However, anybody who has ever spent a night outside in the desert is forced to revise this oversimplification at the latest at the end of the night when it becomes very cool and sleeping bags become soaked by dew. One quickly realizes that the air contains a considerable amount of moisture and that air humidity is somehow connected with air temperature.

Discussion of air humidity means in most cases a presentation of figures for relative air humidity. This value is termed relative because it gives the actual value of air humidity as a fraction of the maximally possible content of humidity (in the form of water vapour) of the air. This humidity at saturation, mostly expressed as saturation vapour pressure of moist air, depends on the air temperature as shown in Fig. 2.8.

For example, at an air temperature of 20 °C the saturation water vapour partial pressure e_s is 2.337 kPa (Fig. 2.8, P). The actual moisture in the air is expressed as actual water vapour partial pressure e_a; suppose that it is 1.1877 kPa in our example (Fig. 2.8, P_1). Thus the moisture content at 20 °C air temperature is, according to the formula

$$rh = \frac{e_a}{e_s} 100 \, (\%) \tag{2.1}$$

equal to 50.8% relative air humidity (rh). The saturation deficit, for which in plant ecology the abbreviation VPD (vapour pressure deficit) is in general use, is for standard pressure conditions calculated after the formula

$$\text{VPD} = e_s - e_a \, (\text{Pa}) \tag{2.2}$$

Whereas the value of the relative air humidity stands for the amount of water that is actually borne by the air, the term VPD characterizes the remaining capacity of the air for water vapour. The saturation deficit or VPD is a very important concept for plant ecophysiologists; its value determines to a large extent the magnitude of evaporation from a wet surface and therefore also the transpiration of a plant (see Section 4.2).

If saturated air (100% rh) is cooled, a fraction of the water vapour of the air is condensed to water droplets and eventually to dew, because air at a lower temperature can only contain a smaller amount of water vapour. In our example, air with a temperature of 20 °C and a relative humidity of 50.8% (Fig. 2.8, P_1) has an actual water vapour partial pressure of

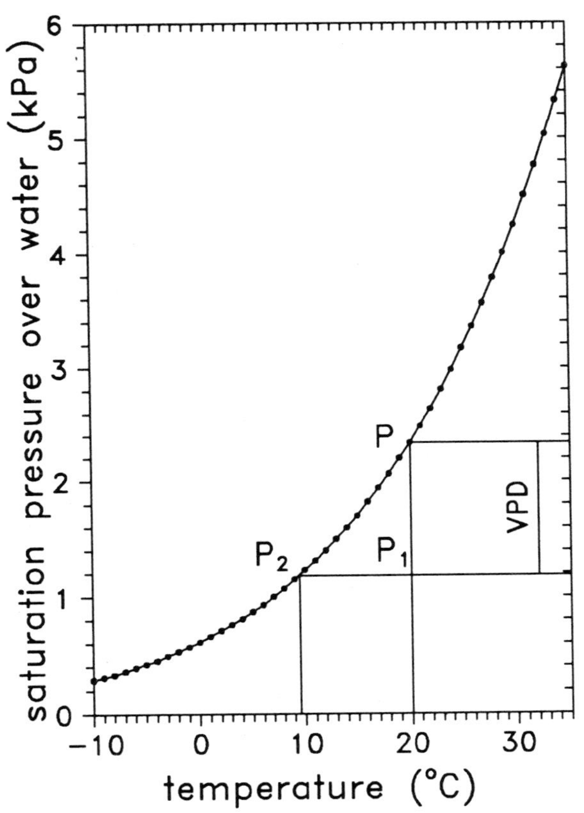

Fig. 2.8 Water vapour saturation pressure over water. P = moisture content of the air at saturation; P_1 = actual moisture content of the air; P_2 = dewpoint temperature of the air; $(P_1/P)100$ (%) = relative air humidity; VPD = water vapour partial pressure deficit.

1.187 kPa. This is equal to the saturation water vapour pressure at 9.5 °C air temperature (Fig. 2.8, P_2). Thus if our air is cooled, e.g. during a clear night, dewfall occurs at 9.5 °C and below. The temperature to which an unsaturated moist air must be cooled down to reach saturation is called dewpoint temperature. It is a figure which, in combination with the value of the actual air temperature, characterizes the actual water content of moist air better than the term relative air humidity.

Air with a relative humidity of 30% is assumed to be dry. Under those conditions a high transpirational water loss by plants is assumed. On the other hand, air with 75% relative humidity is considered fairly wet and a much lower transpiration by plants is expected. However, the assumptions are only valid if the air has the same temperature for both relative humidities, but are completely wrong if the air in both cases is at different temperatures. Let us assume that the air with 30% relative humidity has a temperature of 15 °C. Then the VPD is (according to Fig. 2.8 and equation (2.2)) 1.193 kPa and its dewpoint temperature is equal to −2.45 °C. Taking a temperature of 32 °C for the air with 75% relative humidity, the values for the VPD and the dewpoint are 1.189 kPa and 27 °C respectively. Since the values for the VPD show only a marginal difference, the evaporative demand of the air of 32°C and 75% relative humidity is the same as for the air of 15 °C with a relative humidity of only 30%. This result demonstrates clearly that only the VPD or the dewpoint temperature characterizes the evaporative demand of the ambient air.

Condensation of air moisture does not only occur if the total ambient air is cooled but also if the small boundary layer of air adjacent to a cool surface is cooled below dewpoint temperature. The moisture of the boundary layer air is then condensed on the cold surface. At clear sky conditions during the night the soil surface, rocks, pebbles, leaves of a plant or the outside of a sleeping bag can reach temperatures below the dewpoint of the ambient air. This results from a negative thermal radiation budget (see Section 4.1.4). Depending on the magnitude of the thermal radiation budget of different surfaces the dew formation can be very local and vary in time and amount of dew formed. Little is known to what extent such locally different water vapour condensations contribute to the patchiness of desert vegetation.

The dry soil and sparse vegetation of deserts add little or nothing to the absolute moisture content of the air; this is contrary to the effects of vegetation of humid areas. Therefore the absolute moisture content of the air is not or is only slightly changed during a 24-hour period. In Fig. 2.9 plots of the air temperature and the relative air humidity measured at Numees (Richtersveld, southern Namib desert) on 19 March 1981 are given together with the calculated values for the dewpoint temperatures and the VPDs.

The sky was clear during that day with a moderate wind from the southwest. During the night the net radiation loss cooled the air at the earth's surface down to nearly dewpoint temperature and thus a very high relative air humidity of more than 90% was measured. The evaporative demand (VPD) approached zero. During daytime the high solar energy input heated up the air to 30 °C; a concomitant increase of the VPD to 2.631 kPa and a decrease of the relative humidity to 38% was observed. However, the dewpoint temperature was almost constant during the 24-hour period

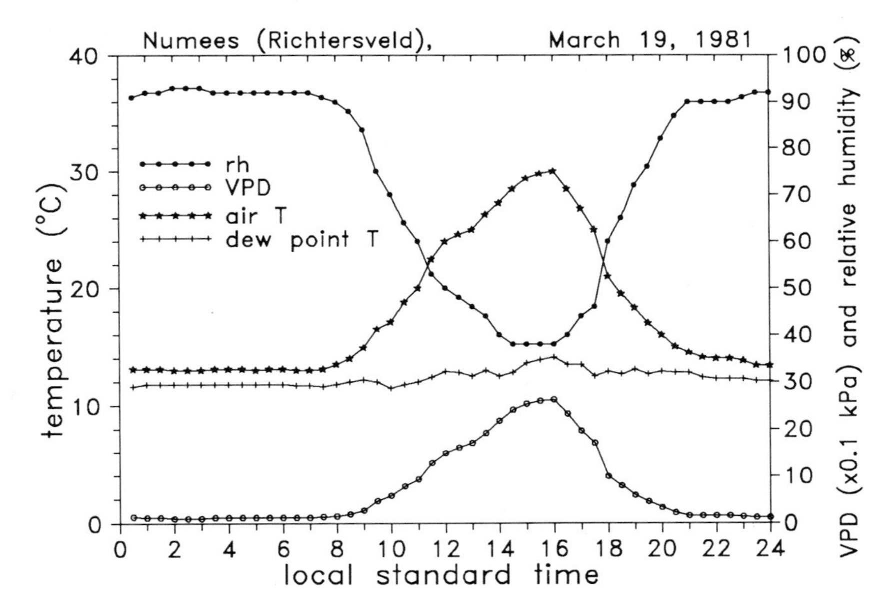

Fig. 2.9 Variation of the temperature (*T*) of the ambient air and the dew point and the resulting values for relative air humidity (rh) and water vapour partial pressure deficit (VPD) under clear sky conditions in the southern Namib desert.

giving evidence that the absolute moisture content of the air did not change substantially and that the changes in the VPD and the relative humidity were only due to changes in air temperature. In the southern Namib the summer is the dry season and in March the sparse vegetation cuts down transpirational water loss to an absolute minimum. The soil was completely dry and therefore even a hot air during the day could not increase the dewpoint of the air by water evaporating from the soil or the vegetation. Under such conditions during clear nights the air temperature can fall below dewpoint temperature and thus cause dewfall whereas during the day the high temperatures are responsible for a low relative air humidity and a high VPD or evaporative demand.

2.1.5 Soil

In most deserts the prevailing impression of the soil is that there seems to be no real soil but only rocks, gravel or sand. Accordingly deserts are often classified as hamada (rock-), serir (gravel-) and erg (sand-desert). Desert soils cannot be compared with soils of more humid regions. They are immature and lack the characteristic profiles. Repeatedly such soils are termed 'non-soils'. Generally, desert soils contain a high amount of inorganic ions, especially calcium. Most soils react in an alkaline way. Some soils are extremely saline and originated from weathering of rocks rich in salt, or accumulated salt from ascending water movements.

Desert soils also have a characteristically low content of dead organic matter (from litterfall, dead roots, etc.) and are dry during most of the year. However, it has been shown that desert soils can house unexpectedly high quantities of microorganisms (Skujins 1984), especially in the rooting horizon and where shed leaves and plant debris are accumulated (e.g. underneath shrubs). This organic matter is decomposed by the microorganisms, firstly to humus and secondly to inorganic substances. The humus fraction of soils in humid climates can exceed 10% whereas soils in arid climates usually have a content of less than 1% in the form of humus. Shreve and Wiggins (1964) reported for the Sonora desert a content of humus of 0.25 to 0.65% and Killian and Feher (1939) for the Sahara gave values of 0.1 to 0.98%. Some of the microorganisms in desert soils, especially those symbiotic with species of the Papilionaceae and Mimosaceae fix atmospheric nitrogen.

After a rainfall in deserts rapid growth of plants, in particular of ephemerals or annuals, demonstrates the ample and instant availability of nitrogen. In humid climates about 98% of the nitrogen of the soil is bound to the organic substances while the inorganic nitrogen is either consumed instantaneously by the vegetation or is lost by drainage during the following rainfall. In deserts the nitrogen-containing ions from microbiological breakdown or from N_2-fixing bacteria accumulate during the dry period. This is the reason why, despite the low content of humus or organic matter, a rainfall can support a quick and vigorous plant growth in deserts.

Mycorrhiza have often been reported from desert plants and in recent times the assumption is strengthened that plants with mycorrhiza are more drought resistant than plants without such symbiotic fungi in the roots.

In arid soils the movement and retention times of water differ significantly from those in humid regions. This is mainly due to differences in soil texture and climate which considerably influence run-off, penetration and evaporation.

In soils with a high water storage capacity (e.g. a high proportion of fine material) a given amount of rain can wet only the uppermost layer, while in a sandy soil with coarse material and low water storage capacity the same amount of precipitation will penetrate much deeper into the soil. If afterwards evaporation dries out the uppermost 2 cm of the soil, this can mean all the water of a loamy soil but only a small fraction of a sandy soil. Consequently, sandy soils are the 'wet' soils in deserts and soils with high amounts of clay are 'dry' soils (Noy-Meir 1973, Walter 1973). Especially in deep sandy soils with no subsurface crusts, the downward moving moisture may be cut off from evaporative forces by the dry layer of sand at the top of the soil. This moisture may be preserved as capillary-bound water at a given depth without further movement. Such moisture has been called ground moisture (Kaiser and Beetz 1919) and is assumed to be responsible for deep-rooting plant growth in many dry areas including the Namib desert.

In relatively flat areas deserts often have a sheet of soil with a foamy structure with air-filled pores near the surface. This sheet can be between a few millimetres and some ten millimetres thick and is covered by a thin crust of tightly packed, somewhat cemented fine soil grains (Volk and Geyger 1970). The origin of this foamy structure is still a topic of debate. The importance of the foamy layer is that it remains absolutely dry after precipitation and prevents the establishment of seedlings. Plants can only encroach on these patches on sites where the foamy structure is interrupted, such as the edges of small flat areas and terraces along rills and runnels or from the edge of a larger boulder.

2.2 Plant life in deserts

Plants of the desert have always fascinated man for several reasons: they exhibit a great variety in growth form, they have strongly contrasting phenologies between seasons or in response to rainfall, they can show large fluctuations in abundance in response to events of rainfall, and they frequently grow in conspicuous patterns. These features often are interpreted as adaptive in the usually harsh and mostly unpredictable desert environment. Features are considered adaptive if they are beneficial to the plants possessing them, e.g. for survival during the unfavourable season or for adequate exploitation of the resources during the short favourable episodes in the deserts. Thus, morphological, anatomical and ecophysiological as well as demographical features may be adaptive. All features together determine the life strategy of the plant.

At present, ecologists try to consider as much as possible the whole spectrum of morpho-anatomical, ecophysiological and demographic plant characters to assess the survival strategies of the species. In desert plants, the kind of reserve tissues by which the species survive the dry periods, the kind of reserves (water or energy), the location of the reserves (above or below ground), the regulation of the transfer from the phase of more or less passive reserve existence to active growth, and the demographic characteristics of various plant phases in particular receive much attention. Based on these characteristics desert plants are commonly classified as follows (Noy-Meir 1973, Evenari 1985b):

– poikilohydrics: These plants maintain their structures under extreme changes in hydration. Undamaged tissues survive extreme and prolonged periods of desiccation in an anabiotic state and then are resistant to extreme heat and cold. They react instantaneously to improved water conditions, take up moisture from rain, dew or directly from moist air, and rapidly switch to metabolic activity. Undamaged they survive repeated biochemical switches from anabiosis to metabolic activity and vice versa. This category of plants includes algae, lichens, some mosses and a small number of ferns and flowering plants.

– homoiohydrous arido-passive plants: These plants maintain no photosynthetically active parts during dry periods. They survive the dry periods with reserves either in seeds (annuals) or in perennial below-ground storage organs (geophytes and hemicryptophytes). Their revival from dormancy to activity may involve a certain time lag after rainfall. Annuals store nutrients and energy in their seeds but hardly any water, while the storage organs of the perennials often contain considerable water reserves as well as nutrients and energy. Activation in these plants is irreversible once triggered, and while the active cycle in the annuals is totally rain-dependent, flowering may be rain-independent in the perennials because of their water storage. Biomass build-up, reproduction, storage, and sometimes also length of the life cycle are highly plastic and variable in these species and correspond to rainfall characteristics. In the annuals extreme variation in germination and establishment in response to rainfall characteristics implies enormous variation in demographic features from year to year. Many of these species build up large seedbanks, and their seeds have long viabilities and often are remarkably heterogeneous in their requirements to be triggered to activity.

– homoiohydrous arido-active plants: These plants maintain some or all of their structures (chamaephytes and phanerophytes) and remain photosynthetically active during dry periods, though they vary in the level of photosynthetic activity maintained. Some species considerably reduce their active above-ground and below-ground biomass during periods of drought or replace wet season leaves by dry season leaves. Others maintain their biomass at nearly constant levels throughout the dry season. Species in this group cover a very wide range of anatomical, morphological and physiological characteristics, which enable them to maintain metabolic activity throughout the year. Very often these plants are unable to respond rapidly to improved moisture conditions and their relative growth rate is usually rather low. Apart from responses to (soil) moisture availability, the plants in this group often also show photoperiodic and thermoperiodic responses in their activity patterns.

Desert plants can have any one of the three photosynthetic pathways, C_3, C_4 or CAM (for details see Section 4.3). The prevalence of plants of each type is clearly correlated with the main climatic features of the sites, but nearly always all types are represented in the species pool (Werger and Ellis 1981, Shmida and Burgess 1988). Thus, relatively short-term climatic fluctuations can sometimes induce remarkable shifts in the predominance of species with each photosynthetic pathway (von Willert *et al.* 1985, Neilson 1986).

Desert succulents cover the whole range from arido-passive annuals through arido-passive perennials to arido-active perennials, and all three major photosynthetic pathways. In the previous chapter we have already dealt with anatomical and morphological features, and the following chapters will particularly outline ecophysiological properties of succulents. Population biological features of these plants are superficially covered, since so little is known about them, particularly of the species of the arid parts of southern Africa, the area on which we focus in this book. However, we shall start with these latter aspects.

2.2.1 Life cycles

In its simplest conception the complete life cycle of a plant comprises all events in the plant's life between germination of the seed or beginning of growth from another reproductive unit and the death of the individual that developed from it after it had produced new seeds or other reproductive units. When applying this concept at a population level, as seems ecologically more valuable, all events should be quantified and evaluated at the population level (Harper 1977). The span of plants' life cycles can vary from days or weeks to centuries. Common to all life cycles is that they start with germination, a process which is inevitably bound to water, although it might additionally be triggered by temperature and light. The most crucial phase of the life cycle follows on germination: it is the establishment of the seedling. Especially during this phase mortality is high.

Harshness of the site (e.g. impenetrable soil surface), unfavourable development of the physical environment (e.g. drought), competition with other seedlings or already established plants, and activities of herbivores can kill seedlings. In desert ecology very little attention has been paid so far to this phase and to the factors and features controlling seedling establishment. Seedling establishment is followed by a phase of vegetative growth and sometimes vegetative reproduction. The next phase is sexual reproduction. In some plants the shift from the vegetative to the sexual reproductive phase can occur only once and then marks the end of its life cycle. In others it occurs repeatedly with a more or less cyclic alternation of vegetative and reproductive growth. In that case the life cycle does not end with seed filling. Although seed dispersal is not strictly part of the life cycle, we would like to add it as the last phase because seed dispersal is predetermined by the plant and consequently part of the maintenance strategy of a plant species. It serves for continuity in occupying the actual site as well as the exploration and invasion of new areas. Plants which flower only once in their life time are called hapaxanths. Those that flower repeatedly are called alloxanths. It is clear that alloxanthous plants must be perennials while hapaxanthous plants can be shortlived (ephemerals, annuals) or longlived (biennials, pauciannuals, perennials). Prominent examples of perennial but hapaxanthic desert plants are several species of *Agave* which can become very old but die upon flowering.

2.2.1.1 Seedbanks and seed germination

Seedbank studies in arid areas are scarce. It is to be expected that seedbanks of many species of arid areas are persistent and huge in comparison with species of other biomes. Seeds of desert plants have to stay viable during long dry periods, often lasting several years, until conditions become suitable for germination. Persistent seedbanks are thus beneficial for the maintenance of the population. Most desert annuals, but not all desert perennials, seem to have persistent seedbanks (Noy-Meir 1973).

A huge seedbank is also advantageous to a species with polymorphic seeds with different periods of dormancy, such that portions of the seedbank are triggered into germination by different environmental factors (heteroblasty). Such species will maintain a portion of their seedbank as viable seeds in the soil when another portion has germinated in response to a favourable environmental factor. In this way the population may survive when sudden unfavourable changes in environmental conditions following germination of a portion of the seedbank kills off the seedlings. Such differentiation in dormancy and trigger factors leading to germination has often been reported for desert plants and has been termed the 'trial-and-error' strategy of establishment. A number of similar examples for desert species are cited by Koller (1972) and Went (1979), while Gutterman (1980/81) discussed the influences of day length during the last few days of seed maturation at the mother plant and of position of ripening seeds in the fruit or on the mother plant upon the development of heteroblasty in some desert species.

Van Rooyen and Grobbelaar (1982) studied the seedbank of the sandy soils in the Hester Malan Nature Reserve near Springbok. They found that the sandy soils of a dry valley, a flat sand plain and a plateau had much higher seed densities than the more stony soils of the hillsides and ridges. The number of viable seeds per square metre in 75 mm of topsoil at the different sites, determined as the number of seedlings germinated under different treatments within 30 days after the start of the treatment, is given in Table 2.2. These values are of the same order of magnitude as those reported from North American deserts by Went (1979). Van Rooyen and Grobbelaar (1982) also found a steep decline in seed density with depth of the soil, except for the sand plain where the layer at 25–50 mm was richest. They suggested that this results from the high degree of disturbance, especially deposition, that occurs in this habitat. As the authors did not find a significant difference between years in the number of viable seeds at any of the sites, they concluded that the species probably have persistent seedbanks that are large in comparison to the annual additions and losses. This is contrary to the findings of Went (1979) in North American deserts.

Van Rooyen and Grobbelaar (1982) reported for a winter rainfall area in the Richtersveld that different temperatures of the soil did not lead to different seedling assemblages. However, they found a tendency for higher rates of germination as environmental temperatures decreased from a constant value of 32 °C to a constant 17 °C, and also at temperatures fluctuating diurnally within this range. Similar results on germination response to temperature were reported for Karoo species by Henrici (1935, 1939). Van Rooyen and Grobbelaar (1982) found that established seedlings neither promoted nor inhibited the germination of remaining seeds in the seedbank. On the other hand, one of the most common sclerophyllous dwarf shrubs of the Karoo, *Chrysocoma tenuifolia*, was reported to possess leachable volatile substances which inhibit germination and seedling growth (Squires and Trollope 1979).

We have only some preliminary observations about the germination patterns of desert plants in their natural habitat. Detailed investigations are still missing.

At the beginning of the summer of 1980/81, in October, we heavily

Table 2.2. *Size of seedbank per square metre in 75 mm of topsoil in the Hester Malan Nature Reserve. (Data taken from Van Rooyen and Grobbelaar 1982.)*

Habitat	Number of viable seeds
sand plain	41 000
plateau	23 000
dry valley	13 750
hillside	9 750
ridge	5 000

irrigated some plots in a winter rainfall area at Numees, Richtersveld, for some days in an attempt to trigger germination. Very few seedlings appeared, however. Among these seedlings were no succulents. Such poor response to irrigation also has been reported for other desert areas (Evenari and Guttermann 1976). Towards the end of that summer, in March, about 25 mm of rain fell during a period of 30 hours. This resulted in a massive germination. We sampled four plots of 0.5 × 0.5 m in each of the following five habitats:

(1) A nearly flat area of homogeneous, deep, fine, sedimental soil material, dominated by *Brownanthus schlichtianus*;
(2) a nearly flat area of shallow, gravelly soil over gneiss and quartzite bedrock, dominated by *Cheiridopsis robusta*;
(3) a west-facing slope of 7° covered by rather thin gravelly and rocky soil on quartzite and gneiss with a very open vegetation of various dwarf shrub species;
(4) a lower slope of 12° exposed to the southeast with a fairly thick gravelly soil of mainly quartzite, and a fair vegetation cover of mainly *Crassula grisea*, *Ceraria fruticulosa* and *Euphorbia gummifera*;
(5) a sandy and gravelly dry river bed without any mature plants.

We monitored these plots for a short time. The results are shown in Table 2.3.

It turned out that the lower slope on quartzite (habitat 4) was by far the richest in seedlings whereas the fine, sedimental flat (1) and the dry river bed (5) were poorest. In all plots, except two on the lower slope (4b and 6c), the number of seedlings per plot started to diminish immediately and in most plots this decrease was drastic. In the course of the year there were only two small additional rainfall events until August 1981. At that time only seven adult plants were counted in all five sites (there were originally 10 123 seedlings) and none of these plants was a perennial succulent.

The more even distribution of rainfall in this area in the 1982 season (see Table 3.2 and Fig. 3.7 in Section 3.2) led to a dense cover of the annual succulent *Mesembryanthemum pellitum* on one of the plains at Numees. Among these *Mesembryanthemum* plants only few specimens of another annual succulent, *Opophytum aquosum*, could be observed. The following year precipitation fell rather late in the winter season and the same plain was then completely covered by *Opophytum aquosum* while only a few specimens of *Mesembryanthemum pellitum* could be detected.

From these observations one can conclude that the response of a given seedbank in the southern Namib can be very complex: number and diversity of seedlings are influenced by the date and the distribution of sufficient precipitation. This remarkable phenomenon may be caused on the one hand by endogenous rhythms of the seeds modifying the portion of the seed population which is prepared for germination and on the other hand by temperature and photoperiodic regimes before and/or during the process of germination. This may have considerable consequences on the qualitative and quantitative composition of a plant community at a given site, especially after a more or less total breakdown of the community in periods of prolonged drought, an event not all that rare in deserts.

Germination patterns of desert succulents from the southern Namib can be very diverse. Dormancy of fresh seeds can prevent the seeds from germinating at an unfavourable time, i.e. towards the end of the rainy season or at unexpected off-season rainfalls. Growth chamber experiments (Ihlenfeldt 1985) showed that after breakdown of dormancy different basic patterns of germination can be distinguished (Fig. 2.10).

Seeds of the perennial, very long-lived and slow-growing succulent *Delosperma pergamentaceum* nearly all germinate with a very short delay, well synchronized in time. However, this behaviour conveys an extreme risk in the case of lack of rain after the first rainfall. The germination pattern of an annual succulent of the genus *Mesembryanthemum* is quite different (Fig. 2.10). The water storing capacity of this species is only moderate, and flowers are only formed towards the end of the rainy season. After a long lag-phase of more than three weeks, germination starts slowly and stops as soon as only a quarter of the seed population has germinated. The remaining seeds are obviously still viable, but not prepared for germination due to dormancy fading away much more slowly. It is obvious that this behaviour diminishes the depletion of the seedbank by avoiding wasteful

Table 2.3. *Number of seedlings in four plots of 0.5 × 0.5 m in each of five habitats (see text) at Numees (Richtersveld, southern Namib desert) sampled on three dates starting five days after a 30-hour period of precipitation (25 mm)*

	28 March			2 April			5 April			
	Grass	Dicots	Total	Grass	Dicots	Total	Grass	Dicots	Total	Mortality (%)
Habitat 1										
plot a	0	26	26	0	9	9	0	0	0	100
plot b	0	2	2	0	0	0	0	0	0	100
plot c	0	26	26	0	7	7	0	5	5	81
plot d	2	20	22	1	8	9	1	2	3	86
Habitat 2										
plot a	31	36	67	37	15	52	28	7	35	48
plot b	30	32	62	16	9	25	19	6	25	60
plot c	5	49	54	5	20	25	3	11	14	74
plot d	6	52	58	6	18	24	6	5	11	81
Habitat 3										
plot a			30			20	0	14	14	53
plot b			29			17	0	12	12	59
plot c			16			6	0	4	4	75
plot d			30			24	0	27	27	10
Habitat 4										
plot a	13	218	231	17	205	222	19	181	200	13
plot b	7	65	72	5	84	89	11	70	81	−13
plot c	16	243	259	52	247	299	52	191	243	6
plot d	4	47	51	7	38	45	7	33	40	22
Habitat 5										
plot a	1	19	20	0	15	15	0	10	10	50
plot b	0	36	36	0	19	19	0	14	14	61
plot c	0	13	13	0	2	2	0	1	1	92
plot d	1	23	24	0	7	7	0	5	5	79

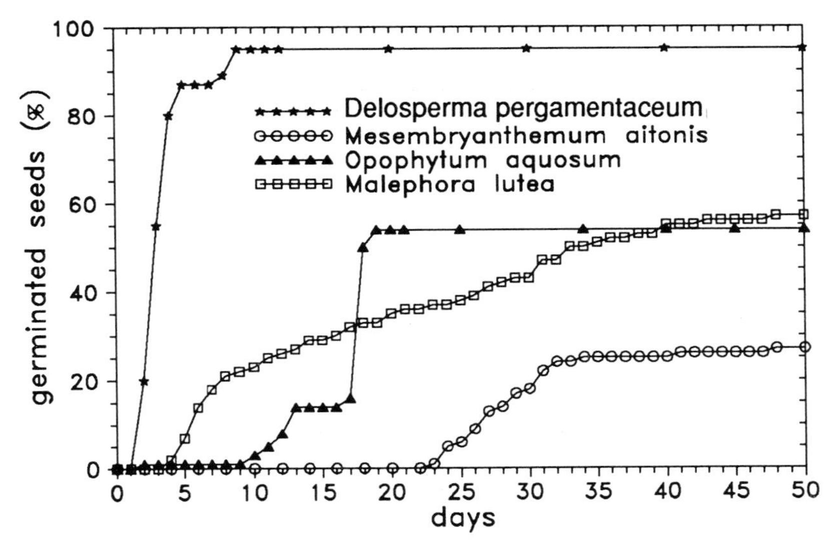

Fig. 2.10 Germination of some succulents in a laboratory experiment. Temperature conditions were 23–26 °C during the light period and 13–16 °C during the night. The number of seeds germinated is given as a percentage of the total number of seeds and is plotted against time in days. *Delosperma pergamentaceum* responded quickly and nearly quantitatively, *Mesembryanthemum aitonis* responded weakly and with a lag-phase of more than 20 days, while *Malephora lutea* and *Opophytum aquosum* behave in an intermediate fashion.

germination in the case of insufficient precipitation or off-seasonal rainfall. Only long-lasting soil moisture will trigger germination which, due to incomplete synchronization, leads to a slower increase in the number of germinated seeds (Fig. 2.10).

A third type of behaviour is demonstrated by the annual *Opophytum aquosum*. In contrast to the previous example, this species is able to develop its first flower very rapidly, and moreover it has an extraordinarily large water storing capacity. The germination behaviour exhibited by this annual succulent is very common also among the perennial but comparatively short-lived species, and it is also known from non-succulent desert plants. This type of germination pattern can be described as biphasic. After a lag-phase of only a few days, there is a first phase of about 10 days during which a certain portion of the seed population germinates. During a second phase another portion germinates quite well synchronized, but there still remains a considerable portion of the seed population which is not yet prepared for germination, thus forming a seedbank for future years. Based on their preparation for germination, the seed population consists of two types of

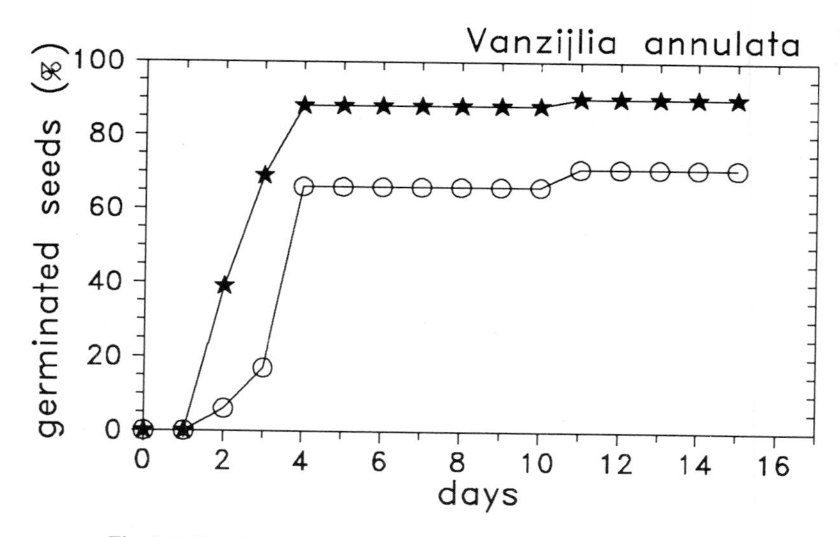

Fig. 2.11 Germination of *Vanzijlia annulata*. Stars give the data obtained from seed of an inland population, circles give those for a coastal population. The lag-phase of germination in the coastal population prevents germination due to the frequent but insufficient precipitation from fog. (Data from Hartmann 1983.)

seeds: fast and slow germinating individuals. In a closely related species from Israel (*Mesembryanthemum nodiflorum*) Gutterman (1980/81) found indications that the different behaviour within one seed population is induced by the relative position of the individual seed in the capsule.

We will discuss later (Chapter 5) how these different germination patterns fit into the whole life strategy of the plants. How exactly the germination behaviour can be adapted to the prevailing environmental conditions of a growing site has been demonstrated for *Vanzijlia annulata* by Hartmann (1983). Inland populations of this species exhibit a germination behaviour similar to that of *Delosperma pergamentaceum* (Fig. 2.11). Populations near the coast experience heavy fogs during the night which can wet the soil temporarily. This soil moisture could trigger germination but is never sufficient for survival of seedlings. By a prolonged lag-phase of three days during which only a few seeds germinate, temporary fog precipitation is discriminated from favourable soil moisture caused by rainfall.

2.2.1.2 Seedling establishment

The total number of perennial species per unit surface area in a desert remains fairly constant over a period of several years. This means that over longer periods there appears to be a one to one exchange of plants either by new ones of the same species or by other species. A successful establishment of seedlings or juvenile perennial plants is a very rare event and the mortality of seedlings of perennial plants is 100% in most years. Only a series of years with exceptionally good rainfall will allow a large-scale establishment as has been assumed to have happened in the Central Namib in 1934 for *Welwitschia mirabilis* (Walter 1936). Similar reactions were also reported for a species of barrel cactus (Jordan and Nobel 1981) and the giant cactus *Carnegiea gigantea* (Steenbergh and Lowe 1977) in the Sonora desert.

The situation is fundamentally different for the short-lived desert plants: the ephemerals, annuals and biennials. In experiments carried out in the laboratory by Van Rooyen and Grobbelaar (1982), by far the highest seedling mortality was found within the first weeks after germination. In field observations at Hester Malan Nature Reserve the same principal researcher (Rösch 1977, Van Rooyen *et al.* 1979a) found a very high mortality rate in newly established plants. In five permanent sampling plots of one square metre observed from April to September 1974, she recorded maximum seedling densities ranging from 115 to 1810 seedlings per square metre. The maximum values occurred in April at the beginning of the observational period. Of the 1810 seedlings in one plot 1558 belonged to one species, *Cotula barbata*. Of all seedlings recorded in the five plots 63.4% reached maturity, which is a slightly higher value than those reported from the North American Mohave desert (Beatley 1967). At each of the five plots between 47.6 and 74.1% of the seedlings survived to maturity. Species differed from 0 to 100% in the survival rate of their seedlings and of all species in all plots 85.5% had individuals which reached maturity.

Very little is known about the seedling establishment of succulents but our own field observations tend to support the basic difference in successful establishment between annuals and perennials outlined above. The environmental factors determining seedling establishment are numerous and range from the amount of rain and its distribution in time, through edaphic factors like soil texture and velocity of downward movement of water in the soil to the damage by herbivores and competition problems with other seedlings or adult plants. Since each of these factors can be crucial, a detailed system analysis is necessary in all studies concerning seedling establishment. Meaningful research requires proper field observations and plenty of time to minimize randomness of results. For a more detailed review of seedling establishment in deserts we refer readers to Evenari *et al.* (1985).

2.2.1.3 Phenological rhythms and reproduction

Phenological rhythms can be determined by water availability, photoperiod, thermoperiod or by combinations of these, with water availability being always the dominating determinant.

Of course, water is necessary for germination. But additional trigger factors for germination can be photo- and/or thermoperiod (see Section 2.2.1.1). The response to these additional factors distinguishes ephemerals from the true annuals as well as drought deciduous species from truly seasonally deciduous species, while evergreen succulents are assumed to respond to available soil moisture at any time of the year.

Flowering is not necessarily coupled to vegetative growth. It can precede, follow or coincide with vegetative growth except for ephemerals and annuals, in which flowering either takes place during intensive vegetative growth or can mark the end of vegetative growth. Most desert species have distinct and limited flowering periods, particularly the geophytes. Nevertheless there are a considerable number of hemicryptophytes, chamaephytes, and phanerophytes in which flowering (and ripening of fruits) (i.e. *Carpobrotus*, etc.) can occur at nearly all times of the year, in response to weather conditions; some species are even capable of flowering more than once a year.

It should be emphasized here once more that succulence makes a plant temporarily independent from external water supply. Therefore succulents are capable of initiating growth and/or reproduction by making use of their endogenous utilizable water. On grounds of this feature we are able to distinguish between seasonally and drought deciduous succulents (see Section 5.2).

Phenological studies on desert plants (including succulent desert plants) are scarce. A brief overview has been given by Ackerman and Bamberg (1974). Phenological data for Namaqualand (southern Namib) have been provided by Van Rooyen *et al.* (1979) and Orshan (1988); Fig. 2.12 gives details for a selection of succulents and geophytes taken from these data.

Orshan (1988) combined morphological and phenological features in his study of plant phenomorphology. Of the 22 species studied in Namaqualand nearly 75% showed a period of vegetative growth during winter and spring. All these plants started flowering in spring during the period of vegetative growth and about one-third continued flowering during summer when vegetative growth had stopped. About a quarter of the species showed growth throughout the year with half of them flowering during winter and spring and the other half during summer and autumn.

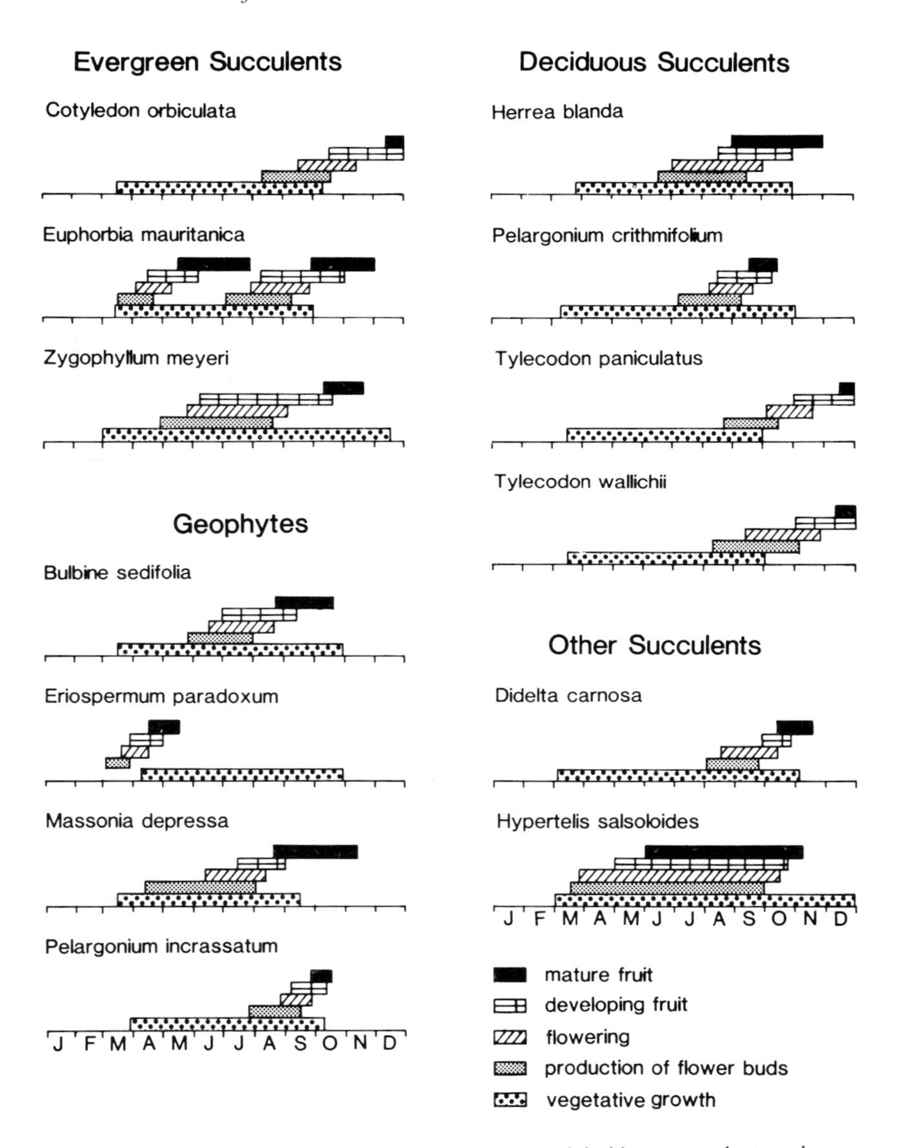

Fig. 2.12 Phenological rhythms of evergreen and deciduous succulents and some geophytes of the Hester Malan Nature Reserve in Namaqualand (South Africa). Data are taken from Van Rooyen *et al.* (1979a, b).

With respect to the succulents of the Namib, vegetative growth is usually restricted to the rainy season (winter or summer respectively, see Chapter 3). Flowering, however, may occur at any time of the year. As far as perennial succulents are concerned, there are three striking features: (1) high investments in the formation of flowers; (2) mass flowering for a limited period; (3) a definite sequence of flowering periods within a given plant community dominated by succulents. All these features are closely related to special demands on the pollination biology in arid regions.

For the majority of the succulents high investments in the formation of flowers are a very striking feature. Flowers are usually large and showy. If the single flowers are comparatively small (as in many *Crassula* species) they are arranged in dense inflorescences. This feature can be interpreted as an adaptation for the strong competition for rare pollinators. Usually flowers of succulents are allotropic, thus being able to make use of a broad spectrum of pollinators. But there are quite a number of well-known exceptions: pollination by birds (ornitophily) in Cactaceae and *Aloe* species, by flies (myophily) in *Stapelia* and related genera, by bats in *Carnegiea*, and by traps for small flies in *Ceropegia*. The large family Mesembryanthemaceae, the dominant family of succulents in the southern Namib, has evolved groups with flowers that are pollinated either by day or night butterflies (*Conophytum*, *Ophthalmophyllum*, certain species of *Sphalmanthus*).

The phenomenon of mass flowering, i.e. a concentrated offer to the pollinators for a limited period by complete synchronization within the population (rarely longer than two weeks), obviously improves the probability of successful pollination, as does the sequence of flowering times within the plant community. The flowers may even be open only for a limited period of the day (Mesembryanthemaceae = 'midday-flowers'), the opening being controlled by light intensity and temperature.

Short flowering periods convey a high risk as during the short period of flowering weather conditions may be unfavourable for pollinators. This is especially true for annual succulents, since these plants have to rely on successful production of seeds for survival as a species. The same applies to many short-lived perennials which have to cope with the risk of not reaching the next favourable season due to premature depletion of their utilizable water. Representatives of these groups therefore exhibit completely contrary flowering strategies. In the annual genera *Dorotheanthus* and *Cleretum* the seedling forms only a small number of leaves and produces a first terminal flower very rapidly. Instead of forming a separate inflorescence, the lateral shoots that are formed later on again produce very rapidly a terminal flower. Here there is no longer a clear distinction between the vegetative phase and the reproductive phase, a feature well-known also from non-succulent short-lived annuals. Flowers are formed (and fruits are produced) as long as the water supply is sufficient. Some annuals (e.g. *Mesembryanthemum crystallinum*), however, have a comparatively long vegetative phase and accumulate as much water as possible; they start flowering at the onset of drought and produce flowers and fruits until all utilizable water has been consumed.

To our knowledge there are no in-depth studies on the influence of water supply on the number of flowers that are formed during one season. But it is obvious that the prerequisite for the formation of flowers and fruits is a sufficient reserve of carbohydrates and water. These carbohydrate reserves have been accumulated during the previous season or seasons in case flowering precedes vegetative growth as in many *Conophytum* species or takes place at the beginning or shortly after the beginning of vegetative growth as in the genus *Argyroderma*. In these cases the flower buds have already been formed at the end of the previous season or during the resting period. In species that flower at the end of the season, these reserves have been accumulated during the current season, and there are some observations indicating that the length of the favourable period influences the number of flowers. This is especially true for those succulent annuals that start flower formation very soon after the establishment of the seedling. In any case, production of carbohydrates is strongly dependent on the amount of available water, and thus formation of flowers is determined by the availability of water.

Up to now, there have been few in-depth studies on the breeding systems of succulent plants. Besides self-incompatibility, which seems to be widespread, numerous cases of self-fertility have been recorded. Sometimes flowers try outbreeding in the beginning of their period of anthesis by protandry, i.e. the male parts of the flowers become functional earlier than the female parts; later they perform self-pollination by changing the position of the anthers which come into contact with the stigmas (*Dorotheanthus*, Ihlenfeldt 1985). Some annuals, in which the pollination chance is especially risky, even switch over to regular self-pollination, at the same time reducing their petals and even performing cleistogamy under favourable weather conditions (*Cleretum papulosum* ssp. *papulosum*).

2.2.1.4 Dissemination and dispersal

Plants in arid areas have to face two completely opposed dispersal requirements. On the one hand resources are limited and it is thus important to provide facilities for wide dispersal in order to reach a great number of favourable sites. On the other hand it is most likely that the habitat of the mother plant is a favourable habitat, and it is thus beneficial to occupy that site again and to refrain from wide dispersal and avoid the risk of wasting a large number of seeds. Consequently, many desert species have developed strongly antitelechoric (dispersal inhibiting) mechanisms. Some species seem to avoid the dilemma by producing two types of inflorescences (amphicarpy), an aerial one on a flower stalk and a morphologically reduced inflorescence growing subterraneously. The two types of inflorescences produce different types of seeds. Examples of such amphicarpic plants are *Enneapogon desvauxii* in southern Africa and *Catenanche lutea* in the Near East. With respect to the mechanisms of dispersal desert plants in general and succulents in particular do not differ from other plants. Usually seeds are dispersed but in some cases whole fruits are the dispersal units.

As in other plants, the diaspores of desert plants in general and succulents in particular can be seeds, fruits (containing one seed, e.g. Asteraceae, or more seeds, e.g. capsules of *Fenestraria*, which are dispersed as 'tumblers', Hartmann 1982), or parts of them (mericarps, e.g. *Herrea*, Ihlenfeldt 1983), or even vegetative parts (special cases of vegetative reproduction, see below). In extreme cases even the complete plant may act as the diaspore. With respect to the dispersal agents, there are also no striking differences compared with plants of other regions, except for endozoochory which is rare in succulents (Cactaceae, Vitaceae), since it demands fleshy fruits, i.e. additional investments of water. In deserts especially the wind (anemochory, e.g. Apocynaceae, Asclepiadaceae, Asteraceae) and – unexpectedly – water (hydrochory, see below) are important dispersal agents, even for diaspores without any special structural adaptations. Wind and running water in many cases provide the facilities for long-distance dispersal, the diaspores being deposited preferably on the lee side of stones and other obstacles. A widespread feature of desert plants is that the diaspores are only released when the conditions are favourable (aestatiphorism), i.e. when rain is falling.

Dispersal systems in desert succulents can be very complicated and complex, as is demonstrated by the family Mesembryanthemaceae. The fruits of this family have evolved a perfect adaptation to desert conditions that is unrivalled by any other family of succulent plants. The mature fruits remain closed. They open only when they are wetted by rain or heavy fog and close again after desiccation (hygrochastic capsules). The seeds are ejected by raindrops hitting the open and flooded capsule (rain ballists, a special case of hydrochory). Thus it is guaranteed that the seeds are only released when conditions are favourable for germination. Moreover, the raindrops disperse the seeds over a certain distance in the vicinity of the mother plant (Van Rooyen *et al.* 1979). Additional complicated structures in advanced fruit types (see Fig. 2.13) prolong the process of seed release, at the same time increasing the distance to which the seeds are dispersed. Some fruits possess special pockets, in which a certain number of seeds become enclosed. These seeds are only released after the complete disintegretion of the fruit, which may take years. For further details of these complicated and fascinating fruits and implications on seed dispersal and seedling establishment in Mesembryanthemaceae we refer to Ihlenfeldt (1971, 1983).

Closely related to dispersal is the ability to reproduce vegetatively (asexually). Asexual reproduction, a technique well-known and often applied in the propagation of succulents, is in the Namib not as important as one would expect. Species with a prostrate habit and the ability to form adventitious roots can disintegrate into separate individuals after the central part of the mother plant has died off (*Cephalophyllum*, *Stapelia*).

The well-known ability of Crassulaceae to regenerate new plants from detached leaves is used as a mode of survival under extreme conditions. In certain *Crassula* species (e.g. *Crassula expansa*) in periods of prolonged drought, the stem becomes very fragile and the plant starts disintegrating. Detached leaves may be blown off by the wind and establish new plants at a different, more suitable site.

Rosette-forming monocotyledons, such as *Aloe* species can develop daughter plants at the base of their stems on adventitious shoots. Among the jointed cacti (Opuntioideae) in American deserts vegetative reproduction from loose joints is quite common. In some species (e.g. *Opuntia bigelovii*) joints detach easily and function as epizoochoric dispersal units.

2.2.2 Vegetation

The floristic composition of the arid zones is determined by their phylogenetic and evolutionary histories. Each area has its own characteristic taxa, often phylogenetically related to the taxa in the adjacent mesic areas. Thus the floristic composition of the arid zones, just like that of the more mesic zones, completely fits into the general plant geographical world pattern, and consequently floristic differences between arid zones within different plant geographical kingdoms are generally large. Some arid areas lie in the border zones between two or three plant geographical kingdoms, e.g. the Sahara and the Middle East desert areas, the Karoo and Kalahari, and the Mexican desert areas. They generally show a clear north–south

Fig. 2.13 Seed capsule of *Cheiridopsis robusta* (a) closed and (b) open after wetting the capsule.

gradient in floristic composition, again in correspondence with the main plant geographical world patterns. Despite the characteristic floristic signature of each of the major arid areas, there are also a number of widespread arid zone weeds, originating from one area but now spread anthropogenically over all or many arid areas, e.g. *Salsola kali*, *Nicotiana glauca* and *Mesembryanthemum nodiflorum*.

The floristic composition of the vegetation types has been studied in detail in some arid areas. Several reviews have been published during the past quarter century, and there is no need to repeat them here in a summary. It is suffice to refer to the many comprehensive chapters in Goodall and Perry (1979, 1981), West (1983) and Evenari *et al.* (1985), and the many references to more detailed accounts given therein.

The physiognomy of the vegetation of the arid areas results from the type, density and distribution pattern of the composing life forms. In arid areas a wide variety of growth forms occurs, covering the full range from trees to annuals, lichens and algae. Also the densities these plants can reach vary widely. As a result arid zone vegetation types range from forest and woodland patches through shrub and dwarf shrublands, marginal savannas, (desert) grasslands and steppes to sparsely vegetated or bare desert surfaces.

The distribution and density of the various life forms is determined by the complex of climatic, hydrographic and edaphic factors. The climatic factors, particularly rainfall characteristics and sub-zero temperatures, play a predominant role in determining the gross vegetational patterns. The hydrographic and edaphic factors determine the patterns on a more detailed level. Furthermore, on the detailed scale, biotic factors such as grazing by large herbivores, rodents or termites can determine the vegetational pattern, as do the competitive interactions between the plants themselves.

Soil texture differently affects moisture availability to plants with different amounts of rainfall: in areas with relatively high rainfall plants have less moisture available in coarse soils than in heavy soils. In areas with low rainfall this is opposite due to the fact that in dry areas evaporation from the upper soil layer, rather than drainage from deeper soil layers, causes the greatest loss of soil moisture. This is called the inverse texture effect. The balance point between the advantage of coarse-textured versus fine-textured soils for plant growth occurs somewhere between the 300 and 500 mm isohyets (Noy-Meir 1973).

Greatly simplified the following correlative patterns emerge. Trees and tall shrubs occur on deep sandy soils, well-drained slopes and footslopes with a coarse and gravelly soil, and along major drainage lines or in oases. Grasses, too, are prevalent on deep sandy soils, while the heaviest clay and silt soils also support grasses, frequently those that form dense turfs with short shoots sprouting from short stolons or rhizomes. Dwarf shrubs predominate on loamy and clayey soils, in salty areas and on calcareous crusts, but they also occur on rocky slopes, intermixed with taller shrubs, occasional trees and grasses. Area-wise, dwarf shrubs and grasses are certainly the most important plant forms of arid zones, covering vast areas in more or less pure or mixed stands. The dwarf shrubs and shrubs of arid areas often have an open canopy structure with a diffuse, scattered branching pattern, perhaps because such a structure may enhance convective heat loss near the soil surface (Shmida and Burgess 1988). Annuals occur throughout the dry areas on various substrates. Their density is strongly related to the temporal and spatial patterns of soil moisture. Annuals seem to be more common in winter rainfall deserts than in summer rainfall deserts, perhaps because rainfall is somewhat more predictable in the former.

Towards the higher rainfall limits of the dry areas the formations dominated by dwarf shrubs, shrubs and trees usually get grassier, except for the narrow gallery forests along major drainage lines.

Though succulents can occur throughout the formations, particularly the stem succulents, they are dominant in areas where the drought is frequently broken by fog or dewfall conditions. In areas with heavy night frosts during the growing season succulents become scarce to absent. Accordingly, predominantly succulent, non-halophytic formations are found in the Namib and parts of Namaqualand and the winter rainfall Karoo in southern Africa, in the dry parts of the Canary Islands, particularly on some of the dry but frequently cloud-covered slopes on the northern halves of the islands, in a narrow coastal zone of southern Morocco and northern Mauretania, and in sections of the dry parts of Peru and Chile. Succulents are also common, and locally even predominant, in some dry areas of Bolivia and northern Argentina, in dry parts along the northern coasts of Venezuela and Colombia, on some West Indian islands, in large parts of Mexico, in Arizona and California (Sonora), in a small part of southwestern Madagascar, on the island of Socotra in the northern Indian Ocean, and locally in Yemen. In Australia some areas carry semi-succulent formations, mainly of Chenopodiaceae, but these seem associated with semi-saline, saline and gypseous soils (Specht 1972, Beadle 1981). Also in the Middle East succulence is principally correlated with saline and gypseous soils.

As moisture availability gets more limited the vegetation becomes more open, changing from steppe-like formations to true desert formations. On soils with the most favourable moisture conditions for plant growth this transition occurs at about the 120 mm isohyet. Areas of lower rainfall and with fairly uniform soil and hydrographic conditions support a sparse vegetation with perennial plants widely scattered. This is called diffuse desert vegetation. Below 70 mm rainfall this vegetation pattern generally cannot be maintained and perennial plants get restricted to wadis, runnels and washes. This is called contracted desert vegetation. Obviously, edaphic conditions that affect moisture availability to plants considerably influence the position on the rainfall gradient of the transition from diffuse to contracted desert vegetation (Shmida 1985, Shmida and Burgess 1988).

In areas of uniform edaphic and hydrographic conditions the distribution of perennial plant individuals may be strongly affected by patterns of vegetative reproduction through stolons or rhizomes and by mutual interactions between plants. The rather regular, circular distribution patterns in *Larrea divaricata* in American deserts resulting from vegetative spread of the clones (Vasek 1980) are well known. Many other species, e.g. *Rhigozum trichotomum* in southern Africa, show a strongly patchy pattern with clear patch fronts slowly moving across the site as a result of rhizomatous growth.

Interactions between plant individuals may be beneficial, as for example when taller plants provide a climatically sheltered zone below their canopies for smaller plants, temporally and locally improve soil moisture conditions around their base due to stem flow and shading, cause a non-uniform distribution of nutrients in the soil as a result of locally increased litter fall and decomposition, or protect smaller plants from large herbivores by their chemical or mechanical grazing defences. Interactions may also be competitive, the intensity of which is also plant size dependent. An example will be given in Section 3.3.

Throughout arid areas the hydrographic pattern leading to redistribution of rain water due to run-off and run-on strongly determines the pattern of plant growth. Many detailed vegetation maps of arid areas closely reflect the hydrographic patterns of these areas (e.g. Fig. 8.23 in Werger 1985). Run-off considerably reduces the available moisture at a site, whereas run-on can increase water availability to several times the annual rainfall. Accordingly, dense patches of perennial vegetation can persist in areas which are climatically too dry for such vegetation.

Geomorphological processes on gentle, homogeneous desert slopes may initiate marked patterns in the distribution of water in the topsoil and lead to conspicuous striped patterns in vegetation. Such patterns have been described from many arid parts of the world. The patches seem to move gradually upslope through establishment of pioneer species at the upper fronts of the patches (Cornet *et al.* 1988).

3

The Namib desert

The precise location of the southern boundary of the Namib is still a matter of controversy in scientific literature. Several authors put it at the Orange River (Fig. 3.1), which is merely a political boundary. Geomorphologically the areas at both sides of the river do not dramatically differ except for the coastal area of huge sand dunes which do not cross the river. South of the river the dunes are not very impressive but sandy plains continue to occur in a rather wide coastal zone till south of Port Nolloth, and then taper into a narrow coastal belt. Further inland follow gravelly desert plains in wide valleys or broad mountain basins at both sides of the Orange River. Their southernmost occurrence lies considerably south of the Olifants River. Also the rugged mountain ranges of granite, quartzite and metamorphic rock and the narrow dry river valleys are similar at both sides and do not justify consideration of the Orange River as a boundary. In the north they start in the area around Aus and become increasingly common towards the south, also continuing to about the area of the Olifants River or even beyond.

There is no climatological evidence for considering the Orange River the southern boundary of the Namib desert. It is true that the south receives rainfall predominantly in winter, as compared to summer rainfall in the north. The transition zone between predominant winter and summer rainfall crosses the coastline somewhat to the north of Lüderitz and then runs in a southern to southeastern direction until about 32° S. The rugged mountain ranges south of Aus are nearly all situated west of the summer rainfall area, and receive scarce precipitation without clear seasonal pattern or during winter.

Biogeographically the Orange River is also no boundary (Werger 1978a). Several species of plants and animals reach their southern boundary somewhere considerably south of the Orange River, while others have their northern distributional limits in the vicinity of Aus (Coetzee 1969,

Merxmüller 1966–1972, Werger 1978b). This latter group of species is distributionally linked to the mountainous habitats occurring from Aus southwards. Some species typical of this habitat continue to occur further northwards in a narrow belt along the escarpment (Volk 1966). South of the Olifants River the admixture with taxa with a typical Cape affinity increases strongly. The Olifants River also delineates the northern boundary of the area where rainfall allows agriculture without irrigation. In our opinion it is therefore appropriate to consider the Olifants River as the southern boundary of the Namib desert, even though it is clear that this delimitation should not be understood as a sharp, dividing line between a desert and non-desert landscape.

This whole area north of the Olifants River up to the vicinity of Aus is dominated by an enormous variety of succulents, mainly members of the Mesembryanthemaceae, Crassulaceae, Geraniaceae, Asteraceae, Euphorbiaceae, Portulacaceae, Liliaceae, Asclepiadaceae and Zygophyllaceae. Although this area receives rain predominantly during winter and in that significantly differs from most of the rest of the Namib which gets erratic summer rain, we see no reason why this area should be excluded from the Namib desert. Rainfall is scarce, and highly unpredictable, with variability coefficients that can exceed ten; fog and hot east wind conditions occur frequently, and the vegetation can be extremely sparse and patchy. We therefore refer to this area as the winter rainfall or southern Namib.

To the east the border of the Namib is generally considered to run through the escarpment mountains, decreasing in altitude with decreasing latitude. In the north it tapers towards the Atlantic Ocean and deep in the summer rainfall area, considerably north of Moçamedes in Angola, the border is a fairly narrow transition zone in which the desert and semi-desert gradually grade into a dry savanna landscape (Mendonça 1961, Barbosa 1970, Werger and Coetzee 1978).

There exist good descriptions of the Namib desert (Walter 1936, 1973, 1985, Werger 1978b, Giess 1981, Walter and Breckle 1984) and we do not want to repeat too much of what has been written in these. We just give a short summary (see Section 3.3) and add some more information on the southern part of the Namib desert, an area exceptionally rich in succulents which display a wide array of different growth forms. This diversity is not found in any other arid area inhabited by succulents. We believe that any attempt to demonstrate and understand the adaptational values of succulents to desert conditions has to take this area into account.

There are three reasons why we deal especially with the succulents of the southern Namib. The first is that we were struck by the extraordinary richness of this group of plants in the southern Namib. The second is that so far so little is known ecophysiologically about the southern Namib desert, and the third is that we have built up a considerable experience in this beautiful part of the world. We start with a detailed overview of those environmental conditions which play a key role for plant life in the southern Namib.

3.1 Geomorphology

The Namib desert stretches along the west coast of southern Africa between about 15° and 32° S latitude from north of Mocamedes in Angola over a length of about 2000 km to the Olifants River in South Africa. A map is presented in Fig. 3.1. It shows that the Namib is only a small strip of land seldom exceeding 150 km in the east–west direction (King 1963).

Because of its long north–south extension several geomorphological subregions can be broadly distinguished on the basis of surface characteristics and climatic differences. South of Moçamedes a sand field along the coast which is up to 50 km wide and almost bare of vegetation indicates the aridity of southern Angola. The sand strip is fringed inland by serir plains, hamada and bare rock outcrops. This area is called 'Northern Namib'. It extends to the lower Swakop River in the south (Goudie 1972, Watson and Lemon 1985).

The 'Middle Namib' is characterized by a 'sea of sand' in the coastal strip bordered by the Kuiseb River in the north and the Orange River in the south. Inland this sea of sand merges into calcrete, gypsum and gravel plains and then grades into the escarpment hills and mountains. The 'Southern (Transitional) Namib' comprises the rocky mountainous land of quartzite, igneous and metamorphic rock that starts in isolated exposures in the vicinity of Aus and increasingly dominates the landscape southwards across the Orange River down to the Olifants River ('hardveld'). To the south of the Orange River the valleys and basins in this mountainous landscape contain flats and plains covered with a thick layer of gravel of the same rock that constitutes the surrounding mountains. The unimpressive, rather narrow undulating coastal sand plains south of the Orange River ('sandveld') should also be viewed as part of the southern Namib. The eastern boundary of this area is again the escarpment of the interior continental plateau.

A west–east transect of the western part of South West Africa at about 26° S latitude is presented in Fig. 3.2. The geological, geomorphological and soil formations in the area near the Orange River (also referred to as the

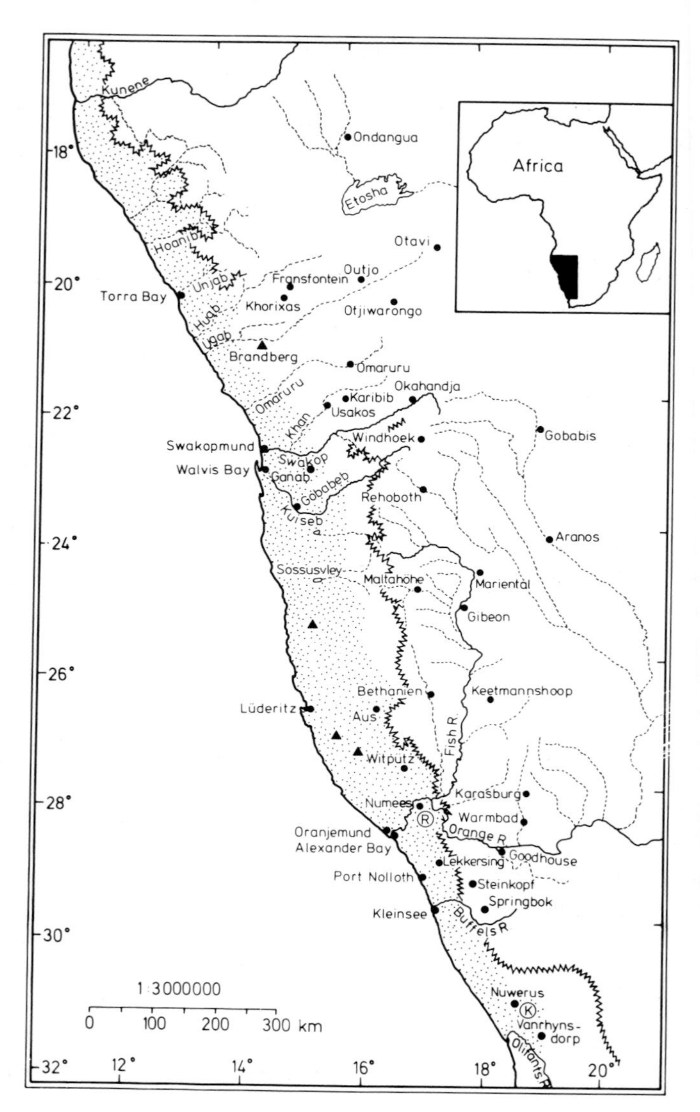

Fig. 3.1 Extension of the Namib desert (dotted) in South West Africa (Namibia) and adjacent parts of Angola and South Africa. The Great Western Escarpment is indicated by the zigzag line. The Richtersveld south of the Orange River and the Knersvlakte between Nuwerus and Vanrhynsdorp in the very south of the Namib desert are indicated by the letters R and K. Mountains are indicated by black triangles. (After Giess 1971 and Leser 1976.)

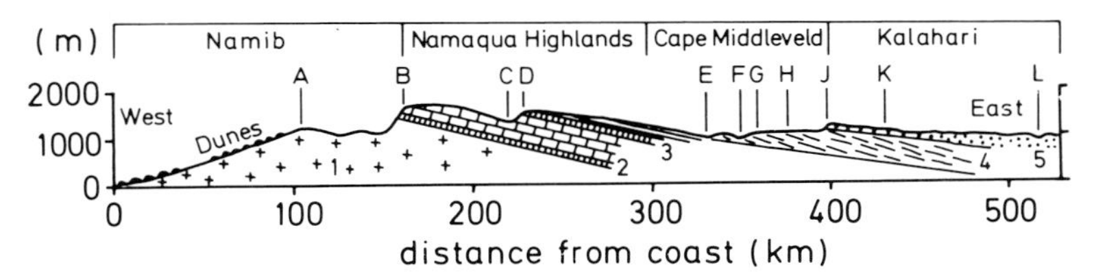

Fig. 3.2 West–east section across the South West African continent at about 26° S latitude from the Namib to the Kalahari after Wellington (1955). The vertical scale indicates height above sea level. 1 = Granite, 2 = Schwarzkalk (limestone, sandstone, shale), 3 = Schwarzrand Quartzite, 4 = Karoo sediments, 5 = Kalahari beds (conglomerate and sand). A = Aus, B = Tiras Mountains (Great Escarpment), C = Konkiep River, D = Schwarzrand, E = Lewer River, F = Great Fish River, G = Gibeon, H = Plain of Gibeon, J = Weissrand, K = Urinanib Plateau, L = Oab River.

Namaqualand coast belt, Richtersveld, Namaland or Bushmanland) are summarized in Table 3.1.

Only the Kunene River in the north and the Orange in the south are permanently running rivers. They cut through the Namib from east to west. Most of the temporary running rivers never reach the ocean but seep into the sand near the coast. Exceptions are the Omaruru, Swakop and rarely also the Kuiseb River.

In the Namib desert physical weathering of rocks predominates but due to frequent fog and dewfall chemical weathering also occurs. True soils with well-developed profiles are absent. The raw mineral soils, mainly sandy and often calcareous or with calcareous crusts, are composed of particles in a very wide range of size classes. Salt crusts are found only in the Westernmost part close to the ocean (Harmse 1978).

3.2 Climate

The Namib desert is divided by a biologically important climatic transition belt which is rather narrow. The belt crosses the coastline north of Lüderitz and runs via Aus in a southeasterly direction separating a northern part receiving summer rain from a southern part that gets winter rain. The narrow strip of land within the belt receives sporadic rainfall without a clear seasonal pattern.

Within the winter rainfall area the climate is rather homogeneous from north to south. This is documented in Figs. 3.3a and 3.3b which compare monthly mean temperature and monthly mean relative air humidity for Numees in the Richtersveld and Kalkgat, a place about 40 km north of Vanrhynsdorp in the Knersvlakte.

Table 3.1. *Summary of the descriptions of the geology, geomorphology and soils of the southern Namib (Richtersveld, South Africa) by several authors*

Geology	Geomorphology	Soil	Author
Primitive Archaean systems, post-Archaean and pre-Karoo eruptive rocks, Precambrian systems	Numerous rock outcrops and plains dissected by intermittently dry rivers	Feebly developed, shallow sandy soils partly covered by a desert pavement of white quartz pebbles and gravel	Wellington (1955)
Precambrian quartzites, dolomites, limestones, conglomerates and various series of the Nama system; Precambrian(?) highly metamorphic rocks of the Abbabis and Kheis system; old granites, gneisses, schists, and granulites of the pre-Karoo system and the azoic complex	Inclined and carved plains with serir, hamada and inselberge (conical mountains); higher mountain region of the great escarpment, vigorously split up at the margins	Rock debris and shallow sand layers; grey and greyish-brown detrital soils	Leser (1976)
		Weakly developed shallow soils: calcareous sands and loams mainly overlying calcrete; lithosols, arenaceous sediments and argillaceous sediments	Harmse (1978)

Summer rain normally derives from thunderstorms while winter rain falls as a soft rain or drizzle. As expected (see Chapter 2) the variability of rainfall in the northern Namib is much higher than in the southern Namib. Additionally, the probability and amount of rain increases with distance from the coast. Consequently the highest uncertainty of rainfall is at the coast. Within 69 years, average annual precipitation at Walvis Bay ranged from 0 to 99.3 mm with a mean of 15 mm. At a distance of 100 km from the coast at Ganab, which also receives summer rain, the mean annual rainfall over a period of eight years was 63.8 mm and varied from 6.1 to 173.5 mm. The gradient from west to east is not only restricted to rainfall but is shown by a great number of climatic factors. A summary of these gradients from the coast inland is given in Fig. 3.4 for a transect through the Central Namib (Lancaster *et al.* 1984). The figure illustrates that for the actual coastline rainfall is not strictly bound to a season but falls more or less erratically, as there can be a rainless period of more than 450 days in a row.

The low precipitation of the Namib results from two effects. Firstly, the easterly trade winds emerging over the warm Indian Ocean lose most of their humidity when rising at the eastern escarpment of the Drakensberge. On their way over the South African continent they dry further over the Kalahari and reach the western escarpment relatively warm and dry. These airmasses sack down onto the low-lying Namib while heating up adiabatically. As they reach the Namib they produce extremely hot and dry winds. These winds are similar to the foehn-winds blowing in the Alps. They are called bergwind in South Africa. These gusty storms frequently lead to heavy sandstorms. The tremendous force of the storm and its ability to carry sand over long distances is demonstrated on METEOSAT-images where sand flags are blown as far as 320 km offshore (Fig. 3.5).

Secondly, the prevailing local southwesterly winds, cooled down by the Benguela Current along the west coast of South Africa, lead to a land inward movement of relatively cold air which is overlain by warmer and consequently lighter air. This inversion prevents convectional rise of the humid air and thus no clouds or thunderstorms can be formed. The cold Benguela Current is also responsible for another characteristic of the Namib: fog. Normally a stable layer of fog lies over the Benguela Current. This is often blown inland by a sea breeze. The advectional fog can penetrate into the Namib as far as 50 km or more before it disappears by heating up over the warmer land surface. For the transect from Walvis Bay to Ganab the number of foggy days as shown in Fig. 3.4 decreases from about 80 to less than 5.

It is still controversial to what extent fog is able to contribute to the annual precipitation and whether the water derived from fog is available to higher plants. As has been already pointed out, the amount of fog precipitation very much depends on the method used to measure it (see Section 2.1.1). The data for fog precipitation given in Fig. 3.4 were obtained with a cylindrical wire-mesh screen. It is doubtful that the plants of the Namib are able to gain that much water from fog. According to Fig. 3.4, fog

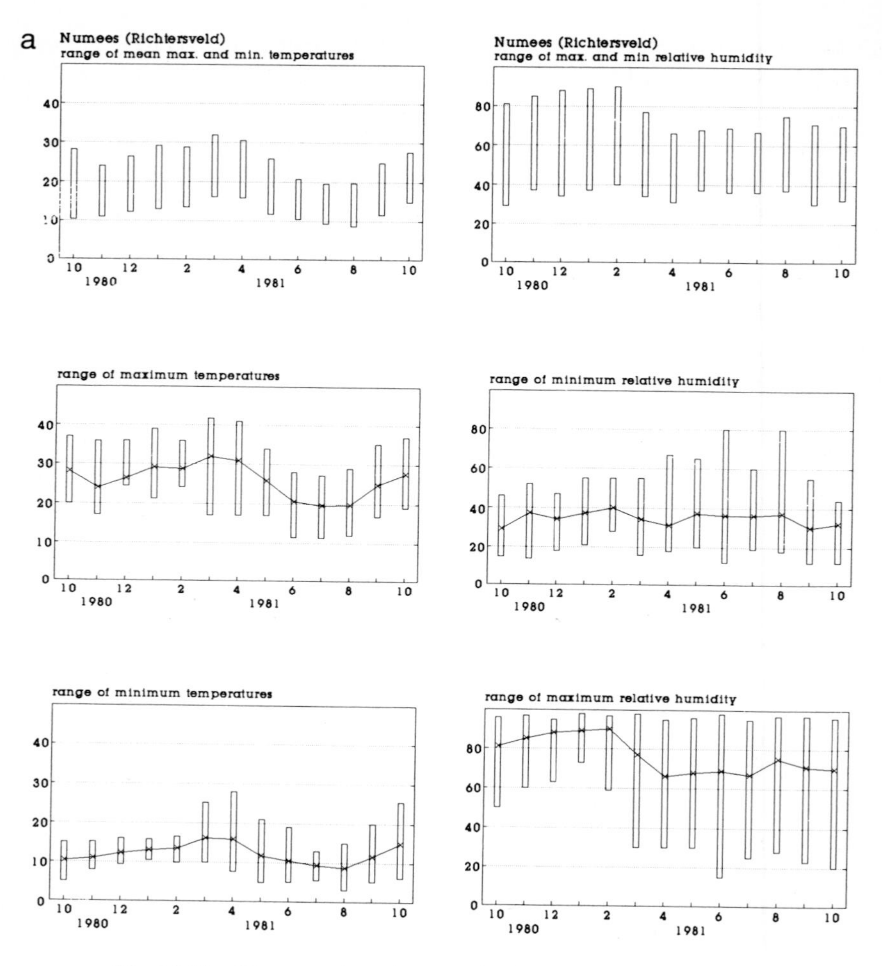

Fig. 3.3 Monthly mean of maximum and minimum air temperature (°C) and relative air humidity, range of maximum temperature and minimum relative air humidity during daytime, and range of minimum temperature and maximum relative air humidity at night for two places in the southern Namib: (a) Numees (Richtersveld) and (b) Kalkgat (Knersvlakte). Monthly means are connected by a solid line.

precipitation can exceed rainfall severalfold. We obtained results from our research site in the Richtersveld, south of the Orange River, at a place about 60 km from the coast. It gets fog frequently throughout the year. An evaluation of METEOSAT images of cloudy and overcast days in the Richtersveld revealed 46 to 86 days per year, while fog over the sea occurred on 56 to 105 days. One should however keep in mind that it is impossible to distinguish clearly between fog and clouds on METEOSAT photographs.

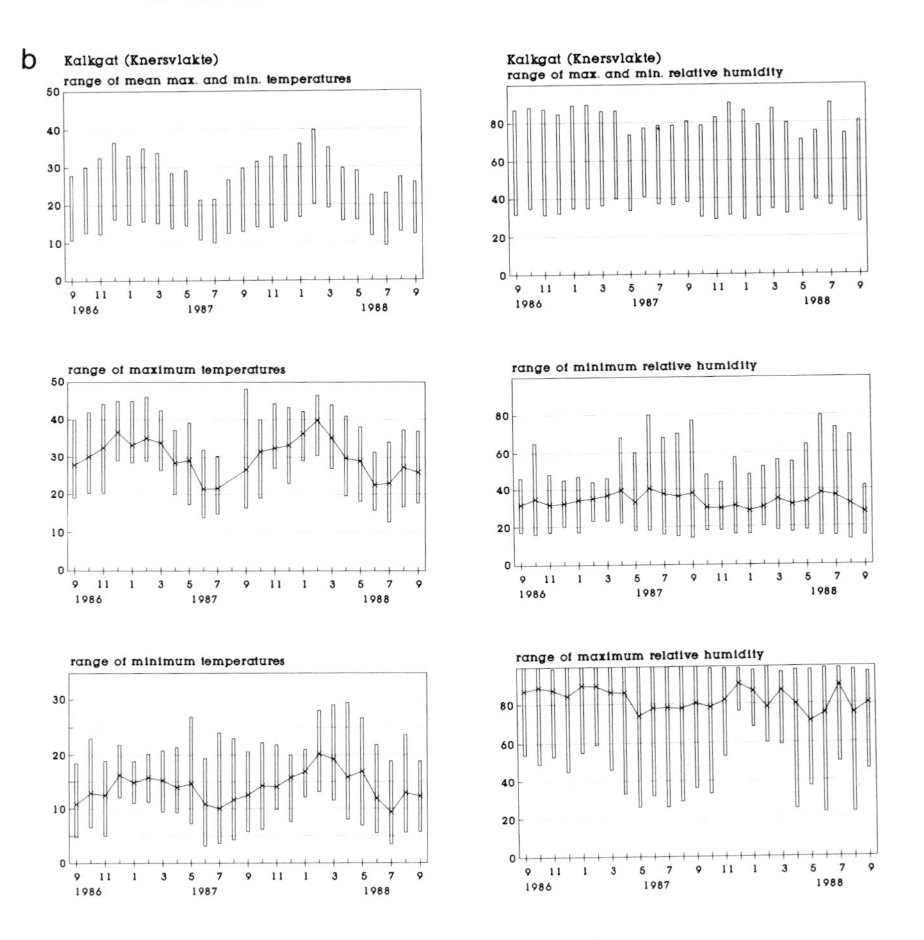

Nevertheless, the data correspond well with those for the Central Namib (Fig. 3.4). With increasing distance from the coast the duration of fog varies considerably. It can last a few minutes – only the passing of a fog patch – but also several days. Normally fog is blown in at night or in the early morning and disappears before noon. This is well documented for the Richtersveld again by the evaluation of METEOSAT images (Fig. 3.6).

The climatic data of the transect given in Fig. 3.4 show that the air temperatures in the Namib increase from west to east, while the relative air humidity values decrease. Due to frequent fog daily and seasonal changes in air temperature and relative air humidity are minimal at the coast and gradually increase further inland. A typical pattern of these gradients is given in Fig. 3.7. Measurements were taken between 11:00 and 12:00 local time during winter on 3 August 1988. A moderate bergwind was blowing

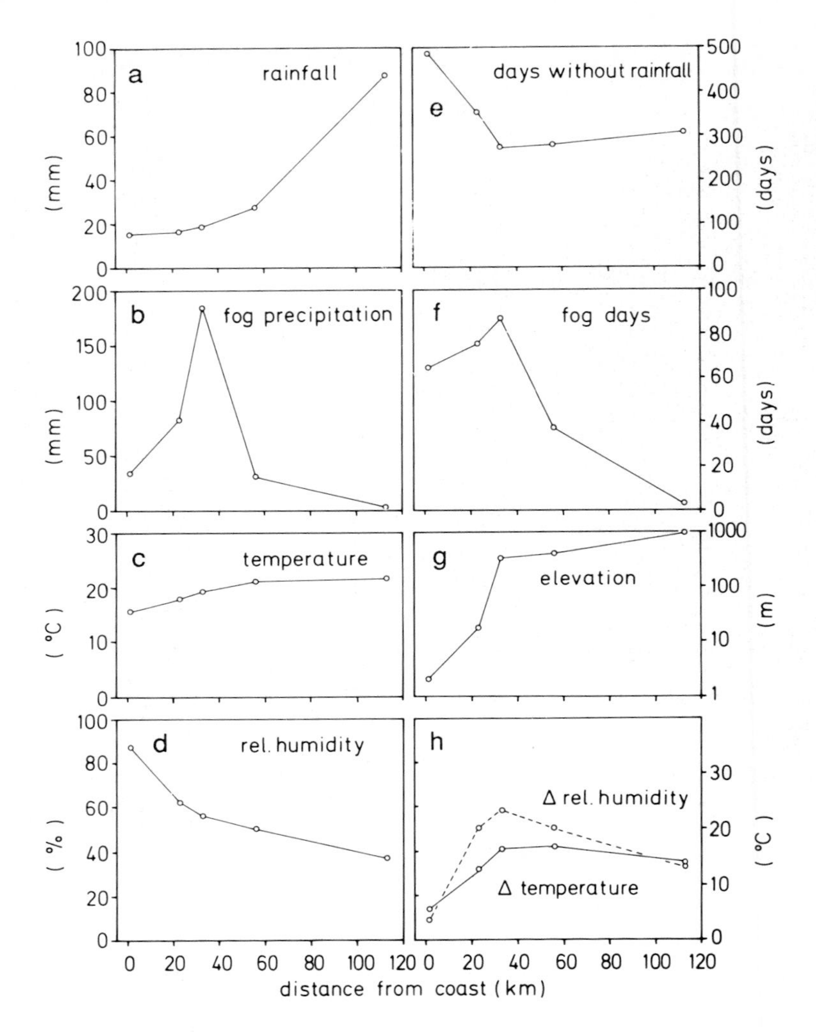

Fig. 3.4 Climatic gradient from the coast inland through the Central Namib desert. Mean annual data of rainfall (a), air temperature (c), and relative air humidity (d) were taken at Pelican Point (distance from coast = 1.5 km). Rooibank (23 km), Swartbank (33 km), Gobabeb (56 km) and Ganab (113 km). Fog precipitation (b) was measured at the same sites except for Pelican Point; instead the data available from Swakopmund were used. The longest period without rainfall (e), the elevation of the weather stations above sea level (g), and the mean annual amplitude (h) of air temperature and relative humidity are also shown in relation to the distance from the coast. Data taken from Lancaster *et al.* 1984.)

Fig. 3.5 Satellite view of South Africa's south and west coast on 2 September 1981. Large dust and sand clouds are blown by easterly winds about 300 km offshore over the Atlantic Ocean. A thick fog line merges the coastline at the border been Namibia and Angola. Further inland the Etosha pan (white ring) and the Okavango swamps (dark delta) are clearly visible. Coastline and land surface contrast is enhanced photographically. (METEOSAT image by courtesy of the European Space Agency (ESA).

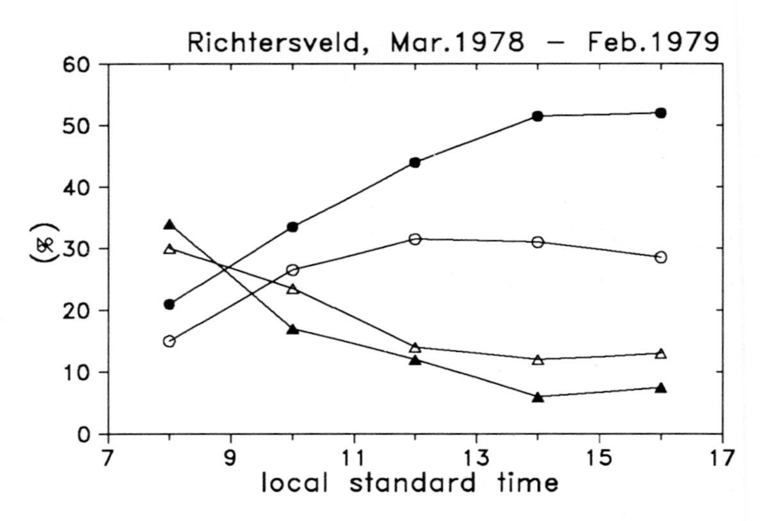

Fig. 3.6 Change in the cloud pattern over the Richtersveld area during the day (average of 344 days). Cloud observations were made from METEOSAT images in the visible channel (0.4–1.1 μm) using four classes of cloud cover: clear sky (●), partly cloudy (○), very cloudy (△) and overcast (▲).

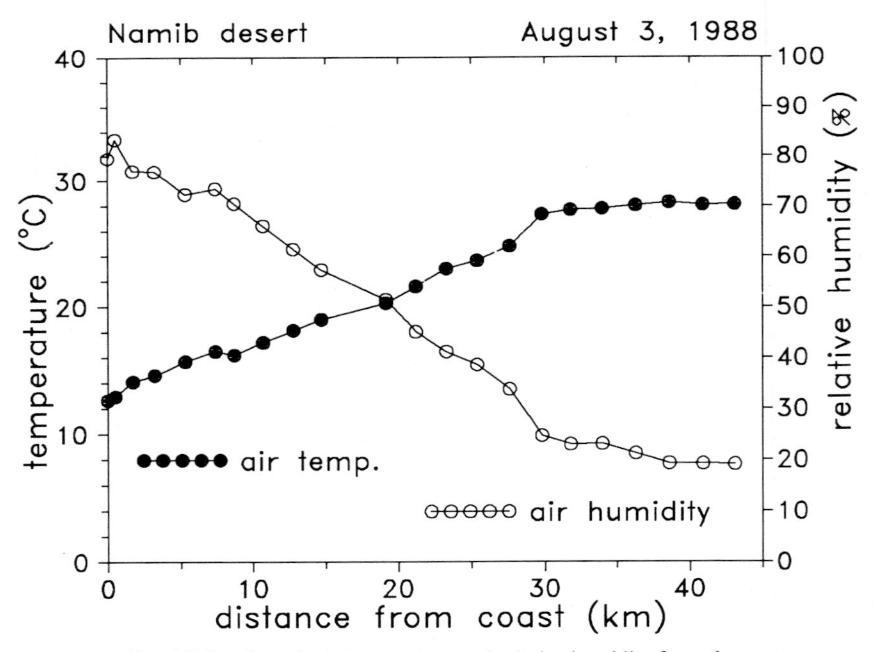

Fig. 3.7 Gradient of air temperature and relative humidity from the coast (Swakopmund) to the interior (Welwitschia Plains). Measurements were taken within one hour between 11:00 and 12:00 along the road from Swakopmund to the Khomas Highland.

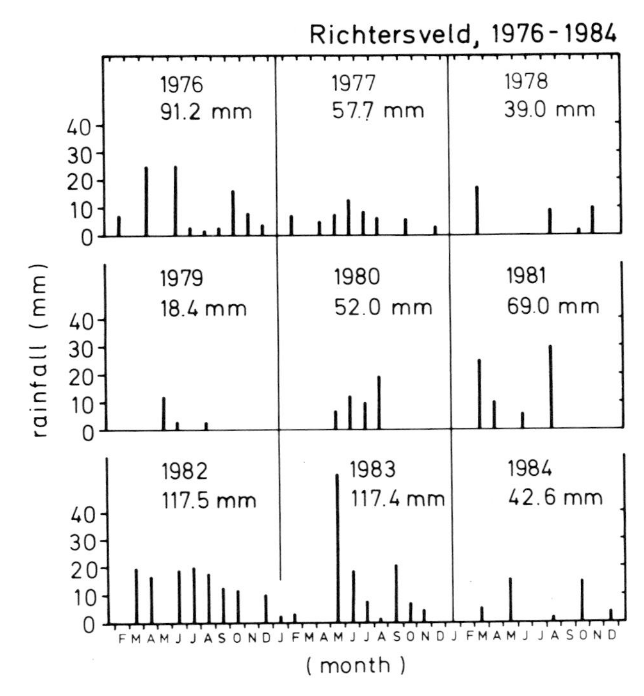

Fig. 3.8 Rainfall pattern at Lekkersing (Richtersveld). (Data by courtesy of the South African Weather Bureau.)

between 30 and 50 km and further inland while a light sea breeze blew from the coast inland. Only the actual coastline was in fog.

3.2.1 The southern Namib

We will now give a more detailed description of the climate of the southern Namib. Rainfall in the southern Namib occurs preferentially in winter from March to October. During winter the stable high pressure belt over South Africa moves to the north so that cyclones, which bring rain to the southwestern Cape region, can extend northwards up to the area of Lüderitz. With increasing distance from the Cape the probability and the amount of rain decreases (Schulze and McGee 1978).

In comparison with the summer rainfall in the Namib the variability coefficient of rainfall is much lower. Over a period of nine years (1976 to 1984) the coefficient is only six. The rainfall pattern of Lekkersing in the Richtersveld is given in Fig. 3.8. The mean for the last 20 years is 64 mm. Fig. 3.8 shows that starting with good rains in 1976, rainfall decreased gradually until 1979 when only 18.4 mm fell but then increased again reaching about double the mean in 1982 and 1983. The 68 mm of rainfall in

Table 3.2. *Rainfall pattern (mm) at Lekkersing (Richtersveld) and at Sendelingsdrif (Orange Valley, Richtersveld) in 1982. (Data by courtesy of the South African Weather Bureau.)*

Date	Sendelingsdrif	Lekkersing
January	—	—
February	—	—
9 March	—	9
10 March	26	9
26 March	—	2
6 April	42	17
May	—	—
4 June	—	8.5
30 June	—	11
1 July	32	14
21 July	—	2.2
22 July	4	4
30 August	11	—
31 August	—	18.5
1 September	—	0.5
9 September	9	—
18 September	4	—
11 October	—	4
13 October	—	5.5
14 October	—	2.5
November	—	—
5 December	—	1.5
12 December	—	6
13 December	—	2
27 December	—	0.7
Total	128	117.5

1981 fell in only 6 days (23 and 24 March, 24 April, 28 June and 28 and 29 August). Table 3.2 shows the rainfall pattern of 1982 for Lekkersing and Sendelingsdrif in the valley of the Orange River, which are only 120 km apart in a north–south direction. This clearly documents the patchiness of rainfall even in the winter rainfall area.

We have seen, in Fig. 3.4, that the amplitude of mean annual air temperature and relative humidity are not high at the coast and increase only slightly with distance from the coast. The southern Namib does not differ significantly from the Central Namib with respect to the annual course of temperature. Fig. 3.9 shows the maximum daytime and minimum night temperature as recorded at Numees in the Richtersveld from 1980 to 1984. During summer the amplitude between the monthly means is about 15 °C but during winter it is only 10 °C. The monthly means of the minimum

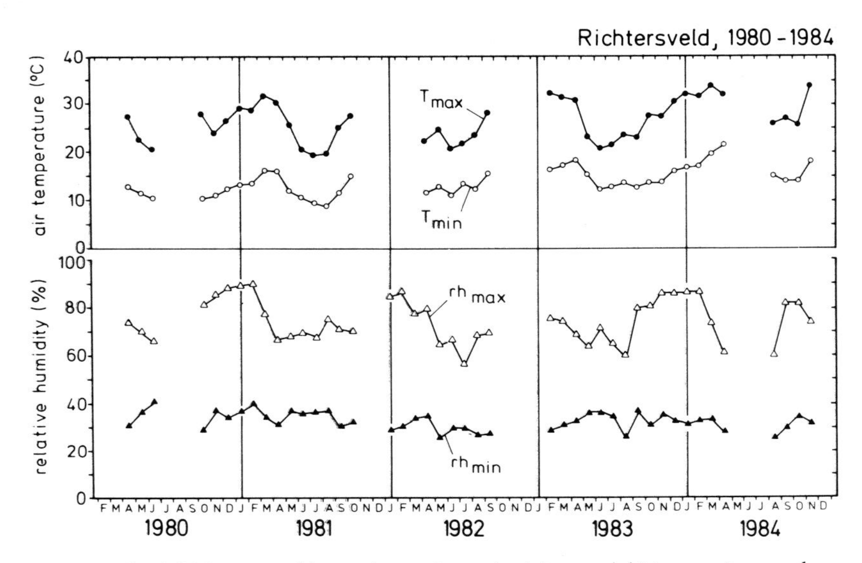

Fig. 3.9 Mean monthly maximum day and minimum night temperature, and mean monthly maximum night and minimum day relative humidity as recorded by a hygrothermograph at Numees (Richtersveld).

night temperature do not change much and even the summer to winter amplitude of the monthly means of the maximum daytime temperature are only 15°C. One noticeable and somewhat unexpected feature can be seen for the monthly means of relative air humidity. Since our humidity sensor could not measure relative humidities below 15%, the lowest values for minimum daytime air humidity are not trustworthy. Although the southern Namib belongs to the winter rainfall area the maximum relative air humidity during the night is much higher in summer than in winter. This, at a first glance, surprising feature has its reason in bergwinds which are very dry and hot fall winds that blow exclusively during winter. In the Richtersveld bergwinds start normally in March and cease in September; from October to February they occur only occasionally. The reason for this is that only in winter can a low pressure cell build up over the ocean. In summer the stable high pressure belt prevents this. Bergwinds can last from one day up to a fortnight. They have a frequency of about one per week and lead to the paradox that the highest night temperatures in the course of a year occur in winter.

Fig. 3.10 shows the maximum and minimum day temperatures together with the lowest and highest relative air humidity in the course of a year. Bergwinds started in March 1983 and became regular events during the

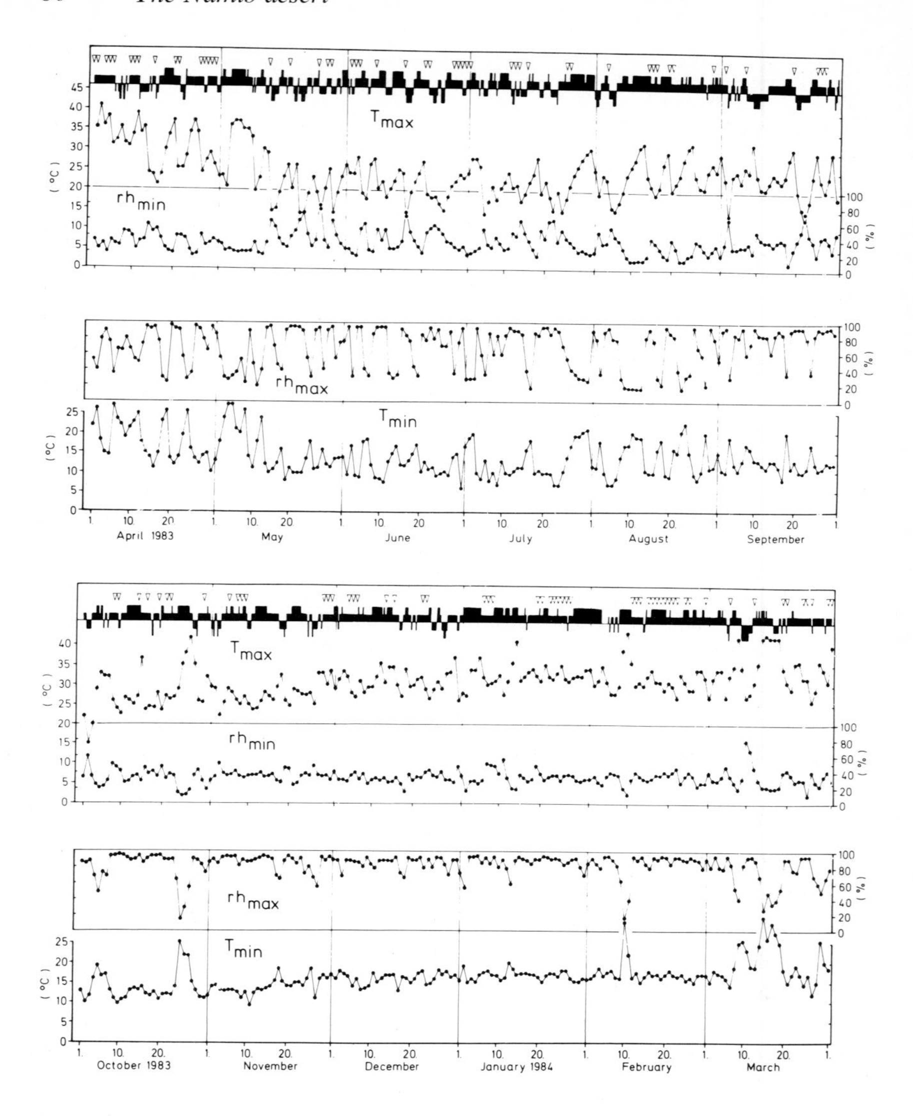

Fig. 3.10 Daily minimum and maximum air temperature (T_{min}, T_{max}) and daily minimum and maximum relative air humidity (rh_{min}, rh_{max}) measured at 50 cm above ground with a Thiess hygrothermograph (90 days recording) in a weather housing at Numees (Richtersveld) from 1 March, 1983 to 31 March 1984. Above the graph of maximum temperature, the cloud pattern over the Richtersveld area is indicated in four classes: upwards large bar = clear sky: upwards small bar = partly cloudy; downwards small bar = very cloudy; downwards large bar = overcast. Days with a thick layer of fog lying over the sea near the coast at the mouth of the Orange River are indicated above the cloud pattern scale (▽). Cloud and fog observations were evaluated from MEOSAT images in the visible (0.4–1.1 μm) and infrared (10.5–12.5 μm) channels.

following months until September. The frequency of bergwinds declined significantly in October with a final heat storm at the end of the month. The following months are characterized by a very stable day and night climate. Bergwinds started again with a very early and solitary event in February 1984 but became regular events in March. Until September they showed a similar pattern as in the previous year. The same pattern of bergwinds occurs in the Knersvlakte and leads to the same consequences as in the Richtersveld. This is well documented in Figs. 3.3a and 3.3b.

The bergwinds which blow only in winter can be seen in Fig. 3.10 in the presentation of maximum air humidity during the night and the corresponding minimum night temperature. Compared with other nights relative air humidity is extremely low during bergwind periods and night air temperature comparatively high. In contrast, during summer night temperature and relative air humidity stay more or less constant (see also Fig. 3.3). It is important to notice that during the rainless summer relative air humidity at night frequently approaches saturation. Due to the facts outlined in Section 2.1.4 this must lead to dewfall during calm and clear nights. This means that there is a reliable source of water in summer and we shall see later (Section 4.3) whether or not this water – though limited – can be utilized by succulents.

During winter there is a steady change between bergwind conditions and foggy or cloudy weather with a higher probability of rainfall. The fact that rainfall is always followed by bergwind is very important. Only an abundant rainfall will supply sufficient water that can reach the rooting horizon of the plants or allow successful germination and seedling establishment. Rainfall of only a few millimetres is of no value because the following bergwind will dry it away more quickly than it can be utilized by the roots; it evaporates before it reaches the rooting horizon.

We started our presentation of the climate of the Namib with annual means and then stepped down gradually to monthly means, and to daily minimum and maximum values. We shall now finish this chapter by giving the daily course of the most important climate factors for plant life: radiation, air temperature and vapour pressure deficit (VPD). As outlined above we must distinguish between three different but typical climate days in the Namib:

(1) the normal day with a cloudless sky and nothing exceptional,
(2) the foggy day with typical advective fog sometimes accompanied by low clouds,
(3) the bergwind day with hot and dry wind throughout day and night.

Representative days of these three types are illustrated in Fig. 3.11.

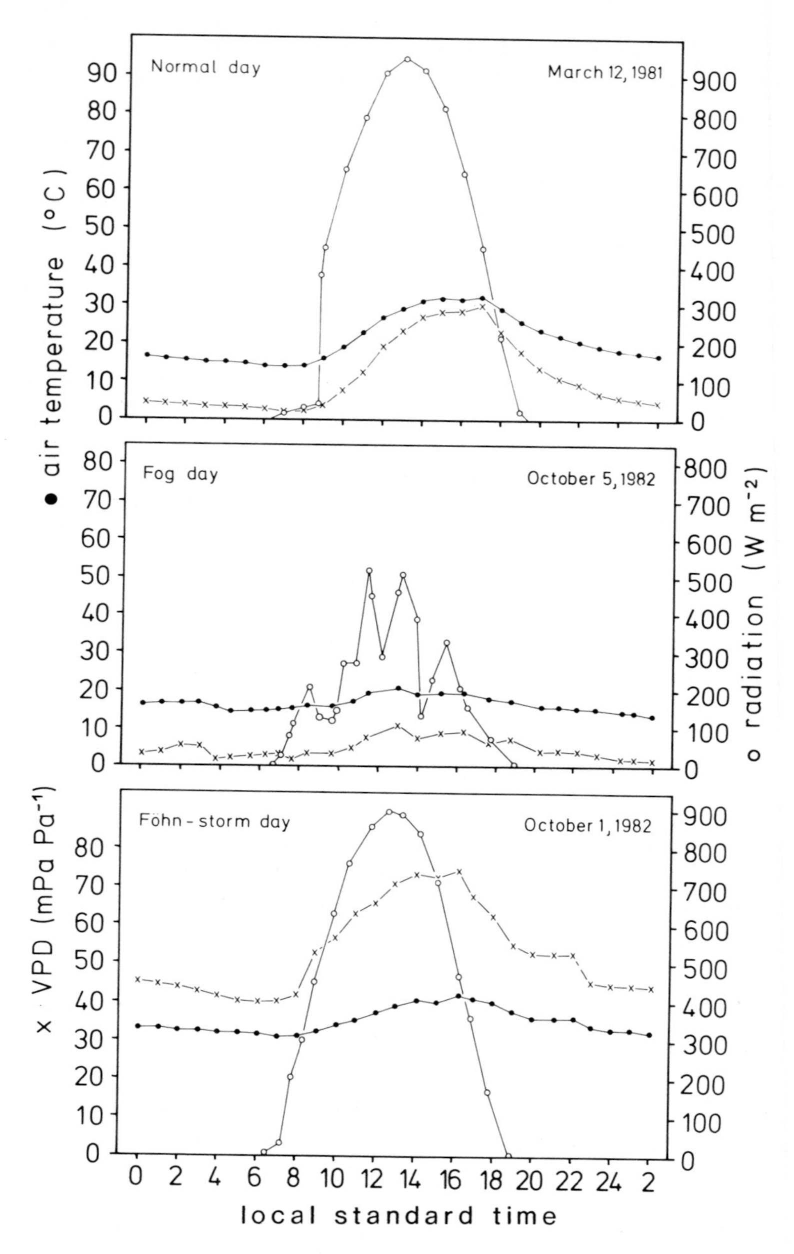

Fig. 3.11 Daily course of air temperature (●), water vapour pressure deficit (x) and global radiation (○) for three typical days measured at 1.80 m above ground by Assmann psychrometer and Licor radiometer at Numees (Richtersveld). The lag time in the radiation curve after sunrise results from shading of a west-facing mountain slope.

In the mountainous areas of the southern Namib the slopes are often steep. Consequently the orientation of the slope plays an important role in determining plant growth. The most pronounced differences are found when east and west-facing slopes are compared. It has been argued that the comparatively luxuriant vegetation on the west-facing slopes is due to a better water supply on these slopes. Advective fog driving in from the west is believed to result in a better water interception on the west-facing slopes than on the east-facing slopes (Jürgens 1986).

A detailed comparison of the microclimate at an east and west-facing slope revealed that radiation and temperature are the most important factors and that they play a crucial role in intensifying the negative effect of each individual factor on photosynthesis and consequently on productivity and plant growth. As soon as plants on east-facing slopes come into shade incident radiation drops to about 10% (Fig. 3.12). This happens between 12:00 and 17:00 in the afternoon when air temperature is still high, especially in winter during bergwind conditions. The available photosynthetically active radiation (PAR) at the prevailing high temperature is insufficient to allow photosynthetic CO_2 uptake to compensate for CO_2 loss by respiration. Hence plants operate below the light compensation point with a significant carbon loss until temperatures drop down after sunset. Plants on the west-facing slope on the contrary experience low light intensities in the morning when, even under bergwind conditions, temperatures are low. Consequently photosynthesis of these plants is able to compensate respiratory CO_2 loss and plants operate above the light compensation point resulting in a considerable carbon gain. These plants come into bright sunlight rather late in the morning and normally temperatures are low to moderate until that moment. During the rest of the day high temperature co-occurs with high radiation allowing the plant to take up CO_2 when water supply is sufficient or at least to operate at the compensation point when water becomes a limiting factor. Consequently plant growth on west-facing slopes is less hampered than on east-facing slopes. Since the storage of water in the plant body is crucial during rainless periods, growth as a result of stored water determines the chance of a plant to survive. As outlined above, this process, additional to a suggested and possibly better water supply, is facilitated on west-facing slopes in comparison with east-facing slopes.

3.3 Plant geography and vegetation

The arid parts of southern Africa have long been recognized as floristically distinct from the surrounding areas. The principal affinity of

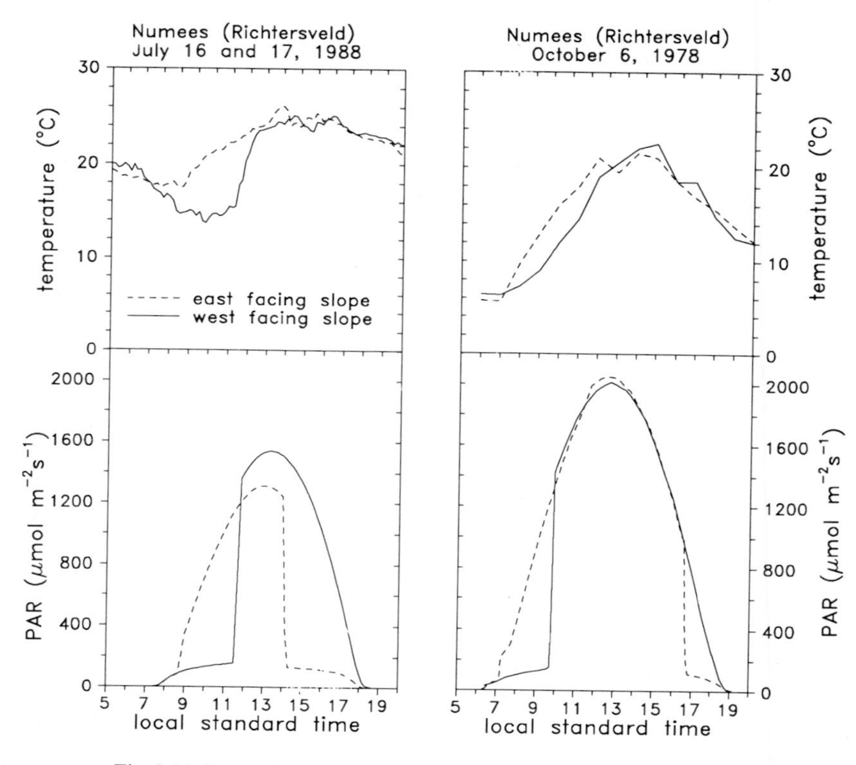

Fig. 3.12 Comparison of air temperature (°C) at 50 cm above ground in the canopy of a succulent bush and of incident solar radiation (PAR, 400–700 mm) in μmol photons m^{-2} s^{-1} on an east-facing (dashed line) and a west-facing (solid line) slope in the mountains at Numees (Richtersveld) on both sides of a ridge. Data in July 1988 were taken during a moderate bergwind condition on two successive days while data in October 1975 were taken on the same 'normal' day.

this flora to the temperate Cape flora in the south or the palaeotropic flora in the east and north has been the subject of some discussion in the literature. Most authors consider the affinity of the flora of arid southern Africa with the palaeotropic flora stronger. Consequently, this dry area most often is regarded as a floristic region, the Karoo–Namib Region, of the Palaeotropic Floristic Kingdom (see Werger 1978a for a review of this literature). Werger (1978a, b) subdivided the Karoo–Namib Region into four domains which are floristically and ecologically distinct and, to a certain level, homogeneous.

(1) The Namib Domain, covering the coastal stretch of Namib desert from Angola to the vicinity of Port Nolloth where the dune area

finds its southern boundary. The land ranges in height from sea level to about 1000 m.

(2) The Western Cape Domain, covering the coastal zone further south as well as the escarpment zone south of Aus, the Tanqua Karoo, and the Little Karoo. This area coincides largely with the winter rainfall and the uniform rainfall areas, but the southern part of the Namib Domain also receives winter rainfall.

(3) The Namaland Domain, covering the escarpment belt and the dry plateau area of Namibia, including the southern Kalahari. This area largely receives summer rainfall but fringes the uniform rainfall area along its southwestern border.

(4) The Karoo Domain, covering the wide plateau area in South Africa. It receives summer rains.

Jürgens (1990) carried out detailed floristic investigations in the restricted area of the coastal belt of southern Namibia and suggested some corrections to the plant geographical picture outlined above. He found that, although there appears to be a distinct vegetation boundary north of Port Nolloth, identified by Werger (1978a, b) as the boundary between the Namib and Western Cape Domains, the succulent flora of the Western Cape Domain continues further north on mountainous 'exclaves' and in a narrow coastal belt to the vicinity of Lüderitz. Therefore Jürgens proposed to include this area in the Western Cape Domain. He further excluded a small, eastern part of the Richtersveld from the Western Cape Domain and included it in the Namaland Domain. Jürgens based his views largely on the distribution pattern of life forms (leaf succulents as against stem succulents) rather than on a complete floristic analysis. Rutherford and Westfall (1986) classified the biomes of southern Africa based on an analysis of distribution patterns of dominant life forms and climatic characteristics. They drew the line between the Desert Biome (Namib) and the Succulent Karoo Biome more or less in the position as suggested by Jürgens (1990) and the line demarcating the Succulent Karoo Biome from the Nama–Karoo Biome (which covers the plant geographical Namaland and Karoo Domains as described above) in the position of the border between the Western Cape, Namaland and Karoo Domains as proposed by Werger (1978a, b).

Within the dry parts of southern Africa several centres of endemism, where a particularly high number of endemic species, usually in one or a few genera, occur aggregated, have been distinguished by various authors. They are outlined in Werger (1978b) and Hilton-Taylor (1987). Jürgens (1986) analysed the distribution patterns of the taxa in the Mesembryanthe-

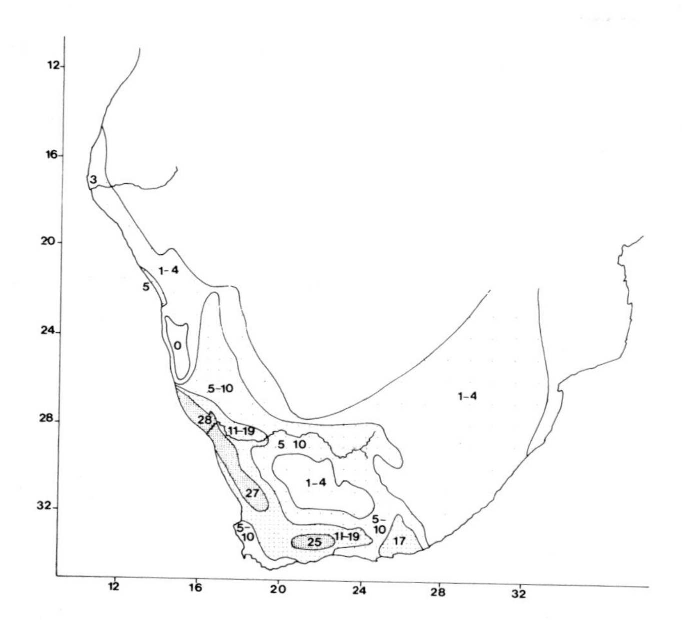

Fig. 3.13 Centres of taxa (genera) of the family of Mesembryanthemaceae. Lines represent isolines of identical frequency of genera per half a degree square of latitude and longitude. Intensely dotted areas mark sites with more than 20 genera. The centre of the Mesembryanthemaceae in the arid parts of the winter rainfall area is obvious as is a secondary centre in the Eastern Cape. (From Jürgens 1990.)

maceae, the richest represented family in the succulent flora of arid southern Africa. He demonstrated that the genera diversity of the family is highest in the winter rainfall area of South Africa and southernmost Namibia in the zone west of the top of the escarpment, as well as in the Little Karoo. In the eastern Cape there is another area of high genera diversity. Towards the summer rainfall area the number of genera rapidly diminishes, though the family is not absent (Fig. 3.13).

Generalized overviews of the vegetation of southern Africa have long been available (Adamson 1938, Acocks 1953, Giess 1971). They show that the vegetation of the Namib Domain consists mainly of very open desert vegetation which becomes denser along the rainfall gradient. It includes desert grassland with a sparse cover of perennial grasses, sometimes intermixed with occasional dwarf shrubs or shrubs in the sandy areas. On more compact soils, or where there is a calcrete bank at or near the surface, dwarf shrubs are more common and following favourable rainy seasons these areas can carry a rather dense vegetation of, mainly annual, grasses and herbs. Along minor drainage lines, in washes and other run-on sites the vegetation gets denser and shrubbier. Here we also find the well-known stands of *Welwitschia mirabilis*, particularly in the Central Namib. Rocky outcrops also support an open vegetation of dwarf shrubs, shrubs and grasses. In the south many of the dwarf shrubs are succulent. Permanent or semi-permanent water courses locally harbour fairly well-developed stands of gallery forest with large trees of *Acacia erioloba* and others. Sites with compact salty soils often support fairly dense vegetation of salt-tolerant grasses or halophytic shrubs. A most peculiar vegetation type in this domain is the well-developed lichen growth locally near the coast.

The Western Domain generally has a denser vegetation than the Namib Domain. Flats, slopes and broken, rocky substrates with a well-developed vegetation of succulent dwarf shrubs and some shrubs prevail. Usually this vegetation is rich in species. Following good rainy seasons these areas are covered with an abundance of colourfully flowering annual herbs, mainly Asteraceae and Brassicaceae, but also species of several other families including some odd, strongly succulent Mesembryanthemaceae. The succulent dwarf shrub vegetation often seems organized in fairly distinct communities, each with their characteristic species, and distributed in a pattern that corresponds to the pattern of hydrological site conditions as determined by topography and substrate characteristics. One such pattern in the Knersvlakte as described by Jürgens (1986) is presented in Fig. 3.14.

On sandy patches, especially near the coast, the shrubby vegetation often is less succulent and contains several sclerophyllous dwarf shrubs and shrubs as well as sclerophyllous grasses. In mountainous areas occasional low trees occur, either succulent or sclerophyllous, and along major drainage lines or water courses the usual riverine vegetation is developed. On the highest tops in this domain, where severe night frosts also occur during the growing season, the vegetation changes drastically, both in physiognomy and in floristics. There a well-developed, narrow-leaved, sclerophyllous dwarf shrub vegetation occurs which has strong affinities with the vegetation of the Karoo Domain.

The vegetation of the Namaland Domain is highly variable. Most common are open stands of tall shrubs, dwarf shrubs, grasses, herbs and occasional trees on the rocky sites along the escarpment and in other broken countryside. Locally, particularly on sites with a better water supply, the woody vegetation can be dense, and temporarily, following good rains, the grass component of the vegetation can be luxurious. There are succulents among the species of this vegetation; stem succulents, in particular, can be locally prominent but leaf succulents are also present. To the east and north these vegetation types gradually change into savanna

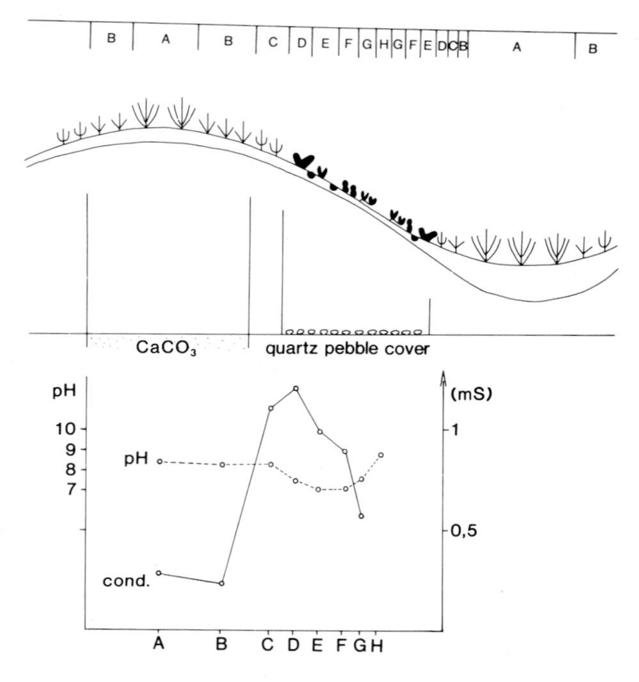

Fig 3.14 Vegetation zonation of a typical plot in the Knersvlakte given in the form of a simplified transect. A catena of diminishing depth of fine textured soil above bedrock causes a series of vegetation units (A–G). Differences in soil properties (quartz pebble content, content of free limestone) as well as pH and conductivity (1:5 extract) are shown in the lower part of the figure. (From Jürgens 1986 and Jürgens pers. comm.)

vegetation with species of *Acacia* and broad-leaved tree species. Also wide, stony desert plains occur in this domain. They carry an open vegetation of shrubs. These can be predominantly stem succulents, such as *Euphorbia gregaria*, or deciduous and mostly bare, semi-sclerophyllous shrubs, such as *Rhigozum trichotomum*. Following favourable rainy seasons these often rather bare plains can show a lush growth of annuals and geophytes. Along major rivers in this domain gallery formations are well developed. Peculiar edaphic conditions determine their own characteristic vegetation types. Most prominent in this respect are the vast stands of desert grassland, grading east and northwards into marginal savanna, on the thick deposits of Kalahari sand. Locally also silt, salt or calcrete pans have formed, each supporting their own characteristic grassy or dwarf shrub communities, usually in a clear zonation around the pans.

In the Karoo Domain narrow-leaved, sclerophyllous dwarf shrublands are prevalent: they cover vast areas of plains and rolling hills, changing in floristic composition with edaphic factors and with the rainfall gradient. To the east grasses become more and more prominent in the vegetation and the dwarf shrublands finally merge into grasslands of savanna affinity. Also on most mountain slopes grasses are more abundant in the Karoo Domain, and shrubs and low trees can be common there. Among the dwarf shrubs and shrubs relatively few species are succulent, but on alkaline soils they can locally achieve dominance. In this wide open landscape drainage lines, rivers and pans all stand out by their deviating, characteristic vegetation types.

Detailed surveys of the vegetation of the arid parts of southern Africa are restricted to limited areas. Most of these studies have frequently been reviewed (e.g. Werger 1978b, 1985, Leistner 1979, Walter 1985, Jürgens 1986, Cowling and Roux 1987) and there is no need to repeat this here. It should suffice to give references to a few detailed vegetation studies of specific interest in that they discuss vegetation types that contain a considerable amount of succulents. Such studies for the Namib Domain include: the area near Moçamedes, Matos and Sousa (1968); the Central Namib, Giess (1968), Robinson (1977) and Moisel and Moll (1981); and the southern Namib, Giess (1974), Robinson and Giess (1974), Coetzee and Werger (1975) and Jürgens (1990). Within the Western Cape Domain very few detailed accounts exist (Compton 1929, Le Roux 1984, Jürgens 1986) and for the adjacent vegetation in the eastern Cape which still contains many stem succulents Van der Walt (1967), Cowling and Campbell (1983) and Hoffman (1989) are relevant. Also the vegetation of the Namaland Domain has enjoyed few detailed studies, except for the southern Kalahari, but there succulents are poorly represented. Relevant studies include Range (1932), Volk and Leippert (1971), Coetzee and Werger (1975), Werger and Coetzee (1977) and Jürgens (1986). In the Karoo Domain succulents play only a minor role in the vegetation.

Jürgens (1986) mapped the vegetation at Numees in the Richtersveld, an area abundant in succulents, and where we did many of our measurements. His map (Fig. 3.15) shows that the vegetation pattern strongly corresponds to the hydrographic pattern, and to a lesser extent to edaphic patterns partly brought about by topography and geology of the substrate. Plants in such desert environments, in Numees as well as in very many other areas in arid southern Africa, often occur strongly clustered, because cracks, fissures and pockets in the rock, as well as the fringes on drainage lines and sites with run-on, favour the establishment and survival of plants.

Some extensive plains and footslopes lack clear drainage patterns at the

3.15 Map of plant communities of Numees. (From Jürgens 1986 and Jürgens pers. comm.)

surface and give the impression of environmental homogeneity. At such sites plants seem to occur diffusely spread over the surface, and often some regularity in the pattern is assumed. Prentice and Werger (1985) studied such a pattern on a homogeneous site on a gently sloping (2°) footslope with a deep soil of sand, silt and some fine gravel at Numees, Richtersveld. The vegetation consisted almost solely of isolated clumps of *Brownanthus schlichtianus* of various sizes. The spatial pattern, analysed by a Monte Carlo method, proved different for clumps of different size. Small plants occurred aggregated, probably as a result of limited seed dispersal from mother plants. Small plants were also aggregated to larger clumps. Medium large plants and clumps had a random distribution, most likely as a result of density dependent mortality among upgrowing plants competing for soil moisture. The largest plants and clumps had a regular spacing. Such an even spacing would minimize the investment in roots for a full exploitation of the soil resources, and thus would be adaptive for plants under long-lasting dry conditions. Under such conditions, which occasionally occur, smaller individuals and individuals with a less favourable balance between investment in roots as against above-ground biomass will have less chance to survive. Similar regular patterns of plant spacing have been reported from other dry areas in the world.

Species diversity in the Numees area is remarkably high compared to other dry areas (Cowling *et al.* 1989). As in other deserts, it seems clearly related to habitat heterogeneity, particularly heterogeneity of soil depth. On rocky sites with strong differences in soil depth over short distances, and where slope and strike of the bedrock determine the abundance of soil-filled fissures and the size of pockets of soil, the species diversity is conspicuously higher than on alluvial or colluvial flats with a deep soil. We determined species diversity in four sites at Numees in 0.1 ha Whittaker-plots (Whittaker 1972, 1977). We found that the clearest difference in species diversity between flats of deep soil dominated by *Brownanthus schlichtianus* and the rocky slopes with strong differences in soil depth and soil pocket size was the much higher rate of increase of species number per unit area (Table 3.3).

Site 1 is a slightly sloping (2° NW) alluvial flat of deep, fine-sedimental soil, dominated by *Brownanthus schlichtianus*. Site 2 is a strongly weathered quartzite and gneiss slope with a slope angle of about 8° W. Soil and boulders are present in large parts of the area but soil depth varies strongly. Site 3 is a steep (40° SE), rocky, quartzite slope with emerging bedrock, fissures and soil pockets. The strike of the bedrock slants parallel to the slope angle and direction. Site 4 is similar to site 3 but 38° west-facing and here the strike of the bedrock is more or less perpendicular to the slope angle.

Table 3.3. *Species diversity in 0.1 ha Whittaker plots at four sites in Numees, Richtersveld. Species diversity is expressed in a least squares regression on the species number and plot size data according to the equation* $S = b + d\log A$, *where* S *is species number and* A *is plot size in* m^2

Site	*b*	*d*
1	2.00	1.00
2	0.58	18.25
3	2.70	17.95
4	8.70	13.34

4

Physiological implications

In Chapter 2 we saw that the climate, mainly the high insolation and the lack of sufficient precipitation, restricts plant life in desert areas. Plants are rooted in their substrate and cannot escape from extreme conditions in their habitat. The evolution of anatomical, morphological and physiological adaptations to arid climate enables some plants to survive in deserts, but often arid areas are not the only habitats where plants that are well adapted to heat and drought stress can grow.

The aim of ecophysiological research on desert plants is to understand what kind of adaptations evolved and evaluate these adaptations with respect to their possible significance for growth or survival of the plant in its desert habitat. At first view, the effectiveness of a plant's adaptations can be evaluated with regard to three primary fields of plant ecophysiology, namely:

energy fluxes
water fluxes
carbon fluxes

However, a detailed analysis reveals multiple interconnections between these three basic flux budgets and also between the numerous singular adaptations. The combination of all characters and functions of the plant relevant for its performance in its habitat characterize the life strategy of a plant species. We will return to this later and first present a detailed description of the fluxes of energy, water and carbon.

Fluxes, either of energy or of mass to and from a plant, usually are summarized in budgets or balances. Frequently the terms budget and balance are used as synonyms. In the subsequent discussions we shall use the term budget for the difference between influxes to the plant and the effluxes from the plant. Consequently, a budget has a value of zero or greater or less than zero. The term balance refers to the effect an actual budget value has on the plant's physiological (and energy) status as a result of a difference between input and output values. If, for example, the water loss by transpiration exceeds the water uptake by the roots, then this resulting negative water budget must be balanced by use of water from the plant's own water reserves. Flux budget minus storage (or plus withdrawal from storage) must be equal to zero. If this demand cannot be met as a result of the physiological status of a plant and its reserves, then the plant cannot survive.

4.1 Energy fluxes

The basis of life on earth is the energy flux from the sun through the energy systems of our planet, the geosphere, back to outer space. The living organisms on earth can, for example by the photosynthesis of green plants, convert a fraction of that flow of physical energy into biochemical energy, which after a shorter or a longer retention time in the biomass of the geosphere is also released to outer space in the degraded energy form of heat.

Two energy flux systems determine the life of a plant, firstly the energy fluxes from and to the plant (including flow, storage and release of heat energy within the plant), which we term the physical energy balance, and secondly the energy turnover in the form of biochemical energy (mainly connected with carbon fluxes) within the plant. The second energy flux system links the single plant by the so-called food chain directly to other living organisms and thus every plant individual is included by its biomass production, consumption and degradation in the overall biochemical energy fluxes of the earth's biosphere. The physical energy balance, however, links the plant's life with the plant's physical environment and for plants living in deserts, with high solar radiation fluxes and a high level of energy in the form of sensible heat (high temperatures of the soil and the ambient air), is a driving force for the evolution of features specialized to allow growth in arid habitats. The ways and extent to which factors of the physical environment influence or restrict plant life in deserts are outlined in the subsequent presentation of single features of the physical energy fluxes.

4.1.1 The physical energy fluxes around a plant

Fig. 4.1 shows the main energy fluxes between the plant and its environment. The plant receives, as already outlined in Section 2.1.2, direct solar radiation, sunlight reflected by clouds or the nearby vegetation and

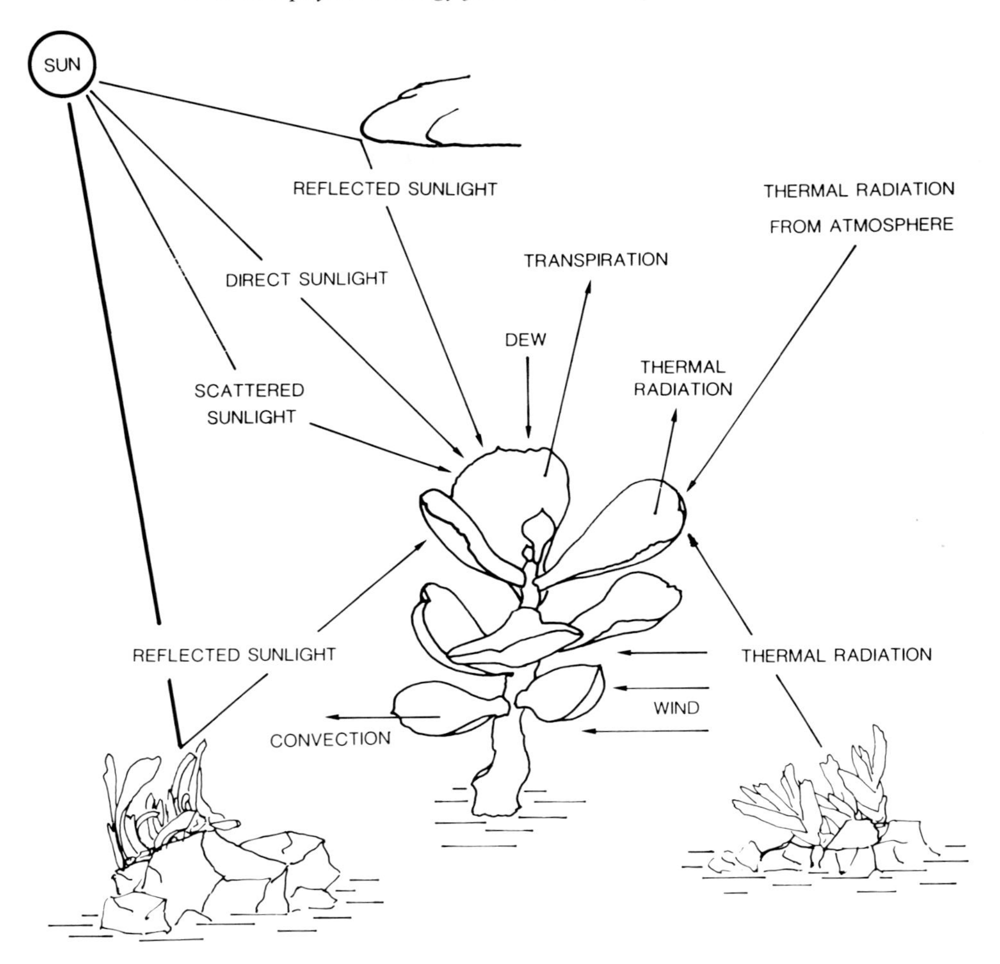

Fig. 4.1 Energy fluxes around a desert plant.

the soil surface, and scattered sunlight from the sky. In addition to this so-called global radiation, the plant receives thermal radiation (longwave infrared radiation) from the atmosphere and the environment. The plant itself emits thermal radiation according to the Stefan–Boltzmann law for blackbody radiation. Energy is also transferred from or to the plant by transpiration or dewfall and by convection. Aside from these main fluxes shown in Fig. 4.1, energy exchange also occurs in connection with the plant's metabolism, heat transport to or from the substrate and short-term heat storage or release connected with changes in the plant's temperature.

Of the radiation energy impinging on the plant, some is reflected or transmitted, while the rest is absorbed and in its major part converted to sensible heat. The absorbed fraction of global radiation G_a and the absorbed thermal radiation R_a are the main energy inputs to the plant. Only under certain circumstances is heat energy generated by respiration, Q_M, a significant quantity, e.g. heat generation in flower stalks of some species of Araceae (Leick 1910, Knutson 1975). The plant dissipates energy by emitted thermal radiation R_e and converts solar energy into biochemical energy Q_P by photosynthesis. Energy flux to and from the plant occurs through convection C and phase transition of water Q_W. Heat energy put into or taken out of storage tissue within the plant is Q_H. In special cases this term Q_H also includes heat energy conducted between the plant and its substrate.

For a given timespan the sum of all these energy quantities must equal zero, therefore

$$G_a + R_a + Q_M - R_e \pm C \pm Q_P \pm Q_W \pm Q_H = 0 \quad (4.1)$$

Different morphological and anatomical features (e.g. inclined position of leaves, epicuticular waxbloom) are considered to be adaptations to prevent excessive leaf temperatures caused by high insolation or are supposed to modify convective energy transfer (e.g. a rough epidermis). However, only a careful consideration of the impact of such structures on the different terms of the physical energy balance of a plant can reveal to what extent morphological or anatomical specializations are proper adaptations.

4.1.2 Optical properties of plants and the soil

The amount of radiation energy absorbed by a surface either of a plant or its surroundings depends on the radiation energy incident on a surface at a particular wavelength and the property of that surface to absorb radiation at the same wavelength. Completely opaque substances reflect radiation only at their surface and no radiation is transmitted. Plant structures, however, are transparent, at least to some extent, for radiation at certain wavelengths. Plants reflect radiation not only at their surface but also within their structures (Fig. 4.2) and even at the plant–air boundary at the surface opposite to the surface where the radiation impinges. Transparent structures can also transmit radiation.

The quotient of radiation energy which is reflected by a leaf and the radiation which is incident on that leaf is termed reflectivity, $r(\lambda)$, and depends on the radiation's wavelength. The similar quotients for the transmitted and the absorbed radiation are referred to as transmissivity, $t(\lambda)$, and absorptivity, $a(\lambda)$, respectively. As the radiation incident on a

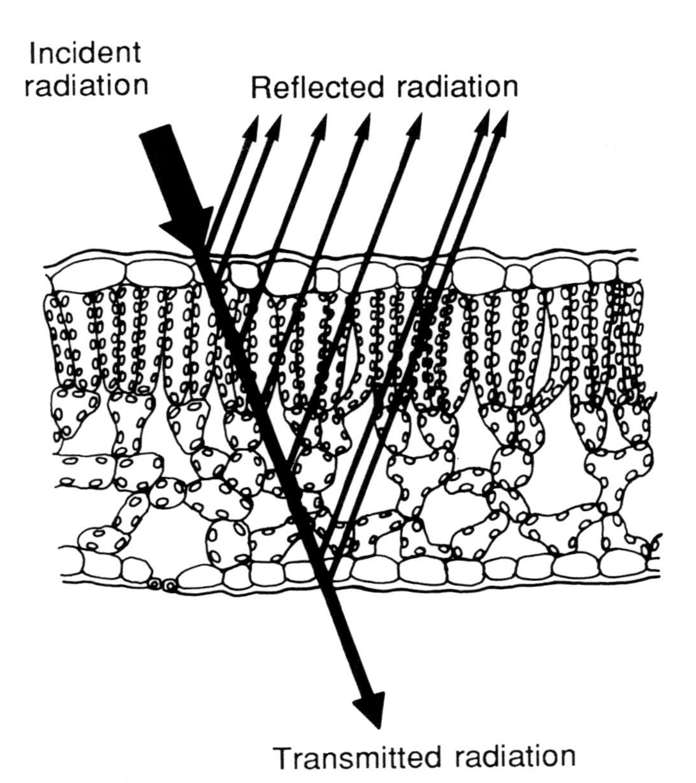

Fig. 4.2 Fractioning of radiation incident on a leaf and origin of reflected radiation (refraction and multiple reflection not considered). Radiation which is neither reflected nor transmitted is absorbed.

structure is either reflected, transmitted or absorbed, the spectral properties – reflectivity, transmissivity and absorptivity – combine into the equation.

$$r(\lambda)+t(\lambda)+a(\lambda)=1 \tag{4.2}$$

By measuring the reflected and the transmitted radiation one can determine $r(\lambda)$ and $t(\lambda)$, and thus $a(\lambda)$ can be calculated. The term absorptivity is *per se* defined for a single wavelength as are reflectivity and transmissivity. Sometimes a mean value for a selected wavelength range is used for comparisons, for instance for mean absorptivities $\bar{a}$ of different species. However, such a value determined by the equation

$$\bar{a}(\lambda_1-\lambda_2)=\frac{\int_{\lambda_1}^{\lambda_2} a(\lambda)\,\mathrm{d}\lambda}{\lambda_2-\lambda_1} \tag{4.3}$$

must not be confounded with the so-called absorptance a_R (or reflectance r_R

or transmittance t_R). The absorptance is the quotient of the absorbed fraction $R_a(\lambda)$ of the radiation $R(\lambda)$ and the radiation incident on the receiver within a specified wavelength range (λ_1 to λ_2), thus

$$R_a(\lambda) = R(\lambda)a(\lambda) \tag{4.4}$$

and

$$a_R = \frac{\int_{\lambda_1}^{\lambda_2} R_a(\lambda)\,d\lambda}{\int_{\lambda_1}^{\lambda_2} R(\lambda)\,d\lambda} \tag{4.5}$$

Values like a_R, which were also termed weighted mean absorptivities, are strictly valid only for the radiation R defined in its spectral distribution of irradiance (W m^{-2} nm^{-1}) and for a selected wavelength range from λ_1 to λ_2. Such weighted mean values are frequently used for the case where the radiation is the global radiation at clear sky conditions with a wavelength range from 300 to 3000 nm or its fraction of photosynthetically active radiation (PAR) from 400 to 700 nm.

The spectral properties of a plant organ can vary considerably during development and senescence and also seasonally, e.g. by colouring during drought periods, but for succulents such information is limited (Eller and Willi 1977, Eller 1982).

Fig. 4.3. shows mean absorptivity values for succulents compiled after data of Gates *et al.* (1965). For ultraviolet radiation below 400 nm wavelength the absorptivity is high. Between 400 and 700 nm (PAR) green plants have high absorptivities resulting from their pigmentation. Multiple scattering of the light rays within the leaf by cell walls and other structures lengthens the light path, and the absorption bands of the pigments, mainly chlorophylls, are broadened and intensified.

At about 700 nm wavelength an abrupt decrease of the absorptivity is the consequence of steep increases in reflectivity and transmissivity. Absorption increases again at wavelengths greater than 1200 nm. The low absorptivities between 700 and 1200 nm are at first surprising but find their explanation in the lack of molecular absorption bands in wavelengths shorter than 1400 nm. This low absorption in the shortwave infrared yields advantages for the plant because most of the infrared solar radiation is reflected or transmitted. By this means the plant can avoid excessively high temperatures which would be unavoidable if a high absorptivity for infrared solar radiation resulted in a high energy input to the plant. Beyond 2500 nm wavelength the reflectivity of plant surfaces is very small, in most cases 2 to 6% (Gates and Tantraporn 1952, Wong and Blevin 1967). Transmissivity is almost nil and consequently an absorptivity of 0.94 to

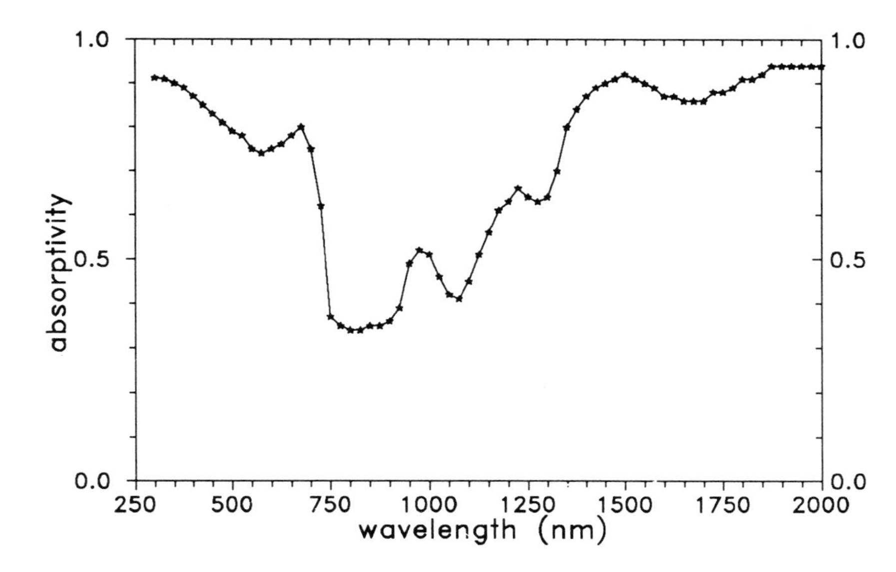

Fig. 4.3 Absorptivity values of succulents (mean values calculated after Gates *et al.* 1965).

0.98 results. Since the impact of global radiation at the earth's surface is restricted to the wavelength range between 300 and 3000 nm, this high absorptivity for wavelengths beyond 3000 nm is unimportant with respect to solar radiation. But a surface can only emit substantial amounts of thermal radiation if its absorptivity for thermal radiation is high, since according to Kirchhoff's radiation law emittance is equal to absorptance at the same wavelength. The high absorptivity values of a plant's surface for wavelengths longer than 2500 nm cause on the one hand a high energy input to the plant from absorbed thermal radiation from the atmosphere or the surroundings, but on the other hand the plant has a substantial energy output by thermal radiation emitted by the plant's surface according to its actual surface temperature.

In Fig. 4.4 the spectral properties of the leaf of *Tylecodon paniculatus* show the interrelationships between reflectivity, transmissivity and absorptivity of a moderately succulent green leaf.

How much of the incident radiation is reflected or transmitted depends on the optical properties of the chemical constituents of the cells and of the cell structures. Transmissionless substances reflect radiation only at their surface whereas the more or less transmitting plant structures reflect radiation not only at the surface where the radiation impinges, but also at their complex internal structures.

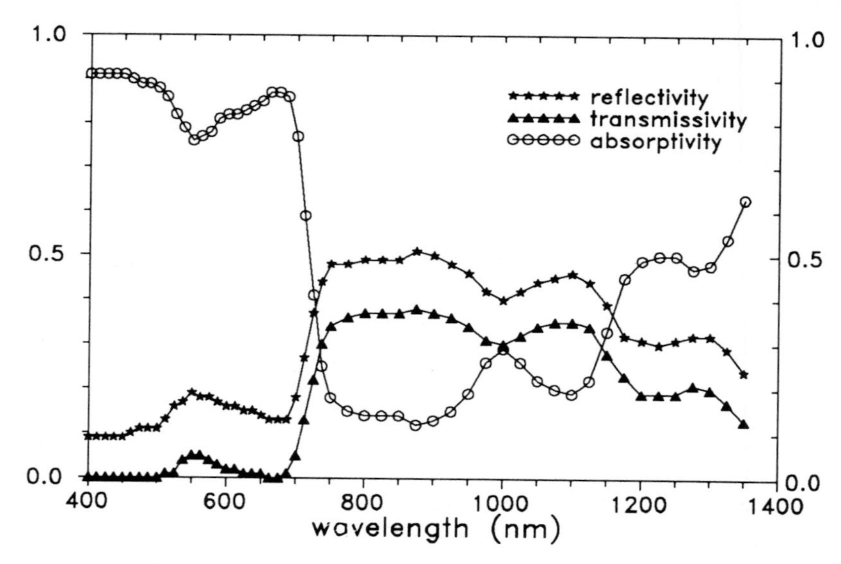

Fig. 4.4 Spectral properties of a leaf of *Tylecodon paniculatus*

The green colour of leaves results only from radiation reflected by the chloroplasts which are situated below the epidermis. The fact that the reflectivity of a plant surface depends not only on its surface structures is also demonstrated by the absorption bands of liquid water, e.g. at 970 nm wavelength, which are clearly visible in the curve for the reflectivity as for the transmissivity of *Tylecodon paniculatus* leaves (Fig. 4.4). This could not be the case if a part of the 'reflected' radiation had not transversed equally thick water layers as the transmitted radiation has.

The chlorophylls absorb little radiation between their strong absorption bands below 500 and above 600 nm wavelength and most of the radiation of the green colour of the light spectrum is either reflected or transmitted by these pigments. The consequence is an increased reflectivity and transmissivity for wavelengths around 550 nm (Fig. 4.4) and thus the absorptivity of a green leaf has low values in this region of the spectrum (the so-called green gap). However, yellow to reddish pigments such as carotenoids, anthocyanins or betalains can alter the absorptivity throughout the wavelength range of PAR (Fig. 4.6).

It seems trivial that for a partially transmitting object like a leaf absorption increases with thickness. Fig. 4.5 compares different leaves of plants growing in the same habitat in the southern Namib desert, and reveals that this generalizing assumption is only valid for infrared solar radiation. No correlation between leaf thickness and absorption exists for

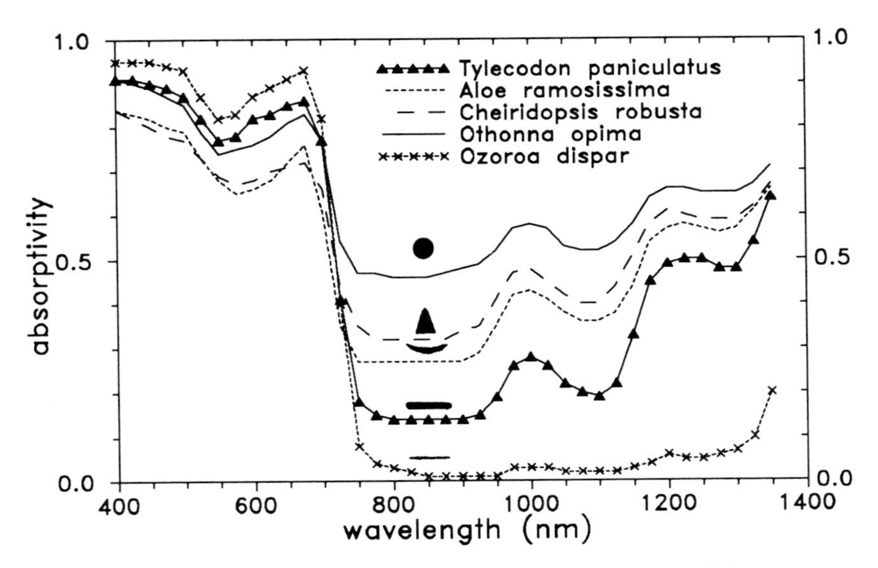

Fig. 4.5 Absorptivities of leaves with different leaf expansion coefficients *S* (surface area in $cm^2\ g^{-1}$ fresh weight). Inserted into the figure are sketches of the leaf cross-sections. The *S*-values are as follows: *Othonna opima* 3.91; *Cheiridopsis robusta* 3.80; *Aloe ramosissima* 5.55; *Tylecodon paniculatus* 8.55; *Ozoroa dispar* 41.67. (From Eller *et al.* 1983.))

radiation up to 750 nm wavelength (Fig. 4.5). The thin leaf of *Ozoroa dispar* absorbs 84%, but the thick leaf of *Cheiridopsis robusta* only 62% of visible radiation.

Colouring, epicuticular structures like wax bloom or trichomes, and the xeromorphy of the epidermis were very often cited to be adaptations of desert plants in order to reduce the input of energy from absorbed solar radiation. Colouring is an eye-catching fact and one tends to believe, for example, that a red leaf must have an absorption that is very different from that of a green one. However, if we are looking on a plant surface we must remember that the human eye detects only that part of radiation that is reflected and by looking at a leaf held against a light source the eye perceives only the transmitted radiation. The absorbed radiation is the fraction of the radiation energy that is neither reflected nor transmitted, and therefore cannot be quantified by looking at a leaf. Moreover, the human eye senses spectral radiation of different colours (e.g. blue, green or red) of the same spectral irradiance value with a different sensitivity. That means human beings cannot even determine by eye the proper amount of visible solar energy that is reflected or transmitted by a plant.

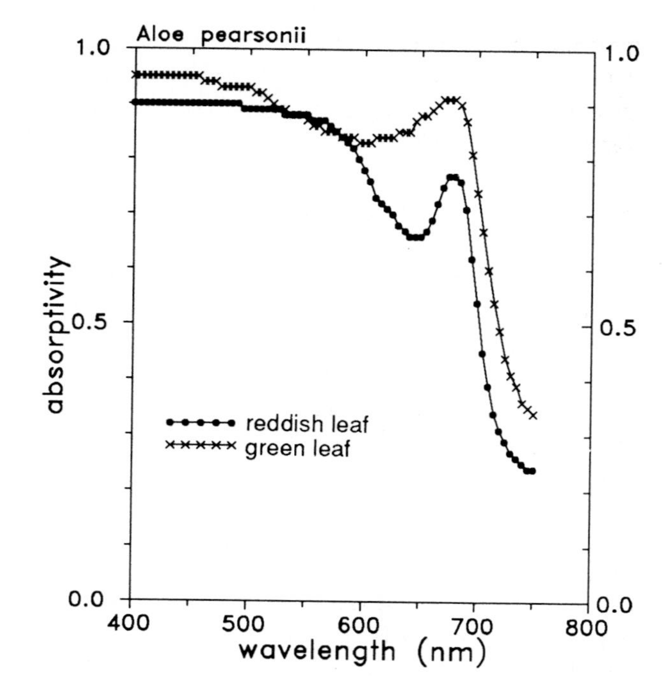

Fig. 4.6 Absorptivities of green and reddish leaves of *Aloe pearsonii*. Measurements were made in the natural habitat at Numees (Richtersveld). Leaf thicknesses at the measured spots were 5.0 mm (reddish) and 4.7 mm (green) respectively.

In Fig. 4.6 the absorptivity values of green and reddish coloured leaves of *Aloe pearsonii* are compared. Aloes often show different colouring of leaves on the same plant. At the top the leaves are fully exposed to the sun and tend to turn reddish whereas the leaves at a lower insertion level are partially shaded and thus remain green. From Fig. 4.6 it becomes evident that in the case of *Aloe pearsonii* colouring induced by light stress lowers the absorptivity for visible solar radiation.

The colour white is not a spectral colour but results from an equally high reflection (or transmission) of all colours of the visible spectrum. Trichomes or wax bloom on a leaf epidermis can turn a leaf white by the total reflection of radiation at the boundary between the optically denser crystals of wax bloom or the cell walls of trichomes and the optically less dense air. A number of studies have concentrated on the effect of wax bloom, leaf hairs or spines on the optical properties of plants growing in arid environments (Gates *et al.* 1965, Cameron 1970, Sinclair and Thomas 1970, Ehleringer *et al.* 1976, Eller and Willi 1977, Ehleringer and Björkman 1978, Eller 1979, Ehleringer 1981).

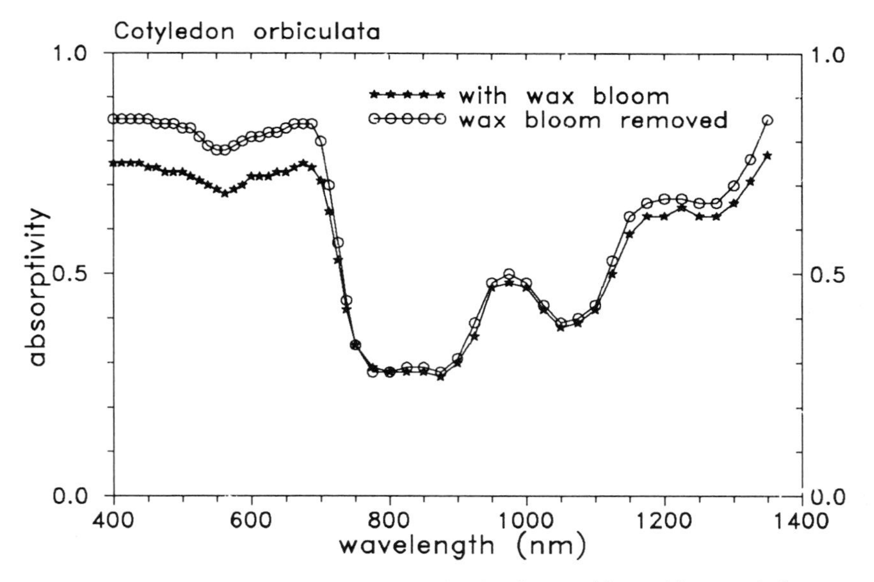

Fig. 4.7 Absorptivity of *Cotyledon orbiculata* leaves with wax bloom and after the wax bloom has been removed.

That a white hair or wax cover increases reflectivity for visible radiation is evident, but a quantification by measurements reveals that the benefit for the plant from decreased energy input by a decreased absorption of solar radiation is much smaller than expected. Fig. 4.7 compares the absorptivities of a leaf of *Cotyledon orbiculata* with wax bloom and after removal of the wax bloom. Although leaves of *Cotyledon orbiculata* do not have a very thick and very white wax layer, absorption in the visible wavelength range from 400 to 750 nm is increased by about 10% if the wax bloom is removed. Sinclair and Thomas (1970) reported for *Cotyledon orbiculata* an increase of 20% for the same wavelengths. Such differences are frequently observed and are either of genetical origin or can result from different microclimatic conditions during leaf expansion that result in different amounts of wax on the epidermis of a leaf. Eller and Willi (1977) showed also that the youngest leaves of *Kalanchoe pumila* had the highest density of wax bloom and therefore the lowest absorptivity for visible radiation. The absorptivity increased with leaf expansion. This can result either from a decrease of the density of the wax bloom, the same amount of wax being spread out on the increasing epidermis area, or a loss of wax crystals due to mechanical action of wind, dew or rainfall.

It is known from investigations with non-succulent leaves (Eller 1979, Gates 1980) that trichomes influence spectral properties in the same way as wax bloom. In contrast to other groups of plants, e.g. the herbs of humid climates, hairy leaves are scarce among succulents. *Senecio medley-woodii* (Asteraceae), *Gibbaeum pubescens* (Mesembryanthemaceae) and *Kalanchoe tomentosa* (Crassulaceae) are some examples of the rare exceptions. A dense cover of spines, however, can also produce alterations in the optical properties similar to those for trichomes (Gates *et al.* 1965). Determinations of the optical properties of the hairy leaves of *Senecio medley-woodii* and the same leaf with the hairs removed give results that are almost identical to the results presented in Fig. 4.7 for the wax-covered leaves of *Cotyledon orbiculata*.

One feature is common to a white wax bloom and a dense cover of white trichomes: both increase reflectivity and as a consequence decrease absorptivity only for visible radiation (400 to 750 nm wavelength) but for infrared radiation with wavelengths larger than 750 nm the changes in the optical properties (Fig. 4.7) are very small or almost nil (see also Section 4.1.4).

Little experimental work has been done so far to establish quantified correlations between the density of a wax bloom or of hairs and the environmental conditions of the habitat of a species. For the non-succulent pubescent *Encelia farinosa* growing in arid regions of California it is known from studies of Ehleringer *et al.* (1976) and Ehleringer and Björkman (1978) that the absorption of the leaves varied by 10% between plants collected at different sites and the absorptivity declined steadily as the season progressed and the average maximum temperature of the ambient air increased. Together with a concomitant reduction of leaf size during the season, this attenuation of the absorption of solar energy yielded a more favourable energy balance and thus reduced leaf temperatures while other environmental conditions became harsher. These researchers also established a strong correlation between the increased pubescence and decreased amount of precipitation received during the current growing season before the sampling date.

Succulents very often have a scleromorphic epidermis and one would expect that a xeromorphic leaf with its thick epidermal cells and a well-developed cuticle would have a low transmission for solar radiation and thus a high absorption compared with a mesomorphic leaf. However, comparisons of leaves with the aim of obtaining evidence for whether or not the degree of xeromorphy determines leaf spectral properties must be made with leaves of the same leaf geometry and size and with leaves of plants of the same growth form from the same or similar habitats. In Fig. 4.8 we give one such example for the leaves of *Delosperma* sp. which has a xeromorphic epidermis and the leaves of *Delosperma cooperi* which has idioblasts on the

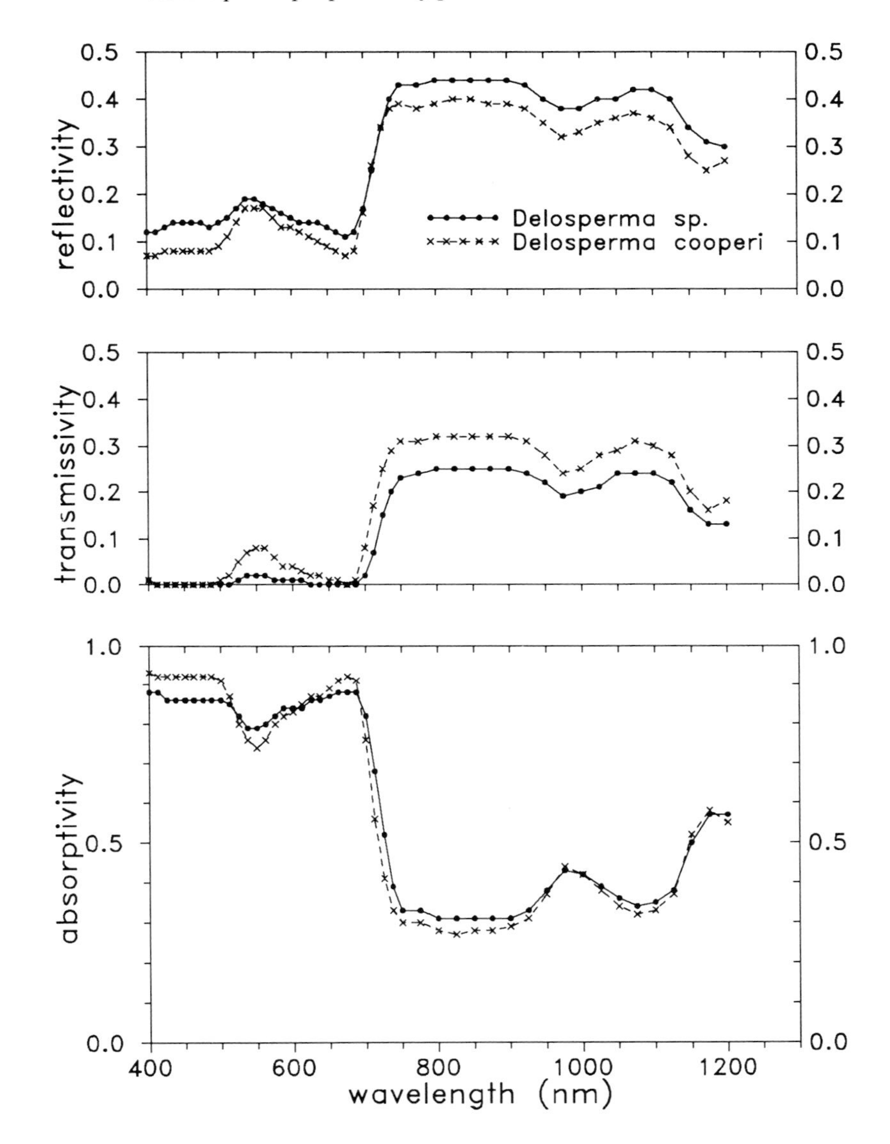

Fig. 4.8 Spectral properties of leaves of *Delosperma* sp. with a xeromorphic epidermis and *Delosperma cooperi* with epidermal idioblasts. (After Tanner and Eller 1986.)

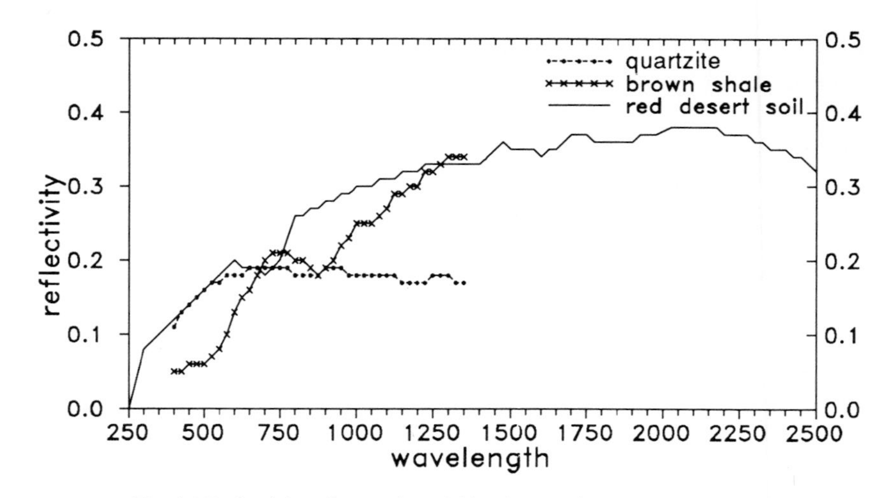

Fig. 4.9 Reflectivity of quartzite pebbles, brown shale and red desert soil. (Data for red desert soil are taken from Leeman *et al.* 1971.)

epidermis. Although the reflectivities and the transmissivities for infrared solar radiation are consistently different for the two leaf types, the absorptivities are nearly identical. The difference in the visible (400 to 750 nm wavelength) spectrum is mainly a consequence of a different pigment content. These results and the more detailed investigations of Tanner and Eller (1986) with leaves of *Delosperma* sp., *Delosperma lehmannii*, *Delosperma cooperi* and *Trichodiadema* sp. aff. *barbatum* proved that neither of the two leaf types and thus the two epidermis types yield advantages with respect to the amount of solar energy absorbed by the leaves.

The plant not only gets radiation from the sky, but also solar radiation that is reflected by its surroundings. The magnitude and the spectral composition of this radiation depends on the reflection power of the environment of a plant. This environment consists either of rocks and soil or of other plants. Compared with a green plant, rocks and soil usually reflect higher fractions of solar radiation in the visible, especially red, radiation (Figs. 4.4, 4.8 and 4.9). One could suppose that a bright white quartzite pebble has the highest reflectivity for visible radiation but Fig. 4.9 does not support this opinion. For the other wavelengths in the visible spectrum the reflectivities are nearly the same. In the wavelength range of the sun's infrared radiation (750 to 1350 nm) quartzite reflects about 15 to 20% less radiation than the red desert soil. So we can assume that radiation input to the plant from solar radiation reflected from the soil is reduced if a plant grows in a habitat with bare soil covered with quartz pebbles instead of uniformly red soil.

The absorptivity of a receiver of radiation is one property that determines the amount of absorbed energy but the other is the actual value of the radiation energy of a given wavelength that impinges on the receiving surface. The subsequent sections will give a survey of the qualities and the magnitude of radiation fluxes that must be taken into consideration to assess the plant's energy balance.

4.1.3 Solar and thermal radiation

The radiation incident on the vegetation or a single plant consists of two groups of radiation fluxes with characteristic wavelength ranges, namely

(1) Solar radiation (or global radiation) with wavelengths between 300 and 3000 nm, very often somewhat imprecisely called 'shortwave' radiation,

and

(2) thermal radiation with wavelengths (for surface temperatures where higher plant life can exist) beyond 2500 nm, often termed 'longwave' radiation.

Direct and scattered sunlight and the solar radiation reflected by clouds are the components of global radiation at the earth's surface. Fig. 4.10 shows the energy distribution of global radiation on a clear day in the Namib desert. The wavelength range of solar radiation at the earth's surface can be subdivided into ultraviolet radiation up to 400 nm, visible radiation (400 to 750 nm) and infrared radiation above 750 nm. Direct solar radiation energy fluxes vary with the time of the day, the season and with geographical latitude (see Fig. 2.5). On its way through the atmosphere the extraterrestrial solar radiation is attenuated and the spectral energy distribution of global radiation at the earth's surface shows absorption bands caused by molecular absorption in the atmosphere by O_2, O_3, CO_2 and especially water vapour.

On its way through the atmosphere some of the nearly perfect parallel solar rays are scattered by the molecules of the atmosphere or by aerosols. This scattered and reflected solar radiation produces the diffuse solar radiation or sky radiation. The actual value of the diffuse solar radiation depends on solar height, the aerosol and water vapour content of the atmosphere and the cloudiness of the sky. In deserts with a low water vapour content of the air this diffuse solar radiation for a clear sky at noon is about 10% of global radiation. For the data presented in Fig. 4.10 the diffuse radiation at noon is 10.4%. The direct solar radiation and also the global radiation is markedly reduced concomitant with the decrease of the

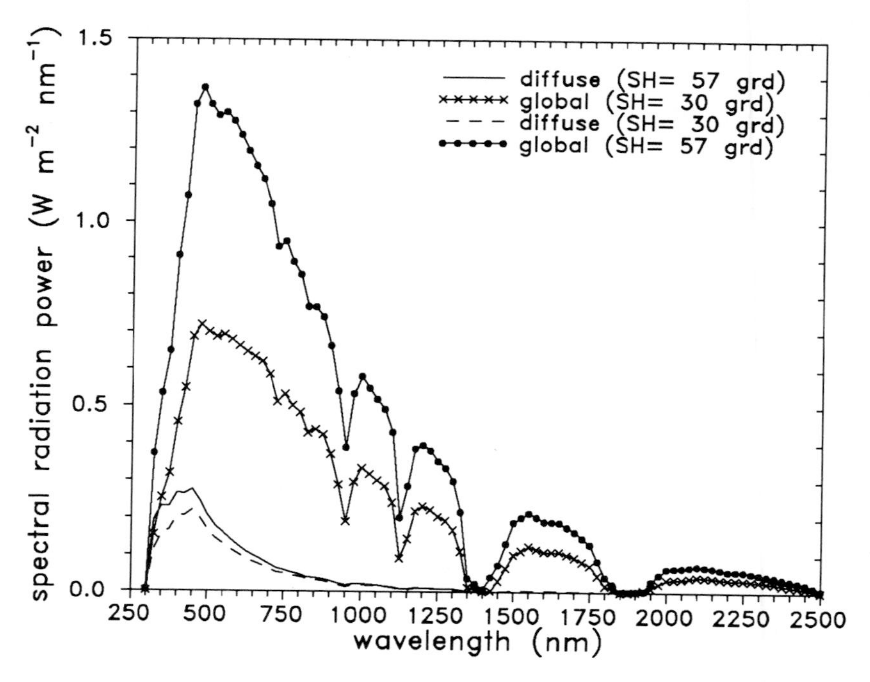

Fig. 4.10 Global radiation and diffuse solar radiation (sky radiation) at 57° (noon) and 30° solar height (SH). Calculated data (after Flach 1986) are given for the Welwitschia Plain in the Central Namib desert for 25 August 1981.

sun's height (SH), e.g. the global radiation decreases from 855 W m^{-2} (SH=57°) to 465 W m^{-2} (SH=30°) . However, the decrease of the sky radiation is much smaller and the fraction of the global radiation that is diffuse radiation increases with decreasing solar height (=15% at SH=30°). The daily pattern of the global radiation and its diffuse component presented in Fig. 4.11 is characteristic for a clear day and emphasizes that the diffuse component of global radiation proportionally can be very high after sunrise and prior to sunset. Moderate cloudiness, mainly if the clouds are white cumuli, can increase global radiation considerably through sunlight reflected by clouds. At such weather conditions the irradiance can reach peak values of 1200 W m^{-2} and every ecophysiologist must bear in mind that solar energy input to the plant can be substantially higher with a sky with some white clouds than with a completely clear sky.

The different wavelength ranges of solar radiation have different consequences for the plant's physiology. Ultraviolet (UV) radiation can destroy organic molecules and disturb growth on the cellular level. However, the

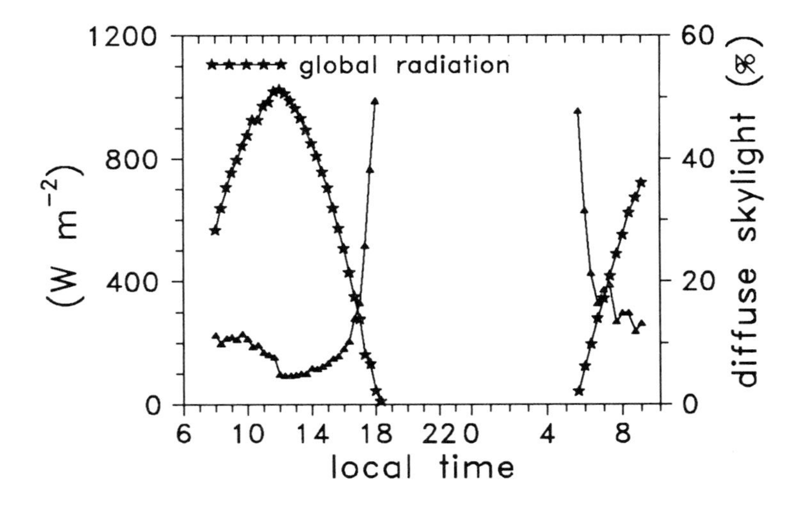

Fig. 4.11 Global radiation and diffuse sky radiation (expressed in percentage of global radiation) measured on the 29/30 November near Pretoria (Republic of South Africa).

energy input into the plant by UV radiation is only 5.5% of the energy of the global radiation shown in Fig. 4.10 (SH = 57°) and is low compared with the visible (48.5%) and the infrared (46%) fraction of solar radiation. The infrared radiation absorbed by the plant is completely converted to heat.

The photosynthetic apparatus of higher plants can only make use of the solar energy in the waveband from 400 to 700 nm wavelength, the so-called photosynthetic active radiation (PAR), where the pigments, mainly the chlorophylls, absorb radiation. The visible radiation not used in photosynthesis is degraded to heat. Measurements by Stanhill and Fuchs (1977) made under high insolation conditions as prevailing in deserts revealed that the fraction of PAR (in W m^{-2}) of global radiation can be taken as 0.5 for solar elevations greater than 10°. However the proper quantum flux (PAR in μmol m^{-2} s^{-1}) should be estimated either from direct measurements with an appropriate quantum sensor or by a calculation based on measurements of the spectral energy distribution of solar radiation. It is known to everybody that the sky turns golden at sunrise and sunset, but the human eye overestimates the changes in the spectral distribution of global radiation energy. However, for very low solar elevations the PAR fraction of global radiation changes significantly (for a review see Stanhill and Fuchs 1977) and for such conditions measured actual values of PAR should be used for calculations or for comparison with other values of PAR.

Surfaces of plants or the ground emit thermal radiation proportional to their true surface temperature. This radiation (J_R) has a spectral energy distribution according to the law for blackbody radiation (Stefan–Boltzmann law).

$$J_R = \epsilon \, \sigma \, T^4 \tag{4.6}$$

where σ is the Stefan–Boltzmann constant, ϵ the emissivity and T the absolute temperature of the surface.

Soil, rocks and also plant surfaces emit with an emissivity of about 0.94 to 0.98 instead of 1.0 for an ideal blackbody (Gates and Tantraporn 1952, Van Wijk and Scholte Ubing 1966, Wong and Blevin 1967). At a surface temperature of 15 °C the emitted energy reaches its maximum at 10 μm wavelength and arbitrary limits of 3 and 100 μm are usually taken to define the entire thermal radiation spectrum.

The thermal radiation from the atmosphere (atmospheric reradiation) is emitted by the molecules and particles of the atmosphere, mainly in the lower few kilometres of the troposphere. Water vapour, carbon dioxide and ozone molecules have strong absorption bands in the infrared and these molecules emit radiation according to Kirchoff's radiation law (emissivity equals absorptivity at the same wavelength). That means that the emission of thermal radiation in the atmosphere occurs mainly in the wavelengths of the molecular absorption bands. Therefore the spectral distribution of the thermal radiation from a clear sky deviates in magnitude as well as in its spectral composition substantially from that of blackbody radiation (Fig. 4.12a). The empirically derived formula of Swinbank is suitable for the calculation of the amount of atmospheric reradiation for a clear sky (Monteith 1973, Gates 1980). For a cloudy sky or for overcast conditions the spectrum of atmospheric reradiation approaches that of blackbody radiation (Fig. 4.12b) since the water droplets or ice crystals of the clouds emit nearly like blackbodies. Thus for complete overcast conditions a true blackbody radiation corresponding to ambient air temperature is a better first approximation than empirically modified Swinbank formulas with corrections for cloudiness. For energy balance considerations or radiation budget calculations the atmospheric reradiation should be calculated from measurements with a pyrradiometer that measures total radiation (global radiation plus thermal radiation of the sky) and a pyranometer which measures global radiation only.

4.1.4 The radiation budgets

The radiation budget (radiation fluxes to a plant minus radiation fluxes from a plant) can be subdivided into the two budgets for solar and thermal radiation respectively.

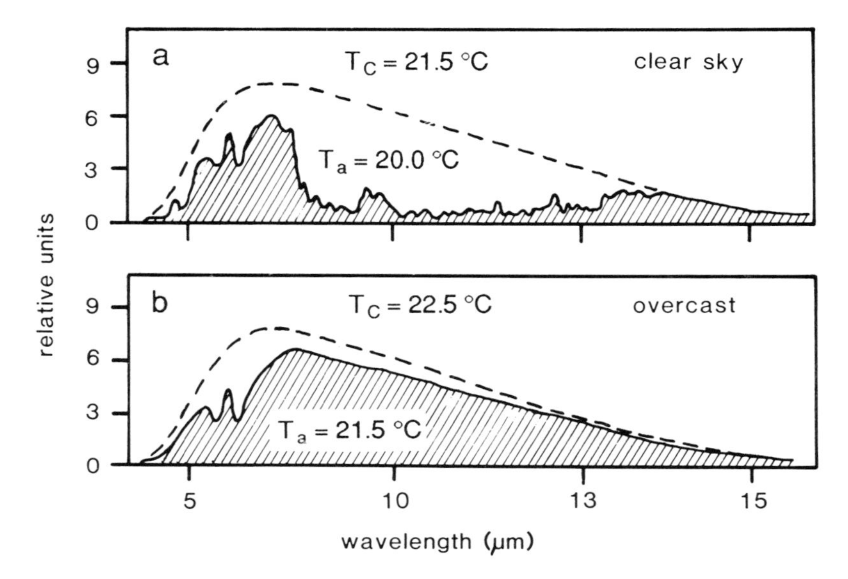

Fig. 4.12 Thermal radiation from the atmosphere for (a) a clear sky and for (b) overcast conditions (hatched areas) compared with true blackbody radiation (dashed lines), selected for matching curves at wavelengths greater than 15μm T_a = temperature of the ambient air; T_c = temperature for which blackbody radiation was calculated. (After Sloan *et al.* 1955.)

The absorbed and the emitted energy from the thermal radiation fluxes constitute the thermal radiation budget of a plant and the energy input to the plant from absorbed solar radiation is determined by the solar radiation budget. The combination of both budgets yields the amount of energy that is transferred to or from a plant by radiation in its entirety. The quality of the radiation source, either the sun or a source of thermal radiation, and the physical properties of an object reflecting, transmitting, absorbing or emitting radiation, determine how and to what extent energy transfer by radiation to and from a plant occurs.

4.1.4.1 The solar radiation budget

The solar radiation consists not only of radiation received by the plant directly but also of solar radiation reflected or scattered in the atmosphere by the plant's environment or by structures of the plant itself. The sum of the radiation fluxes is termed global radiation J_G and its radiant power varies with wavelength $a(\lambda)$. The amount of global radiation absorbed per unit area and unit time by any kind of surface or structure which has an absorptivity of $a(\lambda)$ is given by:

$$J_G\, a_G = \int [J_G(\lambda)\, a(\lambda)]\, d\lambda \qquad (4.7)$$

where a_G is the absorptance for that global radiation.

Radiation incident on a surface is measured in Watts per unit area as is the absorbed, reflected or transmitted fraction of it. By integration of the radiant power per unit area ($W\ m^{-2}$) over a time period (day, year, etc.) one gets the absorbed energy ($W\ s\ m^{-2}$ or $J\ m^{-2}$) for that period. Finally the integration for the whole surface gives the absorbed energy (J) for that surface for the time period considered.

For flat leaves the incident global radiation is different for the upper, J_{Gu} and the lower J_{Gl} surface. Since both surfaces might also have different absorptivities, $a_u(\lambda)$ for the upper and $a_l(\lambda)$ for the lower surface, the energy input from solar radiation to both surfaces of a flat leaf is, in analogy to equation (4.7),

$$J_G\, a_G = J_{Gu}\, a_{Gu} + J_{Gl}\, a_{Gl} \qquad (4.8)$$

The problem gets more complex if not a thin flat, but a thick succulent leaf of regular shape at its cross-section (e.g. triangular or circular) or of irregular cross-section must be considered. In this case an approximation of the surface by plane surface elements must be made and the absorbed energy for each of the surface elements must be calculated separately.

The estimation of the absorbed radiation for such geometrically complex structures becomes very difficult. The absorptivities of the surface elements can be different and the radiation incident on each element changes during daytime. Some plants are known to alter their leaf orientation during the day or track the sun (see Ehleringer and Werk 1986). For succulents this behaviour is most unlikely and has never been observed.

Succulents frequently have their leaves in a more or less upright position often arranged in rosettes, not to forget the ribs and mammillae of the succulent stems of cacti, euphorbias or stapeliads. The cosine law for a collimated radiation like the direct sunlight (J_{sd}) can in this case be used to calculate the direct solar radiation impinging at different angles on the various surface elements:

$$J_{sd}(\alpha) = J_{sd}(0) \cos \alpha \qquad (4.9)$$

where $J_{sd}(0)$ is the irradiance for $\alpha = 0$ and $J_{sd}(\alpha)$ is the irradiance at the angle α. The cosine law is not applicable for diffuse radiation.

For overcast sky and during sunrise and sunset (Fig. 4.10) the diffuse radiation is a major fraction of the global radiation. Consequently radiation reflected from the environment can reach values that must be considered if calculations for surface elements of different exposition are based on radiation measurements with a horizontally exposed sensor. Moreover, the environment of a plant or its own structures can cast shadows on parts of the plant surface (self-shading).

Data for solar radiation presented in Fig. 3.12 for July 1988 at Numees illustrate to what extent the input of solar energy can change within seconds if the irradiation conditions change from shade (cast shadow) to full sunlight. The diffuse sky radiation and the radiation reflected from the environment is about 26% of the global radiation incident on an unobstructed, horizontally exposed receiver (at about 11:30) but can be significantly different for plant surfaces viewing in other directions.

The complexity of shape and exposition of the surface of succulents also raises the question whether or not the spectral properties of a plant's surface depend on the angle of incidence. From investigations by Brandt and Tageeva (1967), Woolley (1971), Farrar and Mapunda (1977) and Yates (1981) with bifacial flat leaves we know that reflection and transmission depend markedly on the angle of incidence of the light rays, but the absorptivity varies only little for angles between zero and approximately 70°. Unfortunately there are no such data available for succulents and we can only suppose that the absorptivity of the surfaces of succulents depends on the angle of incidence in a similar manner as for flat leaves.

All the facts mentioned above make clear that with the exception of fairly flat leaves, for which solar energy input can be calculated using equation (4.8), the estimation of the solar radiation budgets for succulent leaves or stems is a time-consuming task and demands a computer model. Therefore it is not surprising that only a few investigations have so far been made.

The most comprehensive investigation of solar radiation interception using a computer model is the one by Woodhouse *et al.* (1980) for *Agave deserti*. The computed values agree within 19% with the values measured in the field. This modelling was used to quantify the relation between leaf orientation, interception of solar radiation and nocturnal CO_2 fixation. By tilting the whole rosette of *Agave deserti* it could be shown that the number of *Agave* plants on slopes facing steeply north could be related to radiation input during wintertime according to the altered angle of incidence of solar radiation. The same model was used for further comparisons between different species of agaves and revealed that the leaves of the rosette have an orientation so that the sum of intercepted solar radiation throughout the day is at an equal level for the top, middle and old leaves (Nobel 1988).

No similar investigations were made for plants of arid regions of the southern part of Africa but one can assume that the results for agaves and cacti (Nobel 1988) can also be applied to columnar euphorbias or succulents with large rosettes like some aloes.

Very young plants of *Aloe dichotoma* and other giant aloes have their leaves in a more or less upright position whereas the leaves of older but still

Fig. 4.13 (a) Young *Aloe dichotoma* plants near Springbok with either horizontally spread leaves or with the leaves still in an upright position (drawings by Diana Rembges). (b) Columnar cacti (*Cephalocereus hoppenstedtii*) near Tehuacan, Mexico, grow with their stem tops bent to the sun.

unbranched specimens are spread nearly horizontally (Fig. 4.13a). It could be argued that the upright position reduces energy input by direct solar radiation and also from solar radiation reflected by the surroundings. This reduction would lower heat load and could favour establishment and growth of young plants in their natural habitat where a high solar energy input persists throughout the year. The bending of the stem tips towards the sun is nicely shown by *Pachypodium namaquanum* and the columnar cactus *Cephalocereus hoppenstedtii* (Fig. 4.13b) and may be interpreted in a similar way.

Based on field data on the spectral radiation fluxes and the spectral properties of *Aloe dichotoma* collected in 1979 near Springbok (Cape, Republic of South Africa), Flach (1986) computed the daily course of global irradiation for two representative leaves. For the young plant an outermost leaf of the rosette and for the older a horizontally exposed leaf (leaf axis for both $=210°$ azimuth angle, south $=0°$) were chosen. The shape was approximated by three planar surfaces. From Fig. 4.14 it is evident that the peak values of incident global radiation (including solar radiation reflected at that particular site) was by some 40 to 50% lower for the young leaf than for the upper surface of the horizontally exposed leaf. The calculation of the sum of solar energy intercepted during daytime, also taking into consideration the different spectral absorptivities for the different leaf surfaces, proved that the young leaf absorbs per weighted unit leaf area only 62% of the 12.59 MJ m^{-2} per day absorbed by the horizontally exposed leaf of the older plant.

The absorbed solar radiation per unit area according to equations (4.7) or (4.8) is the only inflow of solar energy to a plant and thus is its solar energy budget. Integration of this flux over time and plant surface area results in the term G_a given in equation (4.1). Its value is important for energy balance considerations (see Section 4.1.7) but gives little information for comparisons between plants subject to different solar irradiation. In case the spectral distribution of the radiation incident on plants is the same or at least very similar to that of a clear sky, a weighted mean absorptivity usually referred to as absorptance can be used as outlined in Section 4.1.2.

For green leaves the absorptances for the whole solar spectrum (300 to 3000 nm) vary between 0.5 and 0.6 (Gates 1980, Eller *et al.* 1983). For succulents higher values can result since the absorptivities for infrared solar radiation can be substantially higher (Fig. 4.5). For *Othonna opima* a value for a_G of 71.1% was estimated by Eller *et al.* (1983). On the other hand, white wax bloom or trichomes may reduce the absorptance of succulents for visible radiation. For the energy balance of a plant the entire wavelength

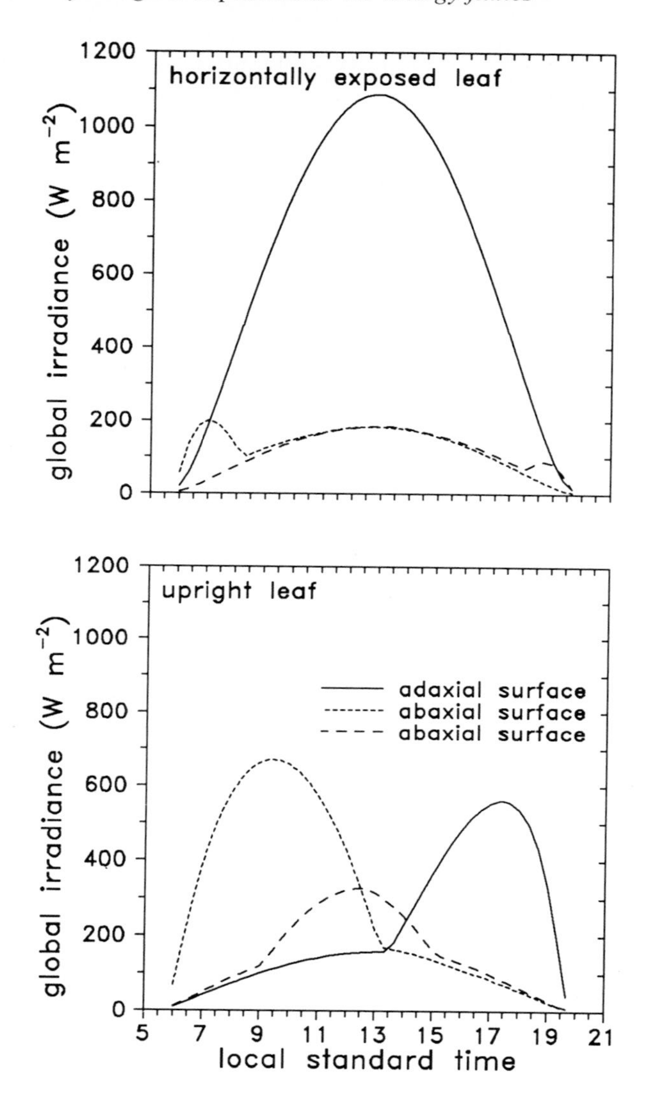

Fig. 4.14 Diurnal course of global irradiation incident on a very young, upright leaf and an older horizontally exposed leaf of a still unbranched *Aloe dichotoma*. The leaves are represented by three planar surfaces: adaxial surface (solid line) and two abaxial surfaces (dashed and dotted lines). Azimuth of leaf axis 210° northeast.

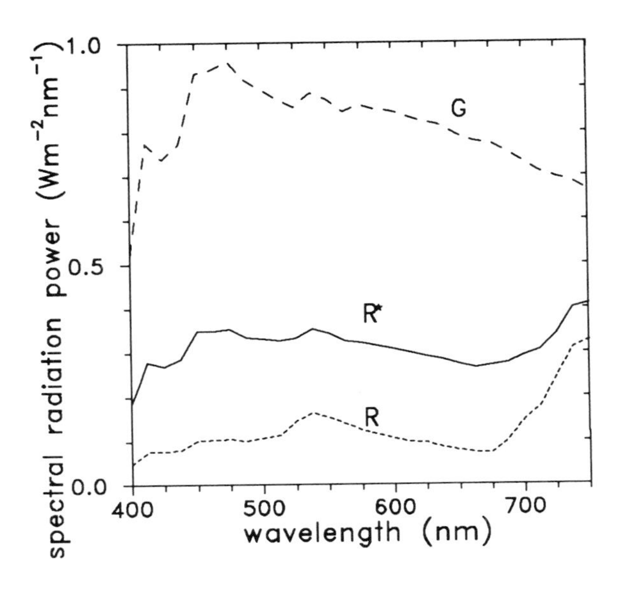

Fig. 4.15 Influence of the wax bloom on the leaves of *Cotyledon orbiculata* on the reflection of global radiation (G). Reflectance (R^*) with wax bloom and after the wax bloom has been removed (R). (Data taken from Eller 1979.)

range of solar radiation is important, but the majority of research where solar radiation is a parameter deals only with the wavelengths of PAR or visible radiation. For such cases the absorptances for either PAR or visible radiation can be calculated using equation (4.5) for that wavelength range.

Hairs, wax, spines or any dead structures superimposed on the epidermis change its optical properties and consequently also the amounts of reflected, transmitted or absorbed solar energy. In Fig. 4.15 the reflected global radiations of a very waxy *Cotyledon orbiculata* leaf and the same leaf after the wax had been removed are shown.

For the visible radiation (400 to 750 nm) the reflectance of the waxy leaf was reduced from 38.5 to only 15.2% after removal of the wax. This resulted in an increase of the absorptance from 59.1 to 80.3%. Eller (1979) showed that the changes for *Kalanchoe pumila* were even more drastic if the bright white wax bloom was removed from their leaves. Reflectance decreased from 42.6 to 9.4%, transmittance changed from 1.6 to 4.2% and consequently absorptance increased from 55.8 to 86.4%. Such figures clearly demonstrate that a wax bloom in habitats with high global irradiance substantially reduces energy input from absorbed solar radiation. Similar changes are known for a dense cover of hairs on non-succulent leaves (Eller 1979) or a dense cover of spines of cacti (Gates *et al.* 1965, Nobel 1988).

A reduced absorption in the visible range of the spectrum implies that PAR at the level of the chloroplasts is also lowered, but unfortunately little information is available about such influences of epicuticular structures. In a study which compared the densely pubescent non-succulent shrub *Encelia farinosa* with the glabrous *Encelia californica*, Ehleringer and Mooney (1978) have shown that with the same irradiance the maximum rates of photosynthesis decreased as pubescence increased. However, if not the radiation incident on the leaves but only the absorbed quanta were considered, then the difference between the pubescent and glabrous leaves disappeared. In the habitat of *Encelia farinosa* its pubescence increases if radiation input increases as the season proceeds. In a modelling analysis it could be demonstrated that the pubescence of *Encelia farinosa* leaves was optimized for maximal carbon gain at given environmental conditions.

One certainly can doubt whether these correlations between the degree of pubescence and the environmental conditions in optimizing the carbon gain of a C_3 plant are transferrable to succulent plants which frequently perform CAM. Further investigations on succulents, including studies of wax layers, are needed, but until such results are available one can only argue that similar correlations as shown for *Encelia farinosa* exist also for succulents. In contrast to *Encelia farinosa* which increases pubescence if irradiance increases, very often the trichomes of leaves are gradually lost during maturation, e.g. in *Senecio medley-woodii*. In most cases also the wax bloom of succulents is thick and dense only on young leaves and the density of the wax decreases with leaf development as has been shown by Eller and Willi (1977) for *Kalanchoe pumila*. In the natural habitat precipitation, wind and abrasion by wind-blown sand and dust removes most of the wax bloom within a short time after the leaves or the growing apex of stems (e.g. in *Hertrichocereus beneckei*) have stopped growth. From these facts one must argue that hairs or a wax bloom mainly reduce solar energy input during the expansion of an organ. This makes sense because in this period a high solar energy input would lead to high temperatures in the growing organ and favour transpirational water loss.

Some species of *Anacampseros* have their small branches more or less covered with paper-like white scales, most pronouncedly found in *Anacampseros papyracea* (Fig. 4.16). Such bright white scales certainly reflect a large fraction of the visible solar radiation, but a quantification is impossible as we lack data on optical properties measured for this kind of plant. Another feature occurs in the Mesembryanthemaceae family. Species of

Fig. 4.16 *Anacampseros papyracea* showing the white scales enveloping its branches.

Conophytum, *Mitrophyllum*, etc. replace their leaves annually but the new leaves develop inside the old coalescent leaf pair (or bases of the leaves as for *Mitrophyllum*). The old leaves dry with proceeding growth of the young leaves until the old ones are reduced to a thin sheet of dry matter. Leaf growth ceases at this stage and *Conophytum* plants in particular then look like a small pile or cushion formed by yellowish or brownish dry globular structures (Fig. 4.17). The new leaves inside the litter of the old ones are hidden from, for example, herbivores, but the yellowish or brownish envelope also can reflect more visible radiation than the green leaves inside. However, the lack of measured data prevents determination of whether this feature substantially reduces solar energy input to the young leaves.

Different coloration of leaves usually does not alter the absorption of leaves to a great extent as was demonstrated in Fig. 4.6 for green and red

Fig. 4.17 Cushion of a *Conophytum* during the dry season. The new leaves are still hidden inside the dry envelope of the litter of the old leaves.

leaves of *Aloe pearsonii*. This statement needs some modification in the case of leaf colour changes in the course of the year which are accompanied by changes of the sun's radiation input. During the growth period, when sufficient water is available, plants of the genus *Argyroderma* are green with a silvery appearance. In the dry period the leaves turn to a yellowish brown. Moreover, the formerly thick leaves shrink due to water loss. Measured optical properties at the end of the rainy season (September) and in mid-summer (December) differ not only in the visible wavelength range, as can be expected from the different coloration, but in a similar magnitude also for the infrared with wavelengths greater than 750 nm (Eller 1982). The absorptance for global radiation at clear sky conditions was 70.0% in September but decreased to 59.7% in December. In the same time period the global irradiance at noon increased from 707 to 902 W m^{-2}, or by

27.5%. If the absorptance of the *Argyroderma* plant had not changed during this timespan an equally high increase of the energy input by absorbed solar radiation would have resulted. The reduction in the absorptance diminished the otherwise unavoidable increase to only 8.6% (Eller 1982). In this case we certainly can deduce that the plant can benefit from the changes in leaf coloration and leaf thickness.

As outlined in this section, the absorptance of the surface of a succulent can vary on a single plant with age, exposition of the surface and also with the season. Hence an estimation of the solar radiation budget succeeds only when the spectral optical properties of the surfaces which receive the sun's radiation are known with sufficient accuracy. Calculations with an arbitrarily chosen value, as were often made for thin green and flat leaves whose absorptances usually are between 50 and 60%, are not recommended for succulents with absorptance values somewhere between 40 and 80% at least if the surface is not dull green.

The major part of the absorbed solar energy is directly converted into heat and only a very small fraction of PAR is consumed by photosynthesis. If the rate of photosynthesis is known the energy used in photosynthesis can easily be calculated by the equation

$$J_P = J_c(\text{net})\ \Delta G_{CO_2} \tag{4.10}$$

where $J_c(\text{net})$ is the rate of photosynthesis in μmol CO_2 m^{-2} s^{-1} and ΔG_{CO_2} is the increase in Gibbs free energy per mole of CO_2 incorporated into carbohydrates which is 479 kJ mol^{-1}. The integration of J_P over a timespan and the assimilation surface area results in the term Q_P of equation (4.1). For a rate of 1 μmol CO_2 m^{-2} s^{-1}, Q_P becomes 0.479 W m^{-2}. Since photosynthesis of succulents ranges between 0 and 10 μmol m^{-2} s^{-1}, the amount of energy dissipated in photosynthesis will not exceed 4.79 W m^{-2} and hence is negligible in energy budget considerations.

4.1.4.2 *The thermal radiation budget*

Thermal radiation differs from solar radiation only in its wavelength range as already outlined in Section 4.1.3. Therefore the equations to calculate absorbed solar radiation, absorptance and also energy input can be used likewise for thermal radiation if the symbols or parameters for solar radiation are replaced by the ones for thermal radiation. Moreover, two qualitative changes make it even easier to estimate the thermal radiation budget. Firstly, the spectral properties of plants and of their environment are nearly wavelength independent and differ only to a small degree. Secondly, thermal radiation is a blackbody radiation as given in equation (4.6) or at least does not differ substantially from its spectral energy distribution. Only for clear sky conditions are there deviations (Fig. 4.12), but the first mentioned fact allows us to use a measured value of thermal radiation from the sky in a manner as if it was a proper blackbody radiation obeying the Stefan–Boltzmann law. Moreover, for plants the absorptances for thermal radiation are very similar and therefore an arbitrarily chosen value for the absorptance of 96% is frequently used and no distinction is made between absorptance for, for example, upper and lower surfaces of a leaf. Kirchhoff's law also states that absorptivity or emissivity and absorptance and emittance are equal for the same wavelength or wavelength range.

Replacing $J_{Gu}(\lambda)$ and $J_{Gl}(\lambda)$ of equation (4.8) for a flat leaf by the spectral thermal radiation for the upper J_{Ru} and the lower J_{Rl} surface we get the energy influx which results from the absorbed thermal radiation. In most cases the spectral distribution of the absorptivity can be replaced by the absorptance value, a_R, for thermal radiation as outlined above. Therefore we can rewrite equation (4.8) for thermal radiation as

$$J_R\, a_R = (J_{Ru} + J_{Rl})\, a_R \tag{4.11}$$

where $J_R\, a_R$ is the thermal radiation absorbed by that leaf.

If J_{Ru} is the thermal radiation from the sky or more precisely the atmosphere then it is determinable by empirical equations (for details see, for example, Monteith 1973) or can be measured by a pyrradiometer. If J_{Rl} is the thermal radiation from the environment of the plant or surfaces of the plant, then this term is either calculated according to equation (4.6) if true surface temperatures are known or it is measured by a pyrradiometer. However, unlike pyranometers or solarimeters which quantify solar radiation and for which calibration is available at different meteorological centres, few facilities exist for the calibration of pyrradiometers and the accuracy of calibration is less than for pyranometers. These difficulties are the reason why ecophysiological field investigations are rarely based on accurate determinations of thermal radiation fluxes.

Unlike the sun's radiation, for which a plant acts only as a receiver, plants not only absorb but also emit thermal radiation. Thermal radiation emitted from the upper, $J_{Ru}(T_u)$, and the lower, $J_{Rl}(T_l)$, surfaces is calculated by inserting into equation (4.6) the emissivity e_R $(=a_R)$ and the surface temperatures of the upper, T_u, and the lower, T_l, surfaces respectively. From this we get the sum of emitted thermal radiation which is

$$J_R(T) = J_{Ru}(T_u) + J_{rl}(T_l) \tag{4.12}$$

Combining equations (4.11) and (4.12) we get the thermal radiation budget

$$J_R(\text{net}) = J_R\, a_R - J_R(T) \tag{4.13}$$

This value is frequently referred to as the net radiation flux but this is only valid in the absence of any global radiation. It should more precisely be termed the net thermal radiation flux.

J_R(net) is relatively small (mostly less than 120 W m^{-2}) compared to the global radiation of up to 1200 W m^{-2}. However, it plays an important role in the plant's physical energy budget because its value is usually negative and represents a loss of energy from the plant. This is commonly the case during night-time but can also occur if the plant grows in the open exposed to a clear sky while the plant at the same time lies in the cast shadow of its environment. Such radiation flux relations are referred to as open shade (Stoutjesdijk 1974). However, if we not only consider the net thermal radiation flux but all its different fluxes, we get a proper impression of the amount of radiative energy exchange which actually exists at the plant's surfaces. Surfaces opposite to each other (e.g. a plant surface opposing another plant surface or a surface of the environment) exchange energy by thermal radiation. The proper amount of this energy exchange depends on the emissivities (=absorptivities) for thermal radiation of those surfaces and of the actual surface temperatures. A plant surface temperature of 30 °C and an emissivity of 0.96 results in 461 W m^{-2} emitted by that surface. If this surface is facing the environment with a surface temperature of 20 °C then 403 W m^{-2} are radiated to the plant. Consequently, the thermal energy flux from the plant is 58 W m^{-2} higher than the flux towards the plant, and a net thermal energy flux from the plant to the environment exists. If the temperatures are reversed (plant 20 °C and the environment 30 °C) then the energy flux from the environment is 58 W m^{-2} higher than that of the plant surface and an energy input to the plant occurs. Especially in deserts with patches bare of vegetation where the soil or nearby rocks are heated up by the sun, the surface temperatures of the surroundings can considerably exceed those of plant surfaces. Such exchanges of energy via thermal radiation fluxes can be crucial for the survival of a plant or the establishment of seedlings.

Thermal radiation from the sky also varies with the temperature of the air at the plant's level but to a much greater extent it varies with the aerosol content of the atmosphere. Fig. 4.18 shows data measured at a *Lithops* habitat during a day with clear sky together with the values for the net thermal radiation flux at the top of the plant, determined according to equations (4.11) to (4.13).

With overcast conditions the water droplets or ice crystals that form the clouds emit thermal radiation like liquid or solid surfaces. Because thermal radiation from a clear sky is always lower than that of a cold surface at the

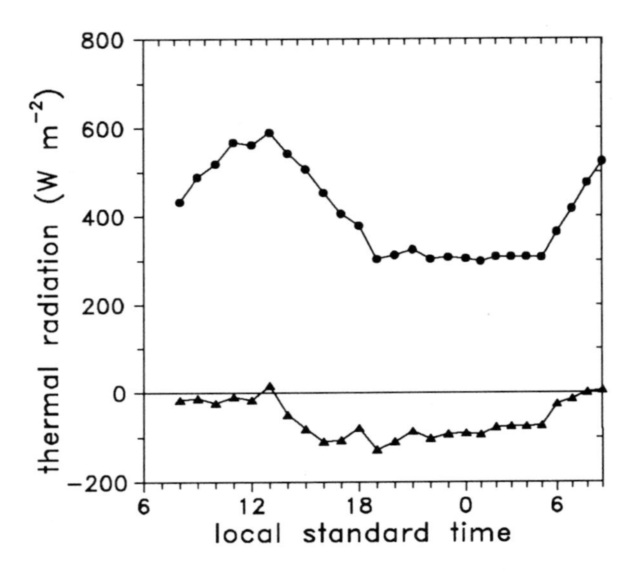

Fig. 4.18 Inflow of thermal radiation from the sky (dots) and net thermal radiation flow (triangles) from the top of a *Lithops* plant. (Data taken from Eller and Grobbelaar 1986.)

temperature of the air near the earth's surface the change from clear sky to overcast conditions significantly alters energy input to the plant. Fig. 4.19 gives an impression of such a change observed during field investigations in the Namib desert. For the timespan of overcast conditions thermal energy influx from the sky is increased and therefore net thermal radiation loss of the plant is approximately six times smaller than during the following night with clear sky conditions. In such cases net thermal energy flow from the plant is usually reduced to almost nil but if the plant has reached a surface temperature below the air temperature net thermal energy output can even change to a net influx of thermal radiation energy to the plant.

Thermal radiation fluxes are very important in arid habitats where a clear sky is frequent. During clear nights a negative net thermal radiation flux cannot only cause a drop of the plant temperatures below the temperature of ambient air but also forces the soil surface and the air adjacent to it to cool down. High relative air humidities and the formation of dew towards the end of the night are primarily a consequence of the energy loss by surfaces through a negative thermal radiation budget of those surfaces.

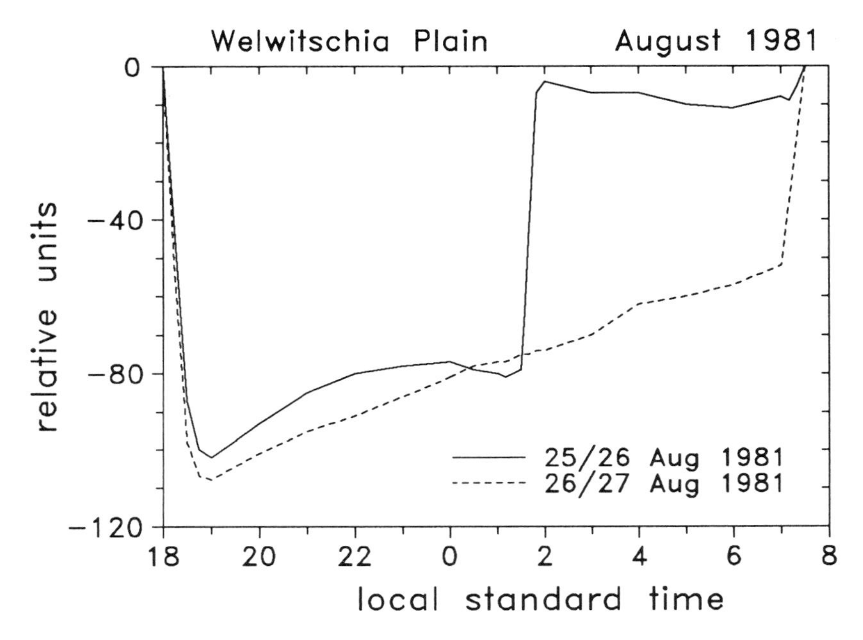

Fig. 4.19 Net thermal radiation fluxes during the night measured with a pyrradiometer. A sudden change from clear to overcast sky occurred during the night of 26 August at 01:30 while prior to that event and during the following night clear sky conditions prevailed.

4.1.4.3 The total radiation budget

The sum of the solar and the thermal radiation budgets result in the net radiation flux in its proper sense. This value could also be termed simply the total radiation budget of a plant's surface. Fig. 4.20 presents the measured and calculated values for the radiation influxes from the sun and the sky and the net radiation energy flux for a horizontally exposed section of a leaf of *Welwitschia mirabilis* as measured in the Welwitschia Plain in the Namib desert by Eller *et al.* (1983). During daytime a high positive influx results from solar radiation. At dawn and dusk global radiation consists only of diffuse skylight and then the energy flow changes from an input to the leaf to a loss from the leaf. As already mentioned above this open shade is a combination of low energy flux from absorbed diffuse and reflected solar radiation combined with a net thermal radiation loss from the plant. If influx is lower than loss the surface temperature of a leaf can drop below air temperature if there is no other energy input to the plant surface, e.g. by convection. Stoutjesdijk (1974) measured (16 September 1971, 11:30,

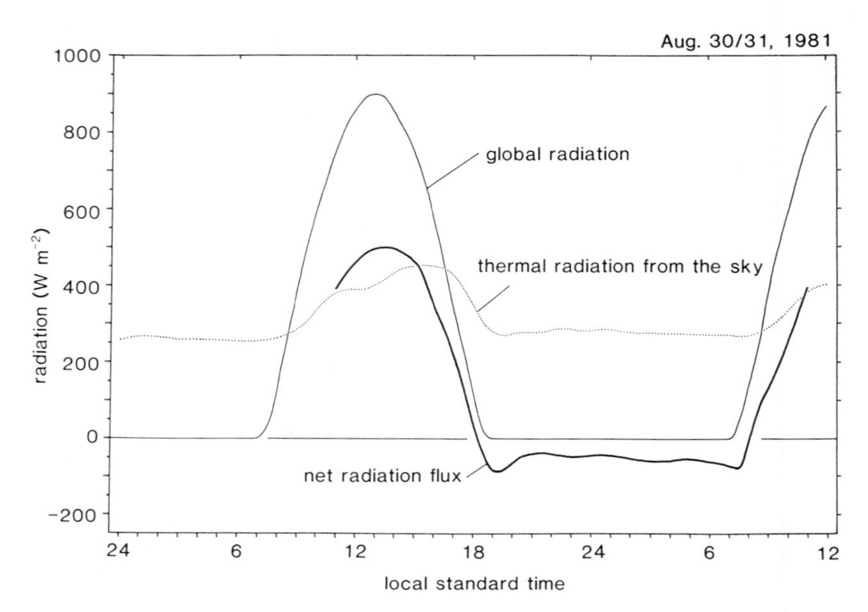

Fig. 4.20 Diurnal course of global radiation, thermal radiation from the sky and the total radiation budget or net radiation flux of the soil surface in the Welwitschia Plain at clear sky conditions. (Data taken from Eller *et al.* 1983b.)

Oostvoorne, The Netherlands) in the open shade a global radiation of 69.8 W m^{-2} and a total net radiation loss of 36.5 W m^{-2}, whereas at sunlit places the global radiation was 580 W m^{-2}. A hawthorn shrub (*Crataegus monogyna*) was at that time still wet from dew and had a surface temperature of 9.1 °C or 5.9 K below the temperature of ambient air. Further data will be given in Section 4.1.7 when we shall discuss the impact of the radiation budgets on the physical energy budget of a plant.

4.1.5 Latent and sensible heat

Heat energy is consumed by phase transitions of substances to a higher energy level, e.g. by evaporation. This heat energy is no longer sensible but by a reversed phase transition is released again. This energy, usually referred to as latent heat, plays a major role in the process of transpiration. Basic textbooks often use the term transpiration cooling, but really mean the consumption of heat energy for the transition of water from the liquid phase to water vapour. The latent heat of water, also termed heat of vaporization of water (L_w), is 2.454 MJ kg^{-1} or 44.21 kJ mol^{-1} (at 20 °C) and varies with the temperature (2.477 MJ kg^{-1} at 10 °C; 2.430 MJ

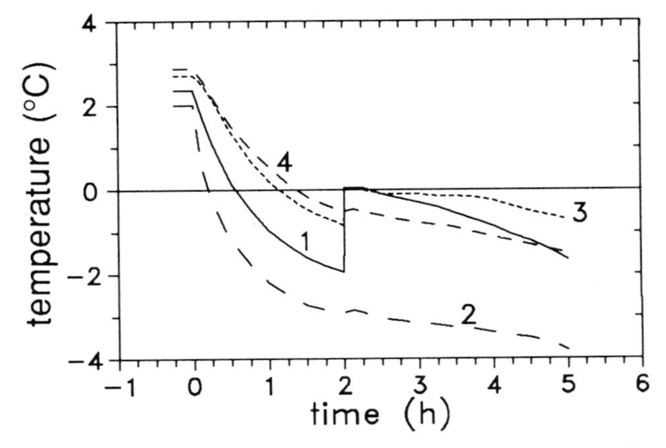

Fig. 4.21 Temperatures of *Lithops turbiniformis* at 1.5 (line 1) and 17.5 (3) mm depth and of the surrounding soil at 1.5 (2) and 10 (4) mm depth. At time = 0 a net thermal radiation loss of 110 W m^{-2} at −0.8 °C ambient air temperature was applied. (Data after Eller and Nipkow 1983.)

kg^{-1} at 30 °C). If the amount of water that changes its phase is known then the energy turnover can be quantified by the equation

$$J_L = J_w(\text{net})\, L_w \tag{4.14}$$

where J_w(net) is the transpirational water loss in mol m^{-2} s^{-1}. By integration over a timespan and the transpiring surface area we get the term Q_w of equation (4.1). We shall give examples of the magnitude of this energy consumed or released by transpiration and how this turnover plays a role in the energy balance of a plant in Section 4.1.7.

Equally important is the energy turnover if water or solutes like the cytoplasma freeze or melt. For water at 0 °C it is 333.7 J g^{-1}. At ambient air or soil temperatures near 0 °C such a release of heat can prevent plants from reaching very low temperatures as was shown by Eller and Nipkow (1983) for *Lithops turbiniformis*. Plants embedded into sandy soil were first exposed to about 3 °C, the temperature of the ambient air, for thermal equilibration at that temperature. Afterwards the plants were exposed to a net thermal radiation loss of 110 W m^{-2} and an ambient air temperature lowered to −0.8 °C. From Fig. 4.21 we can see that the soil and the plant cooled down. Surprisingly *Lithops* did not freeze at 0 °C but supercooled to about −2 °C for the top and −0.8 °C for the bottom of the plant. At these temperatures spontaneous freezing occurred (Fig. 4.21, 2 h) and the released heat of freezing increased plant temperatures to about 0 °C. Despite the continuing net thermal energy loss of 110 W m^{-2} the plant temperatures resumed their decrease only slowly because the plant evidently was not yet completely frozen. Therefore more heat of freezing was released when further cells or parts of tissues froze. Four hours after freezing had started the plant temperatures had dropped again to about −2 °C whereas the soil surface temperature had continuously dropped to −4 °C (Eller and Nipkow 1983).

During clear nights the lowest temperatures of the ambient air and the soil and also of plants occur at dawn. With conditions similar to those outlined above the *Lithops* plant would decrease its body temperature below −2 °C only when the soil surface temperature would have dropped to less than −2 °C at least 5 hours before dawn.

Discussion of freezing of plants raises the question of thermal tolerance of succulents. Eller and Nipkow (1983) supposed that temperatures lower than −3 °C could affect the viability of *Lithops* plants and Nobel (1989) showed that even *Lithops* tissues adapted to cool weather conditions (15 °C during the day, 5 °C at night) were killed at −8 to −9 °C, as was also observed by Eller and Nipkow (1983). Nobel (1989) also found that at temperatures 3 to 4 K higher than the temperatures at which the cells were killed the vitality of the chlorenchyma was already reduced to 50%. Similar values were reported by Nobel (1989) for *Haworthia* but some North American cacti and agaves have temperature tolerances down to −20 °C (Nobel 1988). Macaronesian Semperviva are reported to be resistant against cold down to −10 °C (Lösch and Kappen 1983). For some species of *Delosperma* (e.g. *Delosperma nubigenum* native to Lesotho and adjacent territories) it is known that in cultivation in Europe they survive periods of several days with temperatures lower than −15 °C.

The other extreme is high temperature tolerance. For plants growing in desert habitats one should expect a high degree of heat tolerance. In fact, cacti and agaves can survive temperatures above 60 °C (for a review see Nobel 1988). Eller and Nipkow (1983) reported for *Lithops lesliei* plants collected in the field no injury after the plants had to endure peak temperatures of 56 °C for several consecutive days and Nobel (1989) reported that only temperatures that exceeded 60 °C were lethal for *Lithops* and *Haworthia* species.

Sensible heat is the heat energy content of a substance and is manifested by a temperature which can be sensed or measured by an adequate measuring device. A quantifiable amount of heat energy is needed to raise a substance or a plant tissue one Kelvin in temperature. This energy is called specific heat and its value varies over a broad range for different substances. For water (at 20 °C) it is 4.18 J g^{-1} K^{-1} or expressed as volumetric specific heat (C_v) it is 4.18 MJ m^{-3} K^{-1}. It varies only little with temperature.

Information on the magnitude of the volumetric specific heat of succulent plant tissues is scarce. For succulent organs with a high water content one can assume that its value is close to that of water (Lewis and Nobel 1977, Eller and Grobbelaar 1982) but for scleromorphic epidermal tissues or woody sections of stems it certainly is lower.

A tissue with a volume of V storing or losing a quantity of heat energy (Q_H) is subject to a change (ΔT) of its temperature of

$$\Delta T = \frac{Q_H}{c_v V} \quad (4.15)$$

If in a plant's tissue a surplus of heat energy occurs then the temperature of the tissue increases. Consequently a higher energy inflow by absorbed solar or thermal radiation will result in an increase of the temperature of the plant if this inflow is not balanced by equal outflows in the form of energy dissipation or conversion. One of these outflows is thermal radiation which is in fact a conversion of heat energy into radiation energy. Other flows are the consumption by phase transition as outlined above, the consumption by metabolic processes or photosynthesis and the heat dissipation by convection.

Heat energy flow by conduction can occur, e.g. from the roots to the soil or between plant and soil, if the plant is completely embedded in the soil like *Frithia pulchra*. Laboratory and field investigations on *Lithops* (Eller and Nipkow 1983, Eller and Grobbelaar 1986, Nobel 1989) revealed that the temperature differences between the soil and the plant are small but there exists no clear-cut evidence that in the field this is due to the conduction of substantial amounts of heat energy between plant and soil. On the contrary, the results from Eller and Nipkow's laboratory experiments (1983) as shown in Fig. 4.21 reveal that there was only a minor heat energy flow from the plant to the surrounding soil. During the freezing of the plant the temperature inside the *Lithops* increased at the depth of 17.5 mm by about 2 K whereas the increase of the soil temperature was only fractional (Fig. 4.21). If a substantial conduction of heat existed then this increase should have been greater.

Heat is also conducted within the plant or its tissues and causes changes in their temperatures (according to equation (4.15)). Heat energy flux (J_H) by conduction along a temperature gradient $\delta T/\delta x$ is

$$J_H = k\left(\frac{\delta T}{\delta x}\right) \quad (4.16)$$

where k is the thermal conductivity along the axis x. Knowing volumetric specific heat and thermal conductivity of the tissues of a whole plant one can determine with well-established methods (Dusinberre 1961) the changes in temperature that are caused by thermal energy fluxes to or from that plant (Lewis and Nobel 1977). Summing up for a timespan the heat energy fluxes J_H to and from a part of the plant with a volume V we get Q_H of equations (4.15) and (4.1) respectively. Heat conduction is very important in connection with convective heat energy exchange since convective energy dissipation only works if the air close to the surface of the plant receives heat energy by conduction via the epidermis as will be outlined below.

4.1.6 Convection and wind

The transfer of heat from the surface of a plant to the air is by convection. The essential feature of that convective energy transfer is the transport of energy from the surface by both molecular conduction of heat and gross air movement. That means that one can speak of convection only if there is a movement of air. Heat transfer in the case of absolutely still air, a situation that almost never exists in nature, would be by conduction only.

If the air movement arises from gravity acting on a density gradient in the air, then the process is referred to as free convection. If the air motion is induced externally by wind (or a fan or blower as in plant growth chambers or gas exchange cuvettes) then the process is called forced convection. However, popular use does not make this proper distinction but speaks of free convection if the air speed is lower than 0.3 m s^{-1} (some researchers prefer 0.1 m s^{-1}) and forced convection for faster air motions. Air is a fluid with a defined kinematic viscosity and when a fluid flows along a surface this viscosity produces a drag force so that fluid particles near the surface have a velocity equal to zero whereas the full velocity of gross air movement is reached only at some distance from the solid surface. This transition zone is referred to as the boundary layer. Convective heat transfer occurs only if the temperatures of the surface and the air outside the boundary layer are different. The existence of a temperature profile apart from the velocity profile in the boundary layer is the consequence of this difference in temperature. Boundary layer thickness can vary from a fraction of a millimetre to 10 mm or even more. Both convective heat transfer and the boundary layer depend on two primary types of air flow. In a streamlined flow all particles move in the same direction and follow each other. Such a flow is called laminar flow whereas in a turbulent flow the air movement breaks down in eddies and in irregular motion of air with a prevailing direction of the gross flow.

It is not the aim of this section to give a comprehensive treatment of the theory of convective heat transfer and its application to energy balance calculations but to present some basic information necessary to evaluate the importance of convective heat transfer for plants in arid habitats. Very detailed information is available in biophysical textbooks (e.g. Monteith 1973, Gates 1980) or in engineering textbooks (e.g. Kays 1966).

The heat energy (J_c) transferred between a plant surface and the air per unit time and unit surface area is

$$J_c = h_c(T_a - T_l) \quad (4.17)$$

where h_c is the convective heat transfer coefficient and T_a and T_l are the temperatures of ambient air and the plant surface respectively. Unfortunately, the convective heat transfer coefficient is a rather complex function of different variables. A first set of variables depends on the size and shape of the object (e.g. a leaf or a whole plant) and its orientation. A second set describes physical properties of the air such as density, kinematic viscosity, specific heat and thermal conductivity. Because some of these properties are functions of the temperature, h_c is also a function of the temperature of the air. Finally, h_c also varies with air velocity and the type of air flow, either laminar or turbulent. These rather complex functional relationships are somewhat simplified by the combination of variables in dimensionless groups, and the convective heat transfer from objects of simple shape (e.g. flat plate, cylinder, sphere) can be calculated according to formulas established in aerodynamics (for details see Gates 1980). Integration of J_c over time and plant surface area results in the energy quantity C as given in equation (4.1).

Air flow around a plant structure originates from wind or from locally occurring air movement resulting from density gradients in the air due to local heating or cooling of air pockets. During the day high solar irradiance in deserts nearly always causes air movements around plants that exceed the conditions for free convection. Thus forced convection is the prevailing condition and the dissipation of heat is by forced convective energy exchange. It is important for the physical energy balance of a plant (Section 4.1.7). Fig. 4.22 shows two randomly chosen examples of the daily variation of wind speed measured in the Namib. During the day only very short periods of low wind speeds exist but in the night calm periods prevail (Fig. 4.22b). If there is a net thermal radiation flux during the night then cold layers of air can establish close to the soil surface near a plant.

Plants and/or their organs are rarely of simple shape. Nevertheless, sometimes succulent leaves might be regarded as flat plates or cylinders. Fruits may approach a sphere and succulent stems a cylinder. For such cases different approaches were made to calculate the convective energy transfer (Raschke 1956, Gates 1965, Gates and Papian 1971) but only a

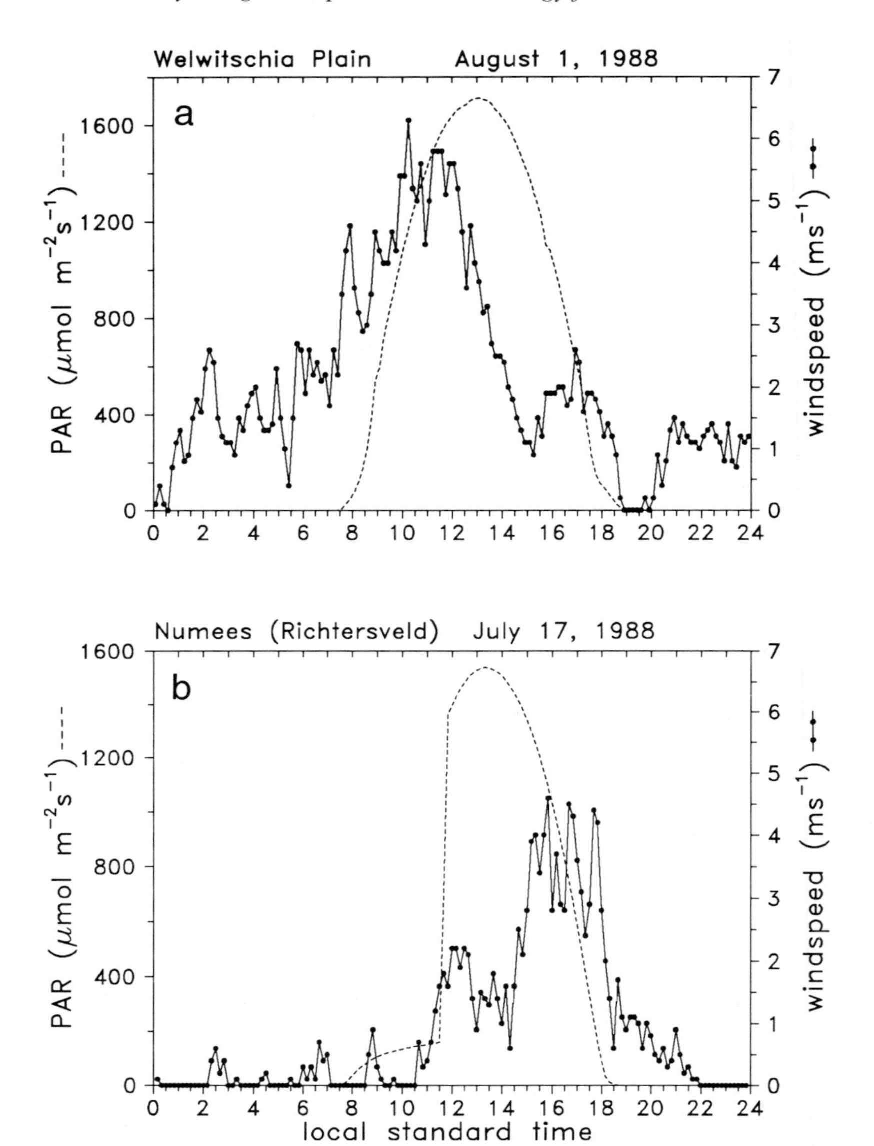

Fig. 4.22 Diurnal variation of wind velocity (mean of a continuous measurement over a period of 10 minutes) at two characteristic places of the Namib desert. (a) Flat plain in the Central Namib 35 km east of Swakopmund; (b) northwest-facing slope in the hills of the southern part of the Namib (Richtersveld). Forced convection exists at windspeeds > 0.3 m s^{-1}.

little work deals with succulents (Pearman 1965, Lewis and Nobel 1977) or with plants of the Namib desert (Schulze *et al.* 1980, Eller *et al.* 1983). Investigations on calculated and measured magnitudes of convective heat transfer coefficients (Parkhurst *et al.* 1968, Parlange *et al.* 1971, Pearman *et al.* 1972, Dixon and Grace 1983) show that h_c measured is 1.5 to 3 times larger than h_c calculated from leaf model values that are also consistent with the magnitude of convective energy transfer estimated from energy balance calculations by Eller and Grobbelaar (1982) and Eller *et al.* (1983).

Investigations of CO_2 gas exchange or transpiration often refer to boundary layer resistance – i.e. the impact of the length of the diffusion path in the transition zone between the stomatal pore and the undisturbed free circulating air. In most cases a boundary layer thickness is chosen that is equal to the boundary layer thickness for convective heat energy transfer.

Boundary layer thickness is a value that depends largely on the assumption made for the type and the magnitude of convective energy transfer. For the case of forced convection with laminar flow over a flat plate the boundary layer thickness δ (in mm) at the distance D (in mm) from the leading edge of the plate at an air velocity of v (in m s^{-1}) is

$$\delta(\text{mm}) = k_\delta \sqrt{\frac{D}{v}} \tag{4.18}$$

where k_δ is 5.6. For most real leaves this is an overestimation since any roughness or irregularities in the shape of the leading edge can act as some kind of spoiler and cause erosion of the boundary layer. Nobel (1975) suggested a value of 4.0 for k_δ. This is probably closer to reality, and is consistent with the results from determinations in field investigations of, for example, Parlange *et al.* (1971) or Pearman *et al.* (1972). For cylinders with a diameter of d (in mm) an effective boundary layer thickness of

$$\delta(\text{mm}) = 5.8 \sqrt{\frac{d}{v}} \tag{4.19}$$

was found by Nobel (1974) and for spheres with a diameter of d (in mm) the boundary layer thickness is calculated with the equation

$$\delta(\text{mm}) = 2.8 \sqrt{\frac{d}{v} + \frac{0.25}{v}} \tag{4.20}$$

as estimated by Nobel (1975). At low air velocities or if free convection prevails one can hardly speak of a boundary layer thickness that is constant over a certain timespan. Eddies and the forming of air plumes rising from the surface of real leaf structures (see, for example Schlieren photographs of Gates and Benedict 1962) make an estimation of the boundary layer thickness a lottery.

Fig. 4.23 Wind damage on a leaf of *Welwitschia mirabilis* due to a sandstorm on 25 July 1985. The photograph was taken on 4 April 1986 in the Welwitschia plains east of Swakopmund. The leaf was turned by the sandstorm and the lower surface was completely abraded by the sand. The edge of the groove from which the leaf emerges is imaged on the leaf and can be seen as a sharp line on the leaf as it grows further.

Convective energy transfer is a feature of rather small dimensions. On the other hand, wind is a feature of large dimensions. Large-scale air displacement determines local and continental climatic conditions but on a reduced scale also has an impact on the performance and the morphology of plants (see, for example, Grace 1975). Wind drag on single leaves or on trees and its results are well known (see Monteith 1973). Succulents with their generally relatively small structures – some tree-size forms of *Euphor-*

bia, *Aloe* or cereoid cacti are exceptions – have not yet been investigated with respect to that topic. That the sturdy succulents are not subject to severe wind drag lacks experimental evidence and is a field for further investigations.

Another impact of air movements or wind has not yet been mentioned. Wind exceeding a certain velocity tends to move particles. In deserts these particles are sand and dust. In Section 3.2 we gave an illustration (Fig. 3.5) of a heavy sandstorm over the central part of the Namib desert on 2 September 1981. As some of the authors were on field research in this area at that date, observations were made on wind damage. Fig. 4.23 gives an impression of the abrasion caused by another sandstorm with peak wind velocities of 30 m s^{-1} (about 120 km h^{-1}). However, such damage seems to be an exception since we could not make similar observations during field observations in the southern Namib (Richtersveld). Labels from cardboard or wood used to identify plants under investigation were worn by sand-blowing wind during the years and had to be renewed, but no visible damage was noticed on the plants. However, leaves of some plants (e.g. *Psammophora nissenii*) were covered with a layer of sand. This can alter energy input as was demonstrated by Eller (1979) for non-succulent leaves. Wind-blown sand also might remove wax bloom or trichomes in considerable quantities and thus alter the absorptance for solar radiation, but no long-term observations have so far been made.

Another influence of wind which can hamper plant life in sandy soils is the existence of moving sand dunes, even if the dunes are not higher than a fraction of a metre. Plants can be buried partially or completely or if the sand is removed by wind the root system is exposed to the dry hot air and solar radiation and the roots can be damaged. These impacts of wind are known to prevent the establishment of a plant cover or at least disfavour plant growth in coastal dunes. The lack of higher plant life in the sand Namib between Walvis Bay and Oranjemund is certainly due to the constant moving of sand in this region.

4.1.7 The plant's physical energy balance

In Section 4.1.1 we presented the formula describing the physical energy status resulting from flows around a plant. In the preceding sections some facts and figures were given for the different terms of equation (4.1) so we should now be able to quantify the interrelationships between the energy quantities of different origin. Subsequently we shall neglect the energy turnover by metabolic processes and photosynthesis and also heat flows to and from the plant by conduction but not heat storage.

Rewriting equation (4.1) for these assumptions we get

$$G_a + R_a - R_e \pm C \pm Q_w = \pm Q_H \qquad (4.21)$$

It is evident that if Q_H is not equal to zero then the resulting energy must be stored in the plant structures or is retrieved from them. Heat energy fluxes to or from a tissue will cause this tissue to heat up or cool down. Within a timespan J_H (equation (4.16)) must alternately become negative and positive for each volume element of a plant's tissue if the plant's temperatures are not to cross the boundaries set by the values of high and low temperature tolerance. Finally we can rewrite equation (4.21) in the more general form of

$$G_a + R_a - R_e \pm C \pm Q_w \pm Q_H = 0 \qquad (4.22)$$

This is the equation for the physical energy balance of a plant. G_a is always an energy input to the plant. Energy exchanged by convection C usually is a dissipation of energy from the plant but can turn into an energy input if the ambient air has a higher temperature than the plant's surface. During dew deposition or if a plant tissue freezes Q_w is an amount of energy that increases the total heat energy content of a plant. In the reversed cases the heat energy content is decreased.

The terms Q_w and Q_H are latent and volumetric specific heat functions of the plant's temperature. We have seen in Section 4.1.4 that the radiation emitted by plant surfaces is also a function of the surface temperature and the same is valid for convective energy exchange. One could suppose that by writing every term of equation (4.22) in its explicit form and doing some mathematical manipulations one should get an equation which allows the calculation of the leaf temperature T_l. Unfortunately there does not exist such an explicit form of equation (4.22) for T_l but leaf temperature is determinable by iterative calculation processes provided all parameters are known (for details see Raschke (1956) or Gates (1965)). For thin and flat leaves such calculations are feasible and yield reasonable values (see Gates and Papian 1971) but for thick succulent leaves of complex geometrical shape such an attempt is questionable.

In most cases such investigations do not aim at the temperature of a plant's organ or its surface which can be measured easily by appropriate sensors like infrared radiation thermometers or thermocouples. The goal is more commonly to get information on the magnitude of the different energy fluxes and their significance for the determination of Q_H and thus ΔT (equation (4.15)). Such calculations also reveal whether supposed adaptations like a wax bloom do significantly alter the energy input by solar radiation. We can determine absorbed and emitted radiation as outlined by

Instantaneous Energy Balance of Welwitschia mirabilis

Input	$J_G\ a_G$		$(J_{Ru} + J_{Rl})\ a_R$
Output	$J_C + J_P + J_H$	J_L	$J_{Ru}(T_u) + J_{Rl}(T_l)$

100 Wm^{-2}

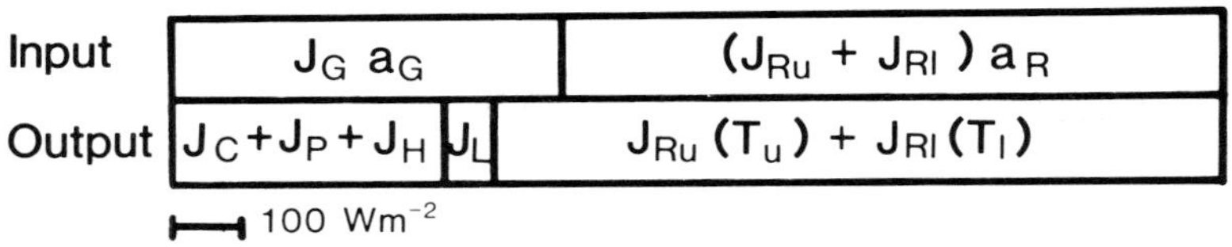

$J_G\ a_G$	511 Wm^{-2}	Absorbed global radiation (upper + lower surface)
$J_{Ru}\ a_R$	417 Wm^{-2}	Absorbed thermal radiation (upper surface)
$J_{Rl}\ a_R$	422 Wm^{-2}	Absorbed thermal radiation (lower surface)
$J_{Ru}(T_u)$	454 Wm^{-2}	Emitted thermal radiation (upper surface)
$J_{Rl}(T_l)$	467 Wm^{-2}	Emitted thermal radiation (lower surface)
J_L	59 Wm^{-2}	Energy dissipation by transpiration
$J_C + J_P + J_H$	370 Wm^{-2}	Energy balance residual

Fig. 4.24 Energy balance of *Welwitschia mirabilis* determined after data collected on 8 September 1981 at 13:30 local standard time in the Welwitschia Plain in the Central Namib desert on a horizontally exposed leaf element, 60 mm from the margin of the leaf. (Data taken from Eller *et al.* 1983b.)

equations (4.8) and (4.13) or the quantity of heat energy fluxes connected with transpiration and heat storage in the tissues according to equations (4.14) and (4.15). It is only a question of the precision and the accuracy of the sensors used to measure the parameters as to what magnitude errors are included in the values resulting from such calculations. However, convective energy exchange is only quantifiable for very simple shapes of plant organs and this only if the aerodynamic parameters can be defined beyond any doubt. The questions whether free or forced convection prevails and whether laminar or turbulent flow exists can rarely be answered for convective energy exchange occurring at a particular situation in the natural habitat of a plant. This holds especially for the complex shapes of succulents and in most cases assumptions and simplifications must be made.

In Fig. 4.24 the instantaneous energy balance of *Welwitschia mirabilis* is shown (Eller *et al.* 1983). As already pointed out in Section 4.1.4.2, the thermal radiation fluxes dominate the energy balance. Summing up input versus output we get a net thermal radiation energy flow of $-82\ W\ m^{-2}$

projected leaf area. That means 16% of the energy input from absorbed solar radiation is dissipated from the plant by emitted thermal radiation exceeding absorbed thermal radiation from both the soil and the atmosphere. Phase transition of transpired water consumes an additional 59 W m^{-2} or 11.5% of absorbed global radiation. The rest, the energy balance residual of 370 W m^{-2} is the energy that flows into heat storage and convection. During the last 30 minutes prior to the measurements that were used for the energy balance calculations given in Fig. 4.24 the temperature of the less than 2-mm-thick *Welwitschia* leaf did not change more than 2 K (Eller *et al.* 1983). Therefore the heat energy influx was less than 10 W m^{-2} during this time. Q_P can be neglected since photosynthesis of *Welwitschia* was only about 1 μmol CO_2 m^{-2} s^{-1} (von Willert *et al.* 1982). Consequently at least 346 W m^{-2} or 67.7% of the absorbed global radiation is dissipated by convective energy exchange. This value seems at a first view very high but Eller *et al.* (1983) showed that such a value is reasonable for the windspeed (mean velocity 1.5 m s^{-1}) measured.

Convection is correlated with wind velocity and if the windspeed decreases the term Q_H must increase and vice versa. The resulting heat flows J_H cause the leaf temperature to change and Fig. 4.25 gives an example of this interrelationship for *Welwitschia mirabilis* in the natural habitat. At 9:50 the windspeed was about 3 m s^{-1} and the leaf temperature of *Welwitschia* was 24 °C and only 1 K higher than the temperature of the ambient air. One hour later the wind velocity was nearly zero and as expected the leaf temperature rose to 35 °C or 6 K above the ambient air temperature. Then a steep increase of windspeed occurred which was paralleled by a significant drop in leaf temperature. The same correlation between wind velocity and leaf temperature is also clearly expressed between 12:00 and 14:00.

At windspeeds below 1 m s^{-1} the leaf temperature of *Welwitschia* increases drastically but this is mainly due to the large size of the more or less horizontally exposed leaves. Low wind velocities or still air do not always mean that the convective energy exchange must be low. In such a case an increase of the leaf temperature or of the ambient air can cause microclimatic turbulences and eddies from raising of hot air pockets. This can cause short-term air displacements close to the leaf with windspeeds characteristic for forced convection as shown by Eller and Grobbelaar (1982) for *Ledebouria ovatifolia*. Such conditions changing between free and forced convection are more efficient if the leaves are small or in an upright position (Vogel 1969) which is frequent with succulents.

Transpiration cooling is only significant for the energy balance if there is a high transpiration rate, a situation that exists for succulents only during the rainy season when sufficient water is available from the soil. The same is valid for annual herbs but for shrubs and trees with deep-reaching roots which can draw water from ground water resources transpiration cooling can be important during the drier season. Lange (1962a, b) for different herbs, shrubs and also for the date-palm (*Phoenix dactylifera*), and Smith (1978) for different perennials of the Sonora desert, reported significant reductions of leaf temperature through heat dissipation by transpiration.

For a rough estimation one can assume that by a transpiration of 1 mmol m^{-2} s^{-1} the energy dissipation is 45 W m^{-2}. This can balance 5 to 10% of the absorbed energy from solar radiation. However, transpiration of succulents usually is lower than 0.5 mmol m^{-2} s^{-1} except after an abundant rainfall event. Much higher transpiration rates are exhibited by nonsucculent annuals, herbs and shrublets (Table 4.4 in Section 4.2.3). In exceptional cases energy dissipations of up to 450 W m^{-2} were estimated according to the peak value of transpiration of 10 mmol m^{-2} s^{-1} exhibited by *Gazania lichtensteinii* during a bergwind day ($\Delta w = 70$ mPa Pa^{-1}) in its natural habitat.

During most of the day the net thermal radiation loss only decreases the heat load caused by absorbed global radiation. At dawn and dusk, under the condition of open shade and during clear nights, this thermal radiation loss is responsible for an energy dissipation from the plant that means that the total radiation budget is negative.

If an energy input from phase transition of water (freezing or dew formation) is not considered the convective energy exchange is the only energy flux that can turn into an energy inflow to the plant. However, this occurs only in cases where the plant's surface is cooler than the ambient air. Under the condition of very low wind velocities or in nearly still air, convective energy exchange tends to be very low and then the thermal radiation loss results in plant temperatures being lower than the air temperature. In Fig. 4.25 such conditions exist before 9:30 and after 19:00. The leaf temperature of the *Welwitschia* leaf lies about 6 K below air temperature since at that time we have nearly still air. After 20:30 an air movement at velocities greater than 0.3 m s^{-1} evidently causes an energy inflow to the plant and results in leaf temperatures that are still below those of the air but the temperature differences are smaller (about 2 to 3 K). That the cooling by thermal radiation loss is most effective if the air movements around the plant are very low can also be seen in the timespan from 5:30 to 7:30 (Fig. 4.25).

Net radiation loss occurs not only from the plant's surface but also from

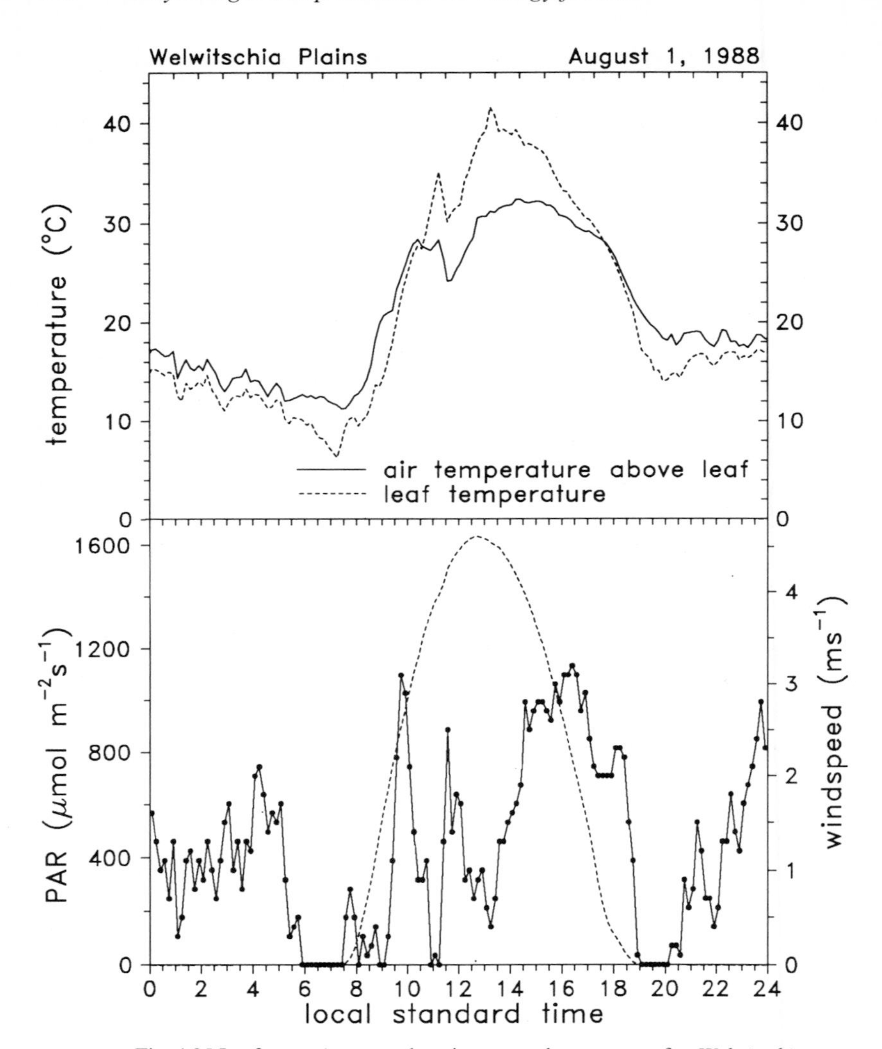

Fig. 4.25 Leaf temperature and environmental parameters for *Welwitschia mirabilis* in its natural habitat.

the soil. In the absence of wind the air adjacent to the soil surface cools down since the soil's surface loses energy and gets cooler. Because the moisture content of the air does not change, a lower air temperature means higher relative air humidities and a lower VPD. Very low values of VPD, dewfall and even the formation of fog are environmental conditions frequent during clear nights in the Namib and similar deserts.

An example of the concomitant changes of air and plant temperatures is given in Fig. 4.26 for a habitat in the southern Namib of *Prenia sladeniana*.

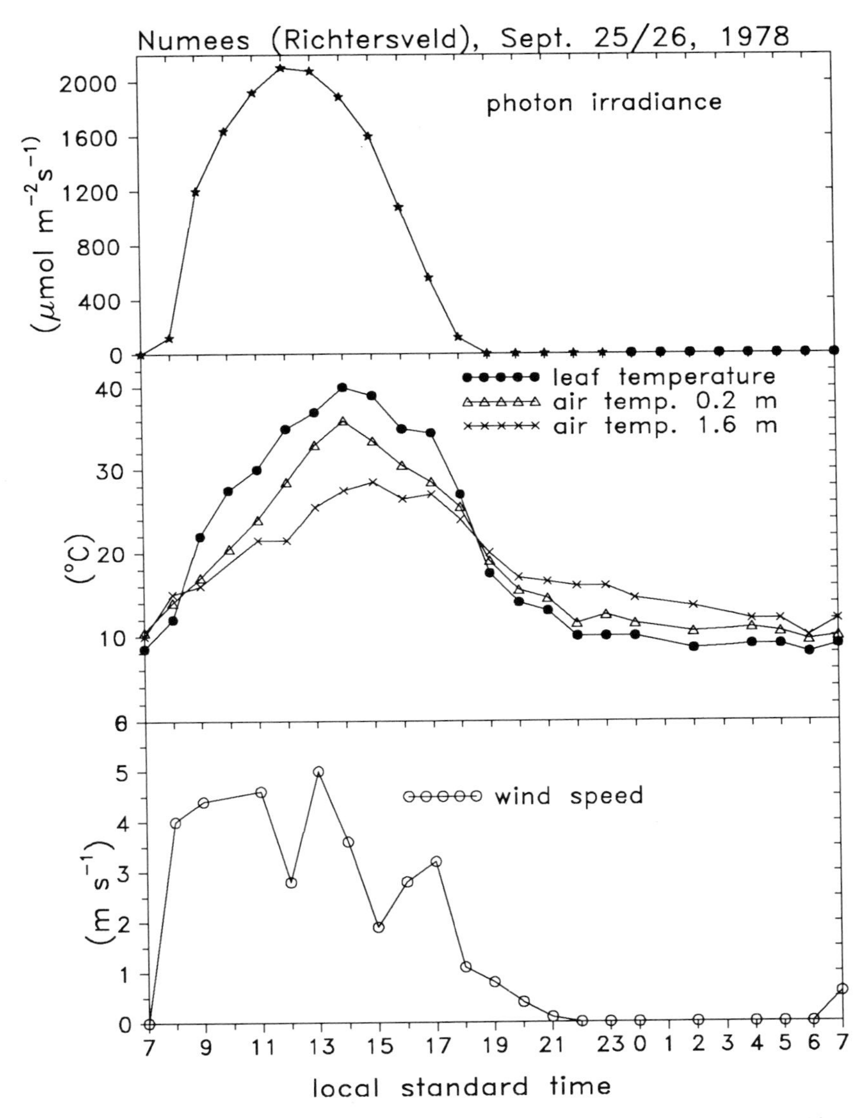

Fig. 4.26 Leaf temperature of *Prenia sladeniana* and temperatures of the ambient air during a day with clear sky and a rather strong wind during the day.

The values given for a height of 1.6 m depict the diurnal temperature fluctuations of the air not directly subject to interactions with soil and plant surfaces. In this habitat the vegetation is only 0.1 to 0.2 m high and thus the temperature given for 0.2 m is the temperature of the air boundary layer adjacent to the soil surface and the ambient air temperature significant for the plant cover. During the daytime the air at 0.2 m is heated up by convection energy exchange from the plant and the soil surface which reach higher temperatures than the air since the plants and the soil absorb considerable fractions of the sun's energy. During the night air movements were low and consequently the plant and the soil cool down as outlined above. At night the plant's temperature lies permanently below the ambient air temperature and the lowering of the latter results mainly from convective energy transfer from the air to the plant. Such a transfer happens only if the temperature of the plant is lower than that of the air adjacent to the plant's surfaces. The emittance of the soil and the plant for thermal radiation are similar and we can assume that there is no significant exchange of energy through thermal radiation between the plant and the soil surface around the plant. This is however not valid during the daytime.

During the day solar radiation is the main influx of energy to the plant and also the soil and other structures. The amount of absorbed global radiation depends on the absorptances which can differ substantially between plant and soil but also between different plants and different soils (Section 4.1.2). Fig. 4.27 compares two habitats of succulents with different soils. The habitat of *Argyroderma delaetii* is a plain densely covered with white quartz pebbles whereas the other habitat with a population of shrublets (*Psilocaulon* sp. and *Ruschia* sp.) has a red sandy desert soil. Irradiance, air temperature and the potential evaporation were the same for both habitats but soil surface temperature in the afternoon was nearly 10 K higher at the habitat with the sandy soil. Leaf temperatures of *Argyroderma* were close to that of ambient air whereas leaf temperatures measured in the habitat with the reddish soil exceeded ambient air temperatures considerably.

The higher leaf temperatures not only result from different absorptances of the plants for solar radiation but must also be due to differences in thermal radiation fluxes. Firstly, red desert soil reflects more infrared solar radiation (Fig. 4.9) than quartz pebbles and, secondly, the maximal value of emitted thermal radiation was about 520 W m^{-2} (emittance 0.96) for the quartz habitat and about 580 W m^{-2} for the other one. So we must assume more than 60 W m^{-2} in higher thermal radiation flow from the soil surface to the plant if it grows on red sandy soil and not on a plain covered with quartz pebbles. However, an absolutely convincing conclusion could only be made if all the energy fluxes could be quantified exactly.

One cannot deny that especially in arid habitats with prevailing clear sky conditions high global irradiance during the day and relatively high net thermal radiation losses during the night tend to give rise to plant surface

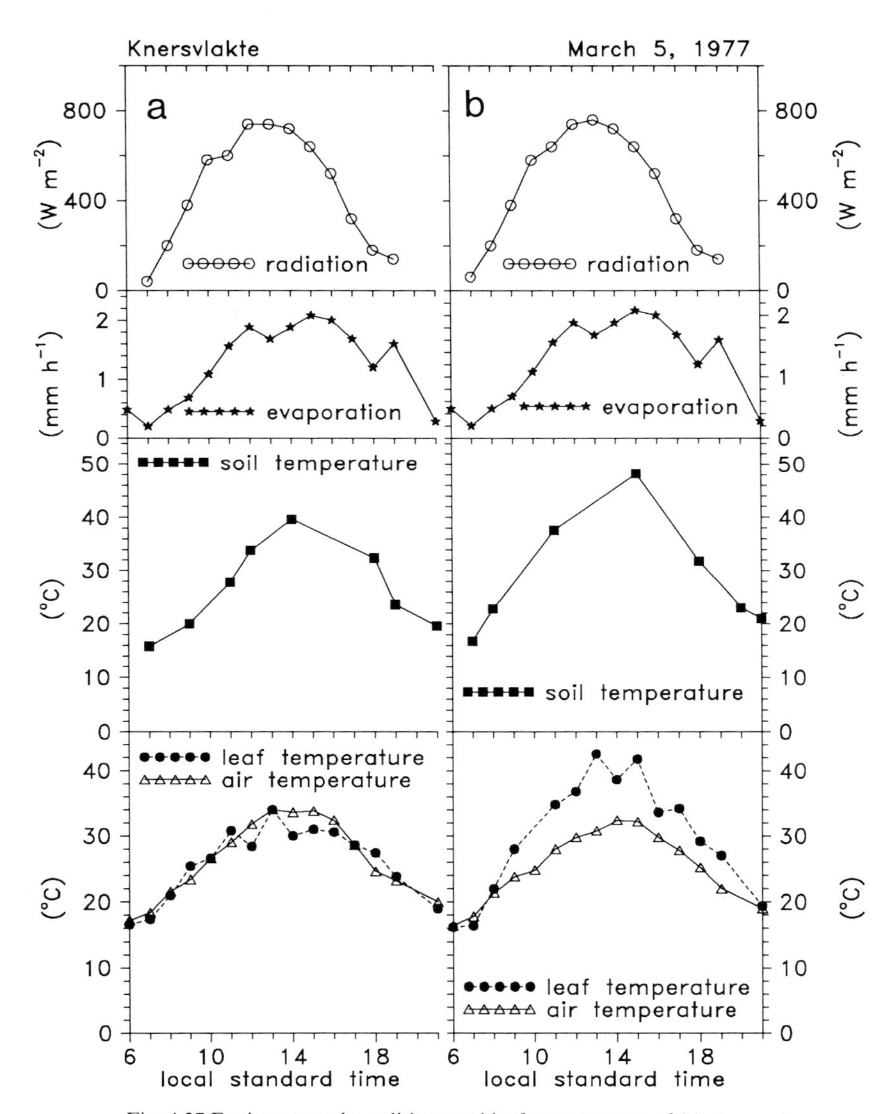

Fig. 4.27 Environmental conditions and leaf temperatures of (a) *Argyroderma delaetii* growing on a plain covered with quartz pebbles and of (b) *Ruschia* sp. growing on an adjacent plain of red sandy soil.

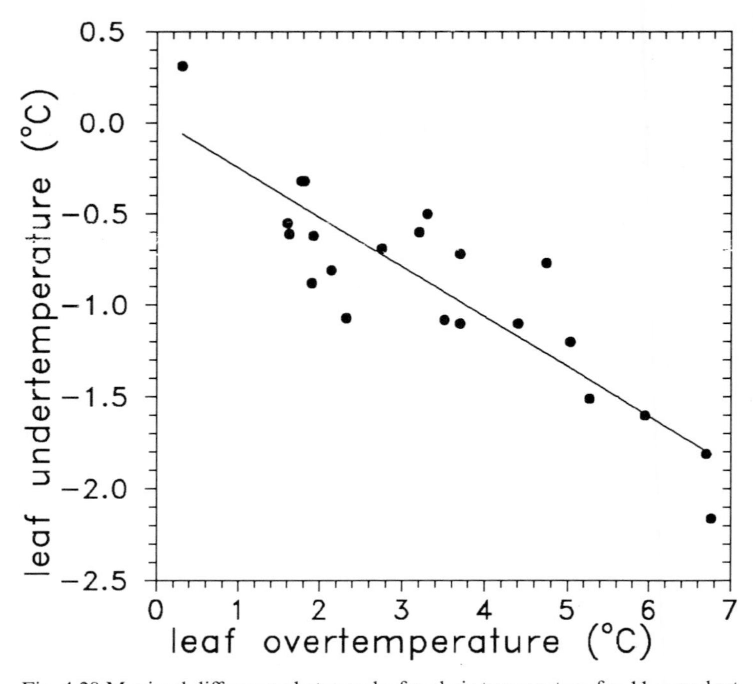

Fig. 4.28 Maximal differences between leaf and air temperature for 11 succulent and one non-succulent species growing at Numees (Richtersveld). Data were collected during several field investigations. For some species data from several days are given. The x-axis gives the overtemperature of the leaves during daytime, the y-axis the undertemperature of the same leaves during the night. The species included were *Aloe ramosissima*, *Brownanthus schlichtianus*, *Cheiridopsis robusta*, *Cotyledon orbiculata*, *Didelta carnosa*, *Drosanthemum* sp., *Mesembryanthemum barklyi*, *Othonna opima*, *Ozoroa dispar*, *Prenia sladeniana*, *Psilocaulon subnodosum* and *Tylecodon paniculatus*.

temperatures that are considerably higher than the air temperature during the daytime and lower during the night. Plants subjected to unobstructed radiation input and output will reach the highest overtemperatures during the day but one can assume that the same plants while facing the cold clear sky at night then have a high loss of thermal radiation. Thus, such plants should have also the largest undertemperatures at night. Highest measured positive and negative differences of leaf temperatures with respect to ambient air are summarized in Fig. 4.28 for 12 succulent and non-succulent plants of a habitat in the southern Namib. The results confirm to a high degree the suppositions made above and at the same time demonstrate how important the diurnal changes in radiation fluxes are for the plant's energy balance and thus plant temperatures.

4.1.8 The significance of succulence for the plant's energy balance

The ecophysiological performance of a succulent is determined by its physiology adapted to optimize or balance fluxes of energy, water and carbon in a way that guarantees the plant's survival. Succulents have evolved water storage tissues which enable them to endure periods of drought. If we presuppose that adaptation to drought is the dominating trend in the evolution of succulents, then the goal to optimize water relations gets priority against the optimization of carbon gain and physical energy fluxes. Starting from this assumption we cannot expect that succulence has only advantages with respect to the plant's energy balance. On the contrary we must expect compromises which favour optimization of water relations.

Considering radiation energy input we firstly have to notice that thick succulent structures in a leaf or a stem cause absorptances for solar radiation which can be substantially higher than those of other types of plants. In Fig. 4.5 we have shown in which way the absorptivity increases with increasing succulence. A thick succulent leaf absorbs more energy from incident radiation than a thinner one. Evidently succulence has negative consequences for a plant if radiation energy input is focused. So we have to look for features that compensate at least a fraction of this disadvantage of succulence. Some features, like trichomes, wax bloom, etc., can attenuate the energy input as outlined in Section 4.1.4. Inclined or upright leaves absorb less solar radiation during the timespan when the sun's radiation is high. Moreover, a reduced leaf size frequently found with shrubby species of the Mesembryanthemaceae favours high convective energy exchange rates as does a more or less upright position of a leaf or a succulent stem. An improved energy dissipation by convection can compensate a fraction of the increased energy input from absorbed radiation.

Very thick cuticles are frequent among some succulents (e.g. aloes) and are considered an adaptation to prevent excessive water loss from the water storing tissues. The disadvantage in this case is that such optically opaque structures tend to absorb a fraction of PAR which then is no longer available for photosynthesis. A concentration of the chloropohylls in a thin chlorenchyma below the epidermis, leaving the proper water storage tissue free of pigmentation, is frequent with succulents (e.g. aloes). However, often relatively bulky leaves have their photosynthetic pigments more or less evenly distributed over the whole cross-section of the leaf or at least pigments are also found within the water storage tissue (e.g. *Pleiospilos*). Eller and Grobbelaar (1982) showed in their investigation on *Ledebouria*

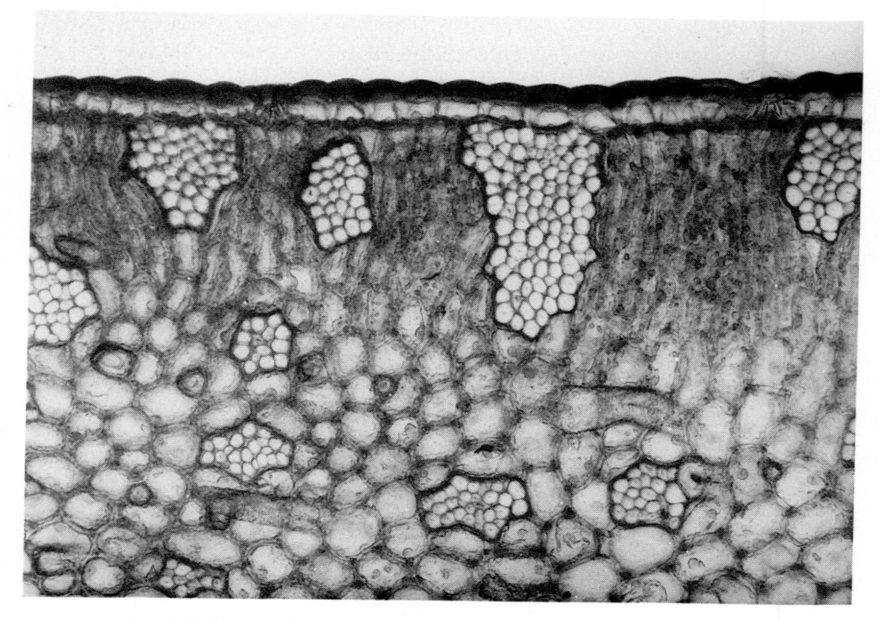

Fig. 4.29 Cross-section of the upper half of the leaf of *Welwitschia mirabilis* showing the large fibre bundles just underneath the epidermis. The fibre bundles run along the axis of the leaf.

ovatifolia, which has moderately succulent leaves, that after 1.8 mm light pathway through the assimilatory tissue the PAR is reduced to almost nil. They proved experimentally that the lack of a sufficient amount of PAR prevented the lower leaf section, despite its pigmentation, contributing to a positive carbon balance of the leaf.

Schanderl (1935) supposed that the scleromorphic fibre bundles often found below the epidermis of xeromorphic leaves act as some kind of light wells or windows through which additional light can penetrate into deeper layers of the pigmented tissue. The famous *Welwitschia mirabilis* of the Namib desert has such fibre bundles (Fig. 4.29) which could act in this way. However, one could also suppose that radiation is totally reflected at the fibre surface opposite to the direction of the light ray incident on that fibre. In such a case the light could, on the one hand, be diffused to those parts of the assimilatory tissue lying between the fibre bundles or, on the other hand, light rays could also take their way back through the epidermis and thus increase the amount of reflected radiation. In fact, *Welwitschia* has reflectivities for infrared solar radiation that are exceptionally high for a thick dull green leaf (Schulze *et al.* 1980). Whether such fibre bundles could also play a role in xeromorphic succulent leaves has not yet been investigated.

Succulents with very thick leaves often show an epidermis with clearly visible dark green dots, e.g. species of the genus *Pleiospilos* or *Conophytum*. These dots are either very large single cells (idioblasts) or groups of cells that are free of pigmentation (Fig. 4.30a). They lie just below the epidermis and certainly act as some kind of light well or microwindow (Schanderl 1935) allowing the light to penetrate deep into the assimilatory tissue. Schanderl (1935) counted 280 such idioblasts per 100 mm^2 on the epidermis of *Pleiospilos bolusii*. The diameter of these microwindows was 0.25 to 3.0 mm. Therefore, about 14 to 20% of the epidermis consisted of microwindows. For *Nananthus albipuncta* and *Cheiridopsis cigarettifera* the fractions were 31% and 43% respectively. Schanderl (1935) also found that the smaller the idioblasts the larger was their frequency and that in leaves with thicker assimilatory parenchyma the idioblasts were bigger and reached deeper into the chlorophyll-bearing tissue.

A different type of microwindow exists in the form of the bladder cells of the idioblast type of epidermis (Ihlenfeldt and Hartmann 1982). That papillae of epidermal cells act physically as a proper lens was demonstrated by Haberlandt (1909). We must suppose that such epidermal idioblasts or bladder-cells act likewise. In leaves with an epidermis of the idioblast type, the subepidermal tissue below the idioblasts is free of chloroplasts or their density is very low (Fig. 4.30b). By this a light well is formed allowing the light focused by the 'idioblast lenses' to penetrate deep into the assimilatory tissue. A modification of these light wells can be found in other succulent leaves such as those of *Ruschia sedoides* which has a xeromorphic type epidermis. Here the small epidermis cells form surface sculptures of lens-like appearance. Below these 'lenses' a light well like in the idioblast type of epidermis is formed. In both cases the light well is not a single large idioblast but a tissue of numerous small cells. Sometimes such non-pigmented zones are quite large and get manifested in dark green stripes as on the leaves of *Senecio corymbiferus* or *Senecio herreianus*.

Several genera of small succulents of the Liliaceae (*Haworthia*, *Bulbine*) or the Mesembryanthemaceae (*Fenestraria*, *Lithops*, etc.) have developed this windowing of the epidermis to perfection. The leaf rosettes of *Haworthia cymbiformis* are partially sunken into the soil. Its leaves are very succulent and only the leaf tip which has a finger-tip-like appearance protrudes from the soil surface. This part of the leaf shows not only achlorophyllous dots but a fasciated surface with large irregular-shaped areas free of pigmentation which are referred to as windows (Marloth

1909). This windowing is maximized if the whole surface of the leaf tip is free of green pigments as with *Lithops* and most characteristic with *Fenestraria* which got its name from that eye-catching feature. These window plants share the feature that they are completely embedded into the soil. The leaves have a more or less upright position with the leaf tips at the soil surface or protruding only a little from the soil. The leaf tips either are rounded and look like finger tips (e.g. *Fenestraria*) or are flattened as with *Haworthia maughanii* or *Bulbine* of which an example is given in Fig. 4.31. The leaves of such plants, or the coalescent leaf pair as for *Lithops*, have the shape of an inverted cone. The greater part of the leaf surface, the cone mantle, is not exposed the direct way to solar radiation since the leaf is embedded into the soil. Thus the assimilatory tissue adjacent to this cone mantle would get insufficient light if the leaf tip absorbed solar radiation like an ordinary green assimilatory tissue. However, during evolution this leaf tip lost its pigmentation almost completely and the solar radiation incident on the window area is transmitted via the colourless water storage tissue to the assimilatory tissue. Therefore the photosynthetically active tissue adjacent to the cone mantle gets light only from inside the leaf and not across the epidermis as can be seen from Fig. 4.30c. From the visible solar radiation (400 to 750 nm wavelength) incident on the window of a *Lithops lesliei* plant 23.0% is transmitted to a depth of 10 mm and 8.6% of the impinging radiation reaches a depth of 20 mm inside the leaf cone (Eller and Grobbelaar, unpublished results). Moreover this light is more or less evenly dispersed over the whole assimilatory tissue since the water storage tissue disperses the solar radiation beam. This diffusion of light might also play an important role in some species of *Peperomia* with a multiple achlorophyllous epidermis especially if the chlorenchyma is V-shaped in cross-section as it is for *Peperomia asperula* or *Peperomia columella* leaves.

All species of *Lithops* or *Ophthalmophyllum* have coloured windows with green parts and some window plants such as *Haworthia maughanii* seem to have dark green pigmented window areas. This green colour, however, does not result from a pigmentation of the window area but is the result of light that penetrates the window and is transmitted across the water storage tissue to the pigmented assimilatory tissue from where a part of this radiation is reflected as at the chlorenchyma of an ordinary green leaf. A fraction of these reflected light rays take their way back across the colourless water storage tissue and back through the achlorophyllous window. Therefore, we do not see radiation reflected from the window but from the assimilatory tissue, or in other words we look through the colourless window on the green chlorenchyma inside the succulent leaf. Additional pigmentation of the windows very often matches with the

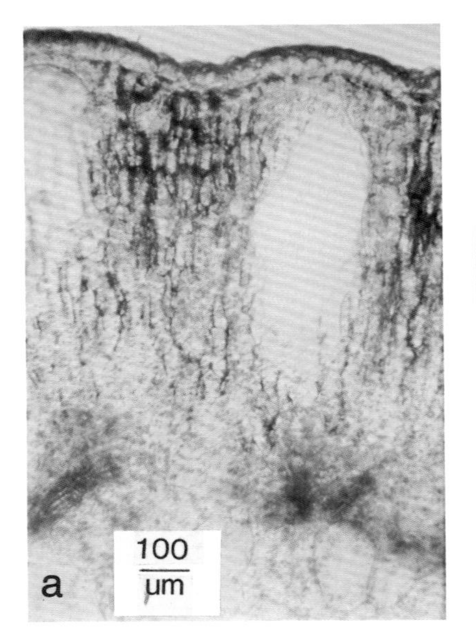

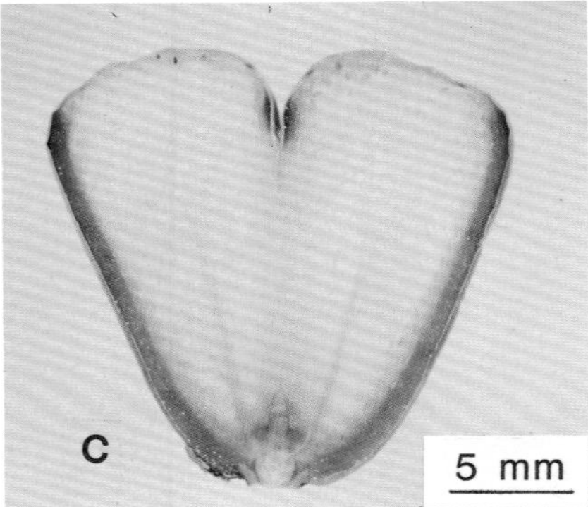

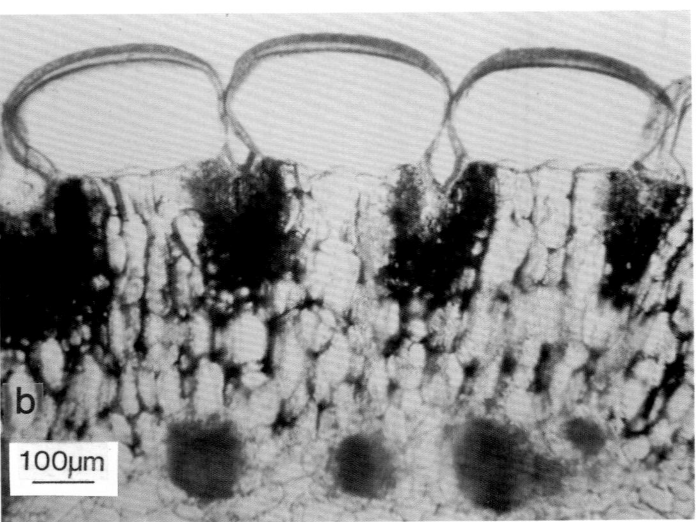

Fig. 4.30 Cross-section through the outer part of the leaves of *Ruschia sedoides* (a) and *Trichodiadema* sp. (b) showing the cell regions below the epidermis which are free of pigmentation and which act as microwindows for light penetration into deeper-lying pigmented cells. (c) Cross-section of the cone-like coalescent leaves of *Lithops karasmontana*. The top of the leaves is free of pigmentation and acts as a macrowindow. The assimilatory pigments adjacent to the cone mantle get the PAR via the window and the colourless water storage tissue.

Fig. 4.31 *Bulbine* cf. *mesembryanthoides* in its natural habitat in the Little Karoo (Republic of South Africa). (a) Shows a plant from which the soil was partially removed to get an impression of the shape of the leaf.

colours of the soil and the pebbles of the plant's habitat and is supposed to act as cryptic coloration (for examples see Cole 1988).

Succulence is, as outlined above, primarily a feature with negative impacts on the energy balance mainly because in general the energy input by absorbed radiation increases with increasing succulence. We mentioned at the beginning of this Chapter that some features like changes in optical properties, leaf exposition and improved convective energy exchange can compensate at least a fraction of this disadvantage. The most interesting fact is, however, that the evolution of succulence gave rise to further adaptations concerning PAR interception as demonstrated above for windowed succulents.

4.2 Water fluxes

A main constituent of a living plant is water. The plant is linked to its environment by two water fluxes, firstly water gains (J_{wg}) through uptake by the roots and by other surfaces (mainly the epidermis), and secondly water loss (J_{wl}), predominantly by transpiration. Gains minus losses constitute the water budget of the whole plant:

$$J_w(\text{net}) = J_{wg} - J_{wl} \tag{4.23}$$

Usually higher plants are rooted in a substrate, predominantly in a soil, and transpirational water loss is to the atmosphere. Consequently, there is a flow of water from the soil through the plant to the atmosphere. This is often referred to as the soil–plant–atmosphere continuum of water flow. Water fluxes approximately obey Fick's first law of diffusion and hence they are determined by driving forces acting over certain distances and by resistances hampering the flow. In order to understand properly the water fluxes we start off with the physical background of the driving forces of water fluxes. We will then deal with the various constituents of the water status of succulents. Their water status and water content is subject to severe alterations by environmental factors.

Transpiration is the main flux of water leaving the plant. There is only a small bank of data referring to this process in succulents. We will add some.

For some plants with specialized epidermal structures, e.g. species of the Bromeliaceae, uptake through the epidermis can exceed or replace uptake by the roots. Whether succulents also have specialized structures for the uptake of water via the epidermis and whether water vapour uptake via the stomata is feasible will be discussed in Section 4.2.4.

In nearly all plants with a small water storage capacity the water status is mainly determined by inward and outward fluxes and the water budget of the whole plant must be positive or at least zero. Poikilohydric plants are an exception in this respect. Succulents with their marked water storage capacity make it necessary that a further quality of water fluxes is considered: the translocation of water from the storage tissue and the refilling of the storage tissue.

We have pointed out repeatedly that utilizable water is that fraction of the stored water which is available for the rest of the plant. This is the most interesting feature of a succulent. We will see how this feature can be quantified and how it determines the water balance of succulents.

4.2.1 The driving forces for water fluxes

As outlined above, the plant is a system with an inflow (uptake by the roots or other surfaces), an outflow (transpiration and other losses), and flows of water within the plant body (flow from one tissue or cell to another, consumption or release in biochemical reactions). For a long time there have been arguments in botanical literature on whether or not water uptake and internal flows are active (consuming biochemical energy) transports or flows along energy gradients. It is now generally accepted that flow of water or water vapour to and from the plant and internal flows of liquid water are driven by water potential gradients. Nowadays every good basic textbook on plant physiology gives a derivation of the water potential based on the fundamentals of thermodynamics (e.g. Nobel 1983).

The phenomena of diffusion or a mass flow of water in plants are the same as in a pure physical system. For both phenomena the driving force is not, as often somewhat imprecisely stated, the water potential but the differences between the water potentials at different parts of a plant or its environment. As other potentials, like the gravitational or the electrical, the water potential describes the energy of a system relative to a chosen level. For the water potential this level is the pure phase of water at standard conditions (273.16 K, 101.3246 kPa). It is equal to a water potential of zero. Field theory teaches that energy is consumed if the potential is increased and vice versa. Thermodynamically speaking, any physical system tends to degrade to a system with less free energy and water obeys this basic thermodynamic law by moving from a high energy level (= water potential) to a lower one.

The representation of Ohm's law

$$J = \frac{U}{R} \tag{4.24}$$

which originates from electric potential theory and which states that the flow J (electric current) is directly proportional to the potential difference U

(electric tension) and inversely proportional to a system depending value R, the resistance, is also applicable to analyse water flow along water potential gradients. As in electricity the driving force is a difference ($\Delta\Psi$) in the water potential of two different objects (Ψ_1, Ψ_2) or two localities within the same object, and the magnitude of the water flux (J_w) is inversely proportional to the resistance ($R_{1,2}$) against water flow from place 1 to place 2 or vice versa. Therefore the equation

$$J_w = \frac{\Psi_1 - \Psi_2}{R_{1,2}} = \frac{\Delta\Psi}{R_{1,2}} \tag{4.25}$$

determines direction and magnitude of water flows to, from and within the plant. Usually the energy of the water potential is expressed in units of pressure (SI unit Pa) and thus $\Delta\Psi$ is a pressure difference. In a plant the water flow which results from such a pressure difference over the transport distance with a certain flow (or diffusion) resistance is the same as the water flow would be through a steel pipe of an irrigation system with the same pressure difference and the same resistance. Therefore the plant and its environment are some kind of a hydraulic system which we can analyse without bothering about thermodynamics.

The equation (4.25) states that J_w is only then equal to zero if $\Delta\Psi$ is zero or $R_{1,2}$ is infinity, a case that never exists in biology. In the second case the two objects (e.g. soil and plant's root) are in an equilibrium with the water potential. Comparisons with a known water potential (e.g. solutions with a defined concentration) or measuring the water potential of one object after the second has reached equilibrium with the first one are widely used methods to determine the unknown water potential of plant organs or tissues (Slavik 1974).

We can assume that in plants water scarcely exists in its pure phase but rather is the major constituent of solutions. The formula describing the water potential (Ψ_c) of a solution, e.g. the cytoplasma of a plant's cell, is

$$\Psi_c = \Psi_M + \Psi_\pi + \Psi_P + \Psi_h \tag{4.26}$$

where Ψ_M is the matric potential, Ψ_π is that part of the water potential which is caused by osmotically active substances (often referred to as osmotic potential), Ψ_p is the pressure potential of the solution (relative to standard pressure conditions), and Ψ_h represents the influence of the gravitational potential (relative to randomly chosen level h_o, e.g. the soil surface or the top of a plant).

The matric potential takes into account the changes of the water potential by solid surfaces interfering with the water or a solution (e.g. at the surface of soil particles, the cell wall, etc.) and also the changes in the water potential of a cell sap by mucilaginous substances. However Ψ_M is only some kind of a convenient term to identify this kind of change of the water potential and one could also include such forces (e.g. capillary forces) into Ψ_h or Ψ_P, or even Ψ_π if this term is not exclusively used to describe the osmotic potential according van't Hoff's law. In most cases we can consider the cytosol a dilute solution and the formula (4.26) is reduced to

$$\Psi_c = \Psi_\pi + \Psi_P + \Psi_h \tag{4.27}$$

The van't Hoff law states that

$$\Psi_\pi = -\mathrm{R}T\sum_i c_i \tag{4.28}$$

where R is the gas constant, T is the absolute temperature of the solution, and c_i is the concentration of diluted substances.

The sum of c_i is expressed in molality and frequently referred to as osmolality. From this formula we can derive that a solution with a sum of c_i equal to unity (= 1 osmol), Ψ_h equal to zero and no interference of a matrix has at standard conditions a water potential of −2.27 MPa. Walter (1960) gives a review of the ranges of osmotic pressure values for different groups of plants. European and North American succulents are listed with values between 0.1 and 2.4 MPa, whereas North American desert xerophytes range between 1.6 and 6 MPa. A detailed description of osmotic values and ion concentration in succulents and non-succulents of one particular site will be presented later.

The pressure potential Ψ_p (pressure relative to the pressure at standard conditions) can be every positive or negative pressure applied to a cell from outside or resulting within the cell from uptake of water. In the latter case it is usually termed turgor or turgor pressure of a cell.

The term Ψ_h is in practice mostly expressed in metres height of a water column, therefore

$$\Psi_h = 9.81\ \Delta h(\mathrm{m})\ (\mathrm{kPa}) \tag{4.29}$$

where Δh is the height relative to the randomly chosen level h_o. For small succulents this term can be neglected. Therefore in most cases the equation for the water potential of a cell or a tissue can be written as

$$\Psi_c = \Psi_\pi + \Psi_P \tag{4.30}$$

With this equation we can describe the water potential of a cell or a tissue but the soil and the air also interfere with plant surfaces. Therefore the water potential of the moist air and the soil are important to know. The water potential of moist air at standard barometric pressure is determined according to Raoult's law

$$\Psi_a = \frac{\mathrm{R}T}{V_w(T)} \ln \frac{e_a}{e_s} \tag{4.31}$$

where R is the gas constant, T is the absolute air temperature, $V_w(T)$ the partial molal volume of water at T, e_s the partial water vapour pressure at saturation and e_a the actual water vapour pressure. Instead of partial pressures the air humidity relations can also be expressed in percentage relative air humidity (rh) since

$$rh(\%) = \frac{e_a}{e_s} 100 \tag{4.32}$$

Inserting $V_w(20\ °C)$ into equation (4.31) we get

$$\Psi_a = 1060\ T \log \frac{e_a}{e_s}\ (\mathrm{kPa}) \tag{4.33}$$

The constant in this equation is based on the partial molal volume of water at 20 °C, but if this formula is used to calculate Ψ_a for temperatures in the range from 0 to 30 °C an error of less than 1% can arise compared with the use of formula (4.31).

Even ecologists are sometimes not familiar with the magnitude of the water potential of moist air. It is also frequently believed that the relative air humidity during rainfall is 100%. However, it rarely exceeds 99% and from equation (4.33) we get for such an air mass at a temperature of 20 °C a water potential of −1.36 MPa. A relative humidity of 98% of similar air would be −2.72 MPa which is lower than a solution of 1 osmolal concentration. At 95% the value (−6.92 MPa) is nearly equal to 3 osmolal and at 90% the water potential of the air, which we still consider very moist, is −14.2 MPa or equal to an osmolality of 6.26 which certainly never exists in a living cell.

There is no formula to calculate soil water potential (in most cases expressed as matric potential) but we can determine it indirectly as already mentioned (Slavik 1974). The amount of water a soil can hold against drainage is termed field capacity. If this amount has been reached then further water falling upon the soil causes the displacement of an equivalent amount of water to deeper soil levels or a run-off. Most freely drained soils in humid climates have a field capacity with a water potential of about −30 kPa, but in the sandy and gravelly soils of deserts this value can be more negative. As with air humidity, human beings have no sensorium for an appropriate determination of soil humidity and therefore estimations without measurements are purely random. After some time a wet soil that dries out looks quite dry but can still have a water potential that is not lower than −0.5 to −1 MPa.

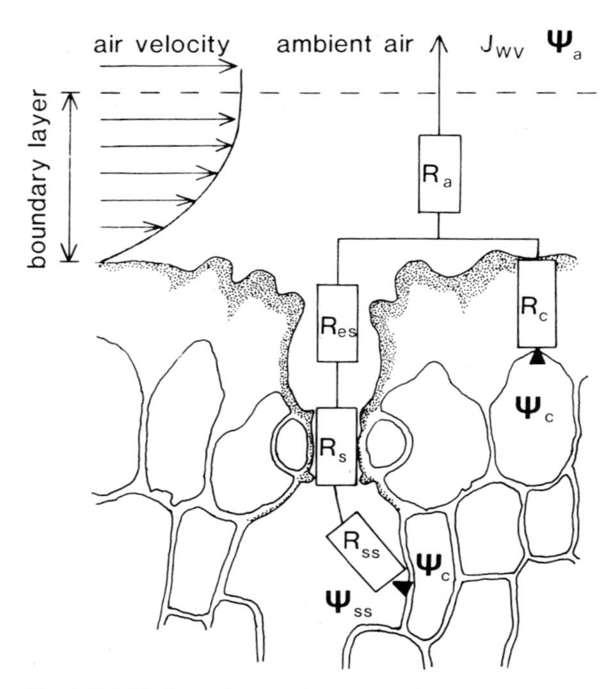

Fig. 4.32 Diffusion resistances in transpiration (J_{wv}). R_a = boundary layer resistance, R_c = transcuticular resistance, R_{es} = epistomatal resistance, R_s = stomatal resistance, R_{ss} = substomatal resistance, Ψ_a = water potential of the ambient air, Ψ_c = water potential of the mesophyll surrounding the substomatal cavity, Ψ_{ss} = water potential of the air in the substomatal cavity.

4.2.1.1 Transpiration

If Ψ_a is the water potential of the ambient air and Ψ_c of the plant's cells in the air spaces of the mesophyll then transpirational water flow J_{wv} is, according to equation (4.25),

$$J_{wv} = \frac{\Psi_a - \Psi_c}{R_{wv}} \tag{4.34}$$

where R_{wv} is the transpiration resistance which describes the modulation of J_{wv} by diffusion resistances in the leaf and the boundary layer of the air adjacent to the leaf (Fig. 4.32). As the direction of the diffusion of water vapour only depends on the sign of the difference $\Psi_a - \Psi_c$ the flow of water vapour is not exclusively a flow from the leaf to the air (= transpiration) but at least theoretically can also occur in the opposite direction. If the water potential of the air in the air space below the stomatal pore (= substomatal cavity) has a lower water potential than the air in the boundary layer then an inflow of water vapour must result. Such a situation would be referred to as 'reversed transpiration'. Whether such conditions can occur or not and which prerequisites must be fulfilled will be discussed in Section 4.2.4.

Transpiration consists of a water loss through the stomatal transpiration, and the loss across the cuticle, the cuticular transpiration. The latter is a diffusion through the outer walls of the epidermal cells and their cuticle. Usually epidermal cells of succulents have thickened outer cell walls with a well-developed cuticle with an epicuticular wax layer and therefore the cuticular resistance R_c is very high. The resistance against diffusion of water vapour through the stomata consists of additive resistances, namely: the substomatal diffusion resistance R_{ss} from the surface of the substomatal cells where the phase transition from liquid water to vapour takes place, the resistance R_s of the stomatal pore and the resistance R_{es} of the epistomatal diffusion path to the bottom of the boundary layer of the air. All these resistances are functions of the geometry, mainly the length, of the diffusion path and are to a great extent genetically predetermined. The plant can only vary R_s by active opening and closing of the stomata. That stomata are sunken into the epidermis is a feature which is widespread among succulents and is considered an adaptation to prevent excessive water loss (see Section 1.2). In fact this sinking of stomata into the epidermis causes an additional diffusion resistance which is absent in mesomorphic leaves.

It is feasible to calculate values of R_{ss}, R_s and R_{es} if the shape of the diffusion path from the mesophyll to the boundary layer is known, and R_a is determinable using the equations for the calculation of boundary layer thickness (Section 4.1.6). The best guideline for such calculations is still the comprehensive work of Bange (1953).

There is no doubt that equation (4.34) only then results in appropriate values if the value of the water potential of the water vapour at the starting point of diffusion is known. In some basic textbooks on botany it is still argued that the air in substomatal cavities is 100% saturated, an assumption that only holds if Ψ_c is equal to zero. If there is no water flux (or diffusion) then water potential theory implies that the layer of water vapour adjacent to the wall of the cell from which diffusion starts must have a water potential (Ψ_{ca}) equal to the water potential in the cell wall (Ψ_{cw}). This one in turn must be in an equilibrium with the cell's cytosol with the water potential Ψ_c. These statements, however, imply that water vapour, cell wall and cytosol are at the same temperature.

A flow of water from cytosol to cell and from cell wall to substomatal cavity can only exist if

$$\Psi_c > \Psi_{cw} > \Psi_{ss} \tag{4.35}$$

If we exclude a 'reversed transpiration' (Section 4.2.4 considers such conditions) then the value of the water potential of the water vapour in the substomatal cavity in the presence of stomatal transpiration is always less than Ψ_c. Only in the absence of stomatal transpiration could the water potential Ψ_{ss} be as high as Ψ_c. However these basic considerations do not include the transition of water from the liquid phase to vapour, a mechanism that consumes energy and thus lowers the ambient temperature. But evaporation occurs only if the water potential of the air in the substomatal cavity at some distance from the evaporating surface is lower than the water potential of that surface. Therefore the assumption that the water potential of the substomatal cavity Ψ_{ss} always is less or equal to Ψ_c is valid. In a first approximation we can assume that $\Psi_{ss} = \Psi_c$.

For a given temperature of moist air Ψ_a is, according to equation (4.33) a function of the absolute water vapour partial pressure $e_a(T_a)$. If Ψ_{ss} is known (e.g. assuming that $\Psi_{ss} = \Psi_c$) and the temperature (T_{ss}) of the moist air inside the substomatal cavity can be quantified (e.g. assuming that T_{ss} is equal to a measured leaf temperature) then the value of $e_a(T_{ss})$ for the air in the substomatal cavity can be calculated by rearranging equation (4.33)

$$e_a(T_{ss}) = e_s(T_{ss})\ 10^{\left(\frac{\Psi_{ss}(\mathrm{kPa})}{1060\ T_{ss}}\right)} \quad (e_a \text{ in Pa if } e_s \text{ in Pa}) \tag{4.36}$$

If ΔW is the water vapour partial pressure difference between the ambient air $e_a(T_a)$ and the substomatal cavity $e_a(T_{ss})$ then we can rewrite equation (4.34) in the form

$$J_{wv} = \Delta W\ g_{wv} \tag{4.37}$$

with g_{wv} being the transpirational conductivity which is the reciprocal value of R_{wv}. The term ΔW must be in the unit Pa Pa^{-1} and with J_{wv} in the units mol $\mathrm{m}^{-2}\ \mathrm{s}^{-1}$ then the conductivity has the same unit. Formula (4.37) has some advantages over the original one (4.34). From the resistance values which constitute R_{wv} only R_a and R_s can vary to a larger extent. The different resistances form a network and R_{wv} can be calculated with the same formula as for electrical resistances. Therefore

$$R_{wv} = R_a + \frac{R_c\,(R_{es} + R_s + R_{ss})}{R_c + R_{es} + R_s + R_{ss}} \tag{4.38}$$

Calculating ΔW from air humidity and the temperatures of the air and the leaf respectively while taking into account Ψ_c and measuring J_{wv}, we can calculate R_{wv}. If we also calculate R_a with the aid of the formulas for boundary layer thickness (Section 4.1.6) then

$$R_{wvl} = R_{wv} - R_a \tag{4.39}$$

where R_{wvl} is the leaf resistance to diffusion of water vapour. Since R_c, R_{es} and R_{ss} vary very little to nil the variations of R_{wv} are a true expression of the variations of R_s or in other words represent the status of control of transpiration by the stomata.

4.2.1.2 *Water uptake from soil*

The flow of water J_{wr} accompanying water uptake from the soil by the roots is

$$J_{wr} = \frac{\Psi_s - \Psi_r}{R_{wr}} \tag{4.40}$$

where Ψ_s and Ψ_r are the water potentials of the soil and the root cells respectively. The term R_{wr} accounts for the uptake resistance in all its manifestations. Water uptake will occur under the same conditions which are valid for transpiration: the water potential of the root cells must be lower than that of the soil. Whether or not liquid water uptake by other surfaces than the rhizodermis is feasible with succulents will be discussed in Section 4.2.4.

4.2.1.3 *Water translocation*

Water flow $J_w(1,2)$ from one tissue to another or from one plant organ to another depends on the difference $(\Psi_1 - \Psi_2)$ and the diffusion resistance $R_{1,2}$ between them (equation (4.25)). The water potential Ψ_1 and Ψ_2 can be expressed by equation (4.30) and we get

$$J_w(1,2) = \frac{\Psi_{\pi_1} + \Psi_{P_1} - \Psi_{\pi_2} - \Psi_{P_2}}{R_{1,2}} \tag{4.41}$$

or

$$J_w(1,2) = \frac{(\Psi_{\pi_1} + \Psi_{\pi 2}) - (\Psi_{P_2} - \Psi_{P_1})}{R_{1,2}} \tag{4.42}$$

Magnitude and direction of $J_w(1,2)$ depend on the values of the differences

$$\Delta\Psi_\pi = \Psi_{\pi_1} - \Psi_{\pi_2} \tag{4.43}$$

and

$$\Delta\Psi_P = \Psi_{P_2} - \Psi_{P_1} \tag{4.44}$$

and thus

$$J_w(1,2) = \frac{1}{R_{1,2}}(\Delta\Psi_\pi - \Delta\Psi_P) \tag{4.45}$$

If $J_w(1,2)$ is positive then water flows from tissue 1 into tissue 2 and vice versa. We will come back to this in Section 4.2.5 as it is a very important feature for succulents.

4.2.2 Water status of succulents

In Section 4.2.1 we saw that under steady state conditions the water status of a cell or a tissue can be expressed by its water potentials following equation (4.30). We now will discuss in more detail matters concerning the water status of succulent plants.

4.2.2.1 Osmotic potential

According to textbooks (e.g. Larcher 1983), the osmotic potentials of succulents lie in the range of −0.4 to −1.8 MPa. A screening among many succulents in the Richtersveld and adjacent regions revealed, however, that this statement is not valid for the Mesembryanthemaceae, the dominant plant family in these areas, nor for succulents of other families. Frequently, and depending on the availability of water in the soil, osmotic potentials below −2.5 MPa were found.

Fig. 4.33 compiles the osmotic potentials of plants in the area of Numees in the Richtersveld. They were measured between 1977 and 1987 irrespective of the season and the amount of rainfall. Ψ_π ranges between −0.1 and −7.8 MPa. Members of the Mesembryanthemaceae and Asteraceae cover the broadest span from higher than −1 down to −6.5 MPa. Both families are characterized by a great diversity of different life forms and anatomical and histological leaf and stem types.

We will now focus on Ψ_π of the succulents in the area around Numees. To do so it is necessary to distinguish accurately between the different seasons, especially between the rainy and the dry season. Fig. 4.34 shows the osmotic potentials for the succulents of five different plant families which were measured in both the wet and the dry season. It is obvious that the range of Ψ_π is much broader in the dry season than in the wet season. This is best shown in succulents of the Euphorbiaceae and Liliaceae. During times of sufficient water supply their osmotic potential varies only between −0.15 and −0.35 MPa whereas in the dry season the span is −0.78 to −2.08.

Fig. 4.34 does not include the Mesembryanthemaceae which we will

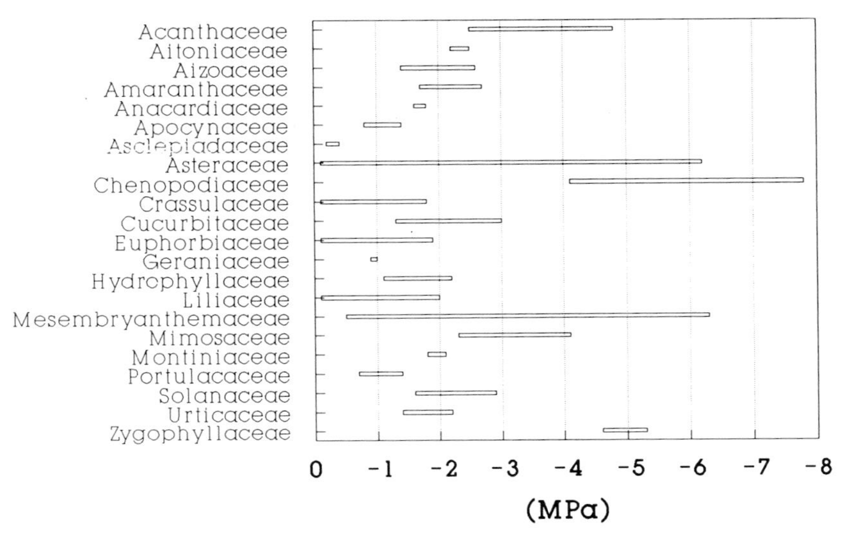

Fig. 4.33 Range of osmotic potentials in succulents and non-succulents growing in the area of Numees (Richtersveld). Measurements of Ψ_π were made during nine stays in the Richtersveld between 1977 and 1988 during different seasons. Number of species per family varies between one and more than forty. Number of samples and replicates ranges from twelve up to more than a thousand.

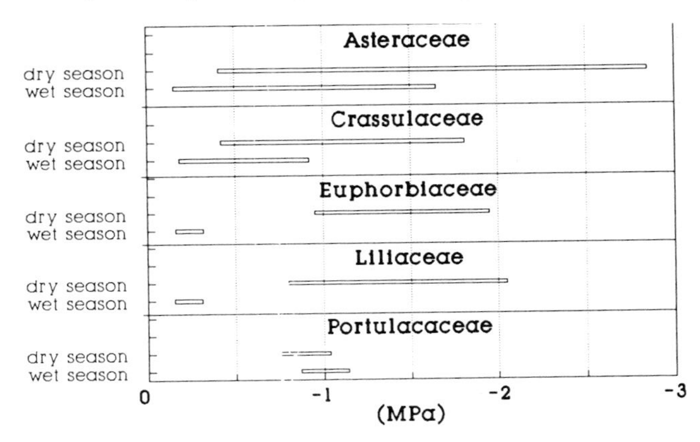

Fig. 4.34 Range of osmotic potentials in succulents of five different families which, apart from the Mesembryanthemaceae, are the most characteristic families with succulent species in the area of Numees (Richtersveld). Asteraceae: *Didelta carnosa*, *Osteospermum microcarpum*, *Othonna herrei*, *O. opima*, *Senecio corymbiferus*, *S. longiflorus*; Crassulaceae: *Adromischus maculatus*, *A. triebneri*, *A.* sp., *Cotyledon orbiculata*, *Crassula* (17 species), Euphorbiaceae: *Euphorbia dregeana*, *E. gariepina*, *E. gummifera*, *E. peltigera*; Liliaceae: *Aloe pearsonii*, *A. pillansii*, *A. ramosissima*; Portulacaceae: *Ceraria fruticulosa*, *C. namaquensis*, *C. pygmaea*. All data obtained during the dry or wet season were pooled. Minimum number of samples is 35.

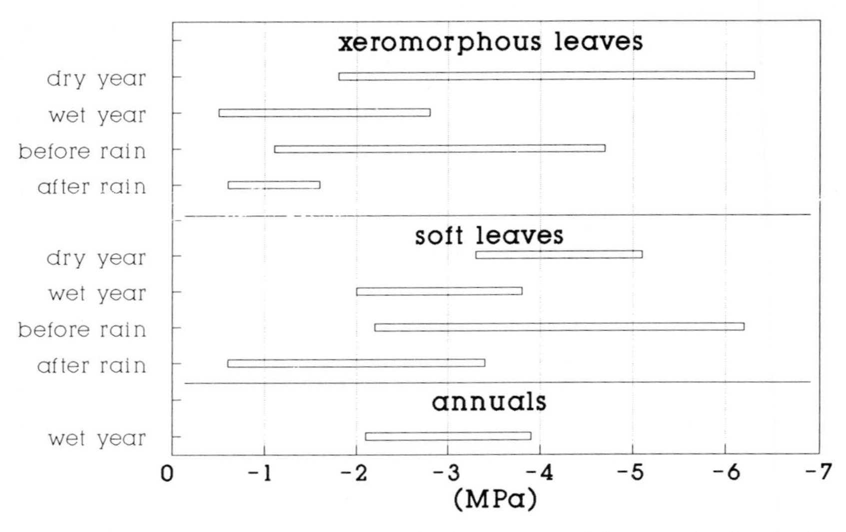

Fig. 4.35 Range of osmotic potentials in succulents of the most abundant family, Mesembryanthemaceae, in the Numees area (Richtersveld). Mesembryanthemaceae were grouped according to their anatomical and histological features into species with xeromorphic leaves (more or less identical with the partial succulent type), species with soft leaves (all-cell succulents) and the true annuals belonging to the all-cell succulent type. Data are given for dry and wet years and also before and after rain. The number of samples was at least 78 and maximally more than a thousand. The number of species was 24 in the group with xeromorphic leaves, 14 in the group with soft leaves and three for the annuals.

present in more detail in Fig. 4.35. This family has the highest species diversity and abundances. Its members belong to either the partially succulent type as characterized by xeromorphic leaves with mostly a central water storage tissue, or the all-cell succulent type lacking xeromorphic features. Both groups, perennial and evergreen, must be separated from the annuals which only occur in years with sufficient winter rain and which normally are absent in late summer.

The osmotic potentials of the Mesembryanthemaceae show the same smaller variation in the wet season as was exhibited by the other succulents. The given grouping of the Mesembryanthemaceae, based on histological criteria, is also reflected by the Ψ_π values in the wet season. Mesembryanthemaceae with a central water storage tissue tend to have higher osmotic potentials (< -1.6 MPa) than the all-cell succulent type species (< -3.5 MPa), whereas the annuals cover the range between -2.1 and -3.9 MPa.

Before we concentrate on the question of which solutes are responsible for the low osmotic potentials and what might be their role in the context of the water status of the succulents, we will first present one more detailed analysis of the osmotic potential.

As shown in Fig. 3.8, rainfall at Numees ran through a pronounced minimum between 1976 and 1983. We were able to follow the effect of the severe drought and the drought relief during many stays in that area. We also experienced an abundant rainfall in March 1981 which marked the end of the prolonged preceding drought. The decrease and increase of the osmotic potentials of eight different succulents, four of which belong to the Mesembryanthemaceae, in the course of the years from 1978 to 1983 is given in Figs. 4.36 and 4.37. The osmotic potentials before and after the rainfall event in March 1981 impressively reflect the degree of desiccation and water deficit in the leaves at the end of the drought period. This is further illustrated in Fig. 4.38 which shows a branch of *Ruschia* sp. prior to and 24 hours after the rain. Even plants that look dead like the one in Fig. 4.38a still have a functioning vascular tissue connecting the roots with the leaves which is able to transport enormous quantities of water to the leaves as soon as water becomes available in the soil.

The main constituent of the osmotic potential of the plants growing in the Numees area is their content of inorganic salts. This is immediately apparent from a rough comparison of the measured osmotic potential values and the calculated osmotic potential values that would be reached on account of the concentrations of the inorganic ions K and Na. We made several thousands of measurements in the period from 1977 to 1987. In Fig. 4.39 we contrast the maximum and minimum osmotic potential values (Fig. 4.33) against the sum of the K and Na concentrations as shown in Figs. 4.41a and 4.41b, irrespective of whether or not these pairs of data really come from the same sample.

Fig. 4.39 also shows the theoretical line that would result if the osmotic potential was determined solely by the concentrations of inorganic ions. Also the regression line for all data points is shown. It is shifted by a constant value and thus runs parallel to the theoretical line. That means that in all the measured species on average 0.9 MPa of the osmotic potentials is accounted for by organic substances. We have no information on what types of organic substances play a role in this respect. In exceptional cases only, e.g. in the annual Mesembryanthemaceae, proline is important (Treichel *et al.* 1984).

As we see from Fig. 4.33, the dominant Mesembryanthemaceae in the Richtersveld cover a wide range of osmotic potentials. In Fig. 4.40 we show the dependence of the osmotic potentials on the concentrations of inorganic

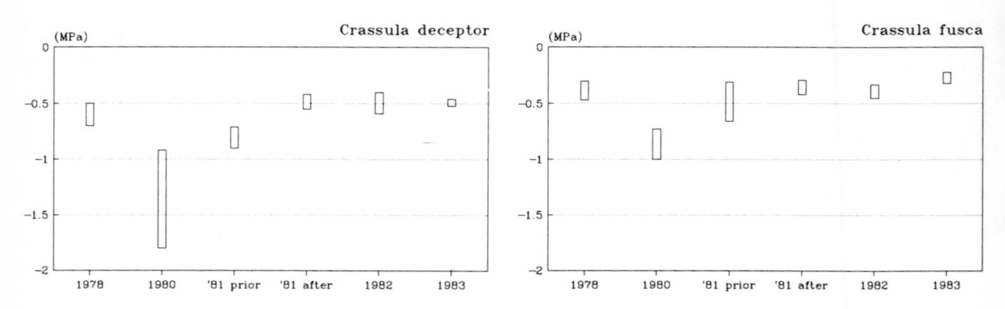

Osmotic potential of four succulents at Numees

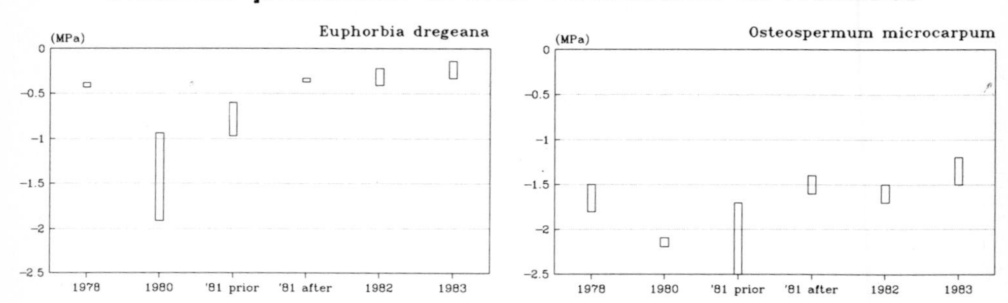

Fig. 4.36 Range of osmotic potentials of four representative succulents in the Numees area (Richtersveld). The graph clearly demonstrates the accidental nature of the data which only depend on the climatic prehistory. The osmotic potentials are directly correlated with the total amount of rainfall, being most negative in years with insufficient rainfall but approaching a maximum value when rainfall is well above the long-term annual average. Lowest number of samples was 36.

ions for this group, separating the all-cell succulent from the partially succulent Mesembryanthemaceae and from the annuals. To allow proper comparison only data points obtained 10 days after an abundant rainfall event when water availability was optimal were plotted.

The plants in the Numees area have very different ion concentrations. We show this in Fig. 4.41a–d for sodium, potassium and chloride, and separately for the dominant Mesembryanthemaceae in Fig. 4.42.

Soil analyses at five different habitats in Numees (Fig. 4.43) show that the availability of ions in the soil can greatly vary even at distances of a few hundred metres. One might expect that the ion composition in the plant mirrors the ion availability in the soil but this is not the case. For example, on the site poorest in sodium and chloride (4) we find species of *Drosanthemum* and *Sphalmanthus* which strongly accumulate salt, while species of *Crassula* with the lowest concentrations of sodium and chloride occur at

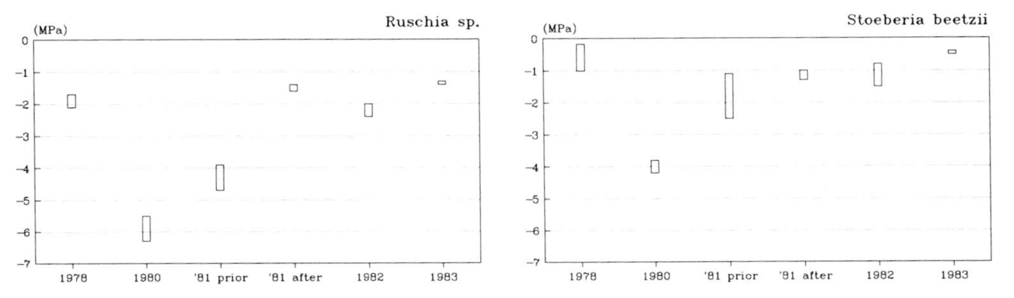

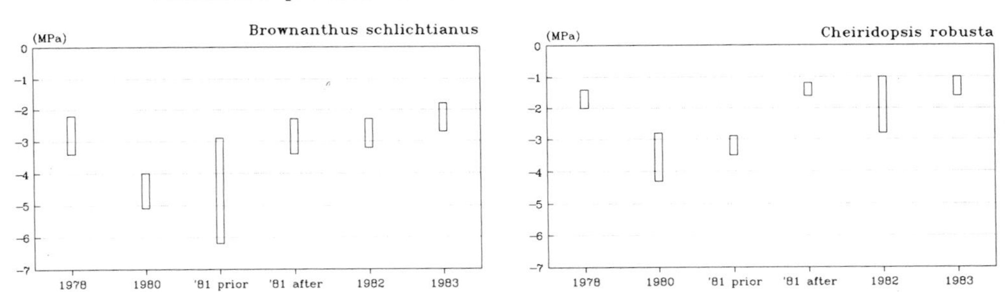

Fig. 4.37 Range of osmotic potentials of four representative succulents of the Mesembryanthemaceae in the Numees area (Richtersveld). *Ruschia* sp. and *Stoeberia beetzii* are woody stemmed evergreen bushes of the partially succulent type; *Brownanthus schlichtianus* is a stem succulent evergreen species with drought deciduous leaves and *Cheiridopsis robusta* is a leaf succulent species with only two annually replaced leaves of the partially succulent type. Again the osmotic potentials reflect the severe drought of 1979–1981 and the tremendous degree of desiccation. Lowest number of samples was 51.

habitat (3) where the availability of inorganic ions is highest. It is unlikely that this pattern reflects an impoverishment of the soil as a result of accumulation in the plant, since the vegetation is too patchy and the ion content in the small quantities of leaf mass is too limited. Crassulaceae, Asteraceae and Mesembryanthemaceae often grow in the immediate vicinity of each other in the same substrate. We therefore presume that the differences are a result of specific – perhaps genetically defined – patterns in ion accumulation. We already came to this conclusion in 1977 (von Willert *et al.* 1979b) based on a comparative study of different species of succulents from a different area, namely plants from the Knersvlakte grown in the field and in the greenhouse.

The Mesembryanthemaceae occupy a special position among the succulents as far as osmotic potentials are concerned, assuming that some

Fig. 4.38 Leaves of *Ruschia* sp. (a) at the end of a prolonged drought and (b) 24 hours after an abundant rainfall at Numees in the Richtersveld.

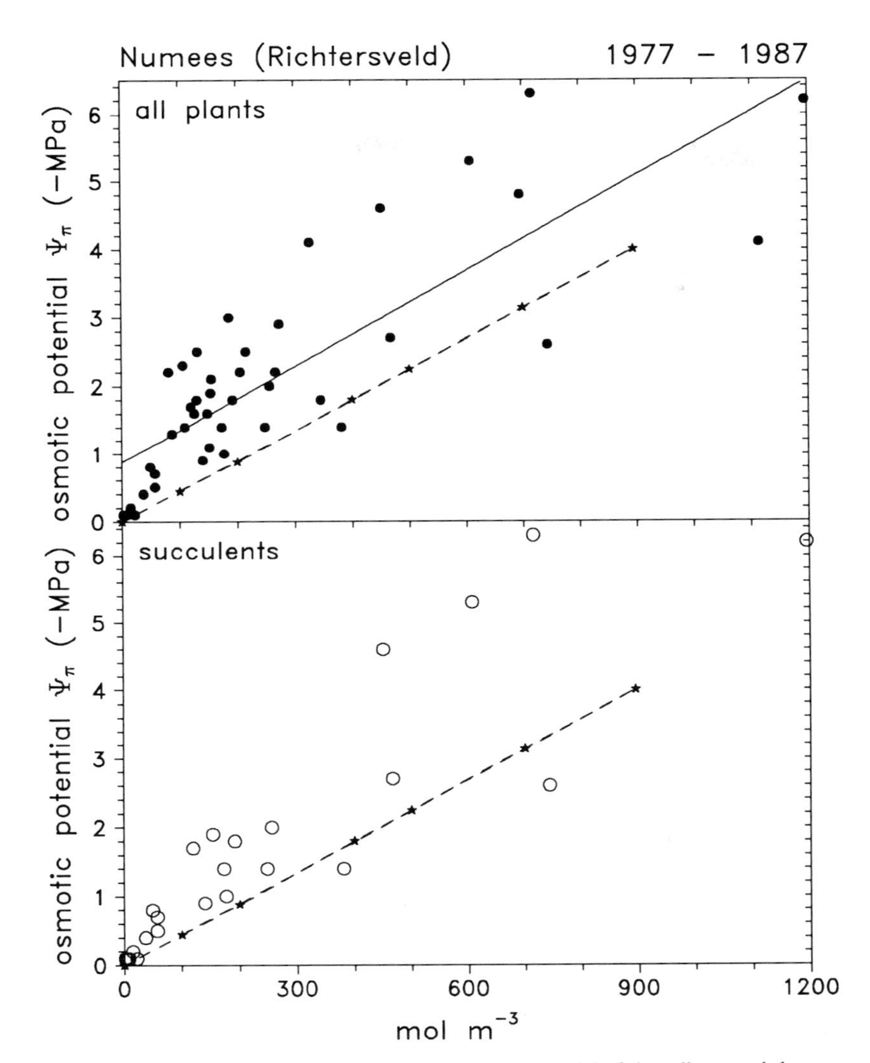

Fig. 4.39 Correlation between the osmotic potential of the cell sap and the sum of the corresponding concentration of sodium and potassium. The mass of several thousand samples over 11 years from 1977 to 1987 was grouped and reduced to maximum and minimum values of the osmotic potentials given in Fig. 4.33 and maximum and minimum values of the Na and K concentrations given in Fig. 4.41, irrespective of whether or not these values really come from the same sample. The upper graph shows the relationship for all succulent and non-succulent species investigated in the Numees area (Richtersveld); the lower graph represents the relationship if only the succulent species are considered. The dashed line and the stars give the theoretical line that would be obtained by a sodium or potassium chloride solution. The solid line in the upper graph is the linear regression line for all values.

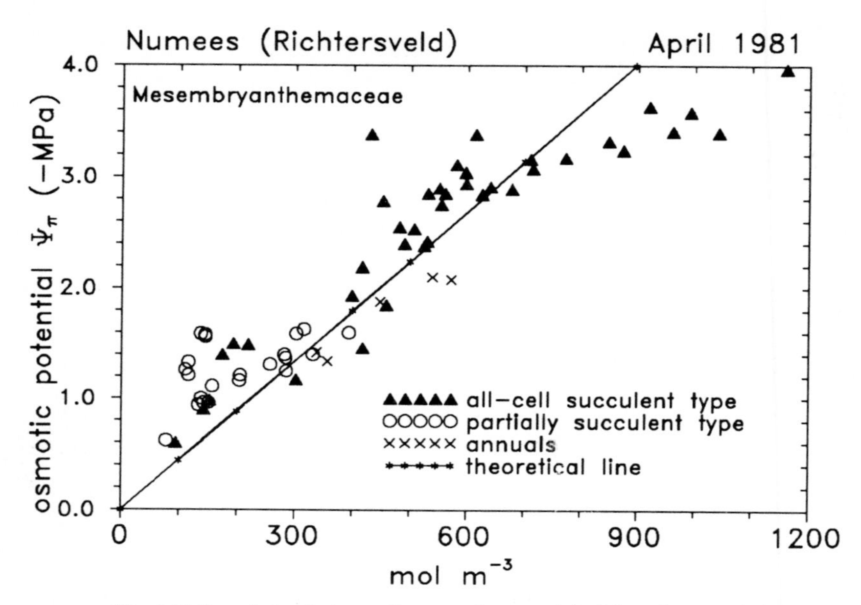

Fig. 4.40 Correlation between the osmotic potential of the cell sap and the sum of the corresponding concentrations of sodium and potassium for the Mesembryanthemaceae growing in the Numees area (Richtersveld) ten days after an abundant rainfall in 1981. Mesembryanthemaceae were grouped according to their histological features and the duration of their life cycle. Osmotic potential and solute concentrations were measured in the same sample and dots either represent different species or different leaves of the same species.

Zygophyllaceae and Asteraceae are disregarded because they are clearly halophytes as indicated by their habitat, and therefore fall outside the scope of this book. Many Mesembryanthemaceae, particularly the annuals, are exceptionally salt tolerant and accumulate considerable quantities of sodium chloride. Nevertheless they are not proper halophytes as they usually grow in non-saline habitats.

4.2.2.2 Turgor potential

Apart from the osmotic potential, a second quantity determines the total water potential of the plant, namely the turgor potential. It can be measured together with the other water potentials by way of a pressure–volume (*P–V*) curve or using a pressure probe that allows direct measurement of the turgor of individual cells. Not all succulents can be measured by pressure bomb and thus cannot be analysed by way of a *P–V* curve. We do not know of any field measurements on succulents. Herppich (1989)

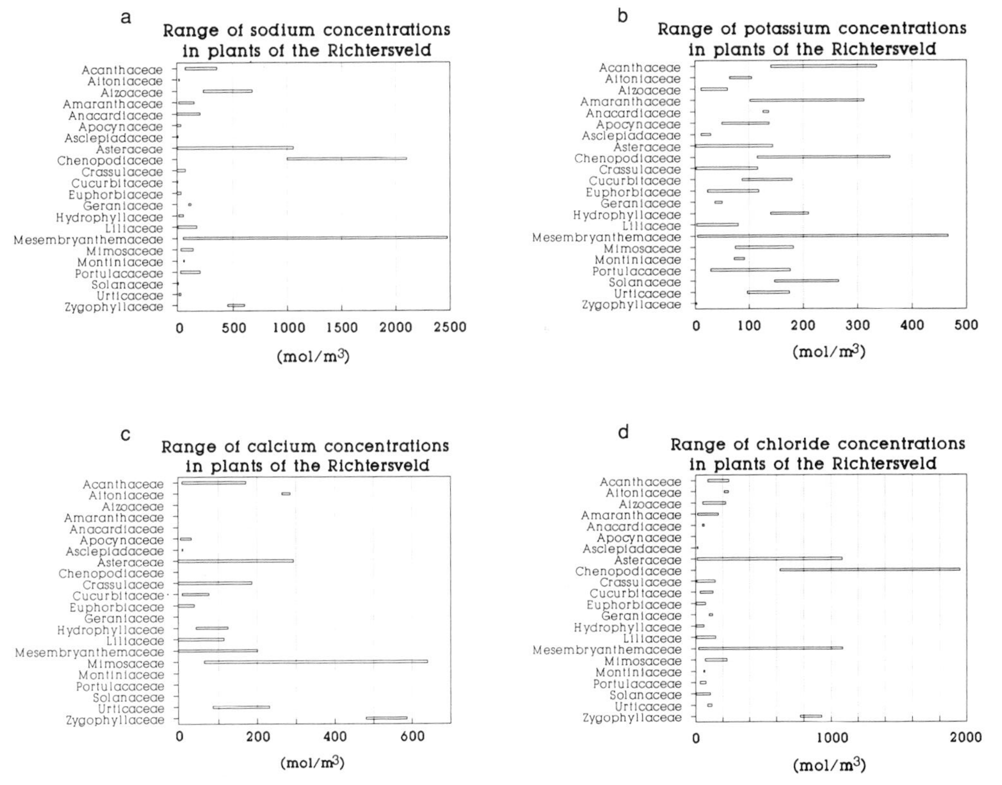

Fig. 4.41 Ranges of sodium, potassium, calcium and chloride concentrations in succulent and non-succulent plants of the Numees area (Richtersveld). For further details see legend Fig. 4.33.

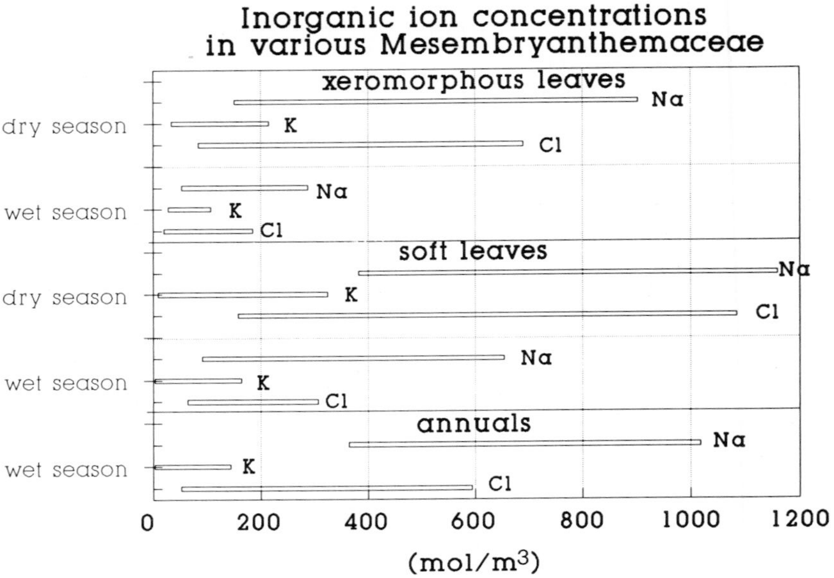

Fig. 4.42 Range of concentrations of sodium, potassium and chloride in the cell sap of the Mesembryanthemaceae growing in the area of Numees (Richtersveld). Data were grouped for wet and dry season except for the annuals. Each bar represents several hundred samples analysed between 1977 and 1987.

applied the *P–V* curve technique in comprehensive laboratory measurements on *Plectranthus marrubioides*, a succulent from arid North Yemen. He obtained values for Ψ_P ranging from 0.18 to 0.23 MPa. Direct measurements using the pressure probe in *Kalanchoe daigremontiana* also gave values for Ψ_P of about 0.2 MPa (Steudle *et al.* 1990), while in *Cereus validus* and *Agave deserti* maximum values of 0.1 MPa were measured (Lüttge and Nobel 1984).

Many data are available from the ephemeral *Mesembryanthemum crystallinum*. Like many Mesembryanthemaceae, this species has epidermal idioblasts or bladder-cells (external water storage). Furthermore, *Mesembryanthemum crystallinum* can fix CO_2 through the C_3 and the CAM mode of photosynthesis (see Section 4.3). This species therefore offers the possibility to investigate the variables determining the plant's water status during the shift from C_3 to CAM. The most recent results show Ψ_P values for the idioblasts between 0.02 and 0.09 MPa during the C_3 mode and between 0.01 and 0.1 MPa during the CAM mode (Rygol *et al.* 1989). The Ψ_P values of the mesophyll cells cover the same range (Rygol *et al.* 1986).

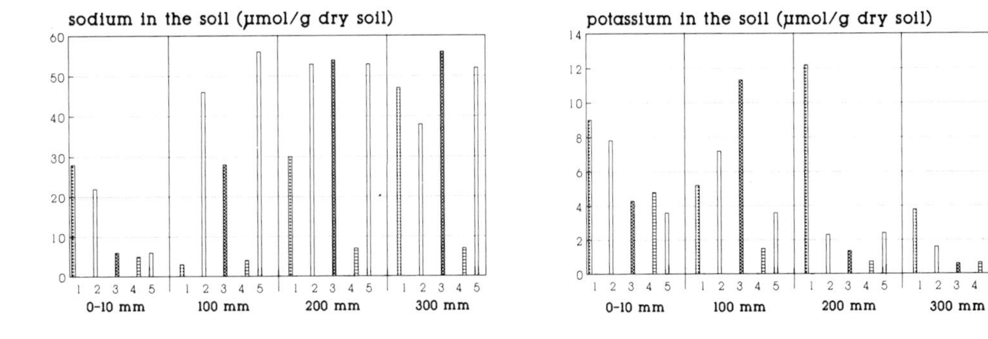

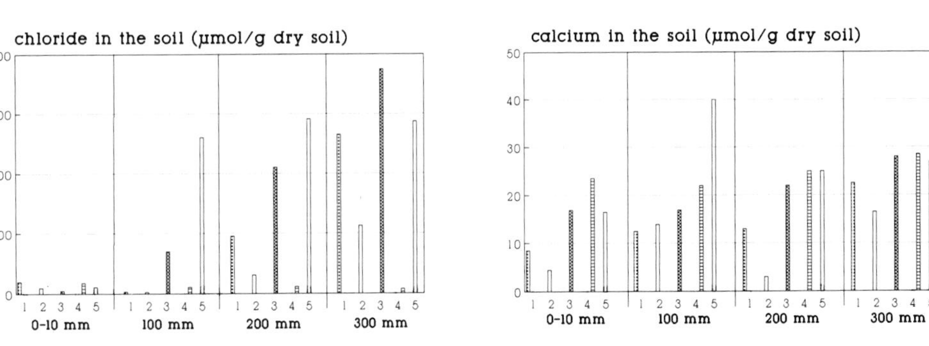

Fig. 4.43 Content and distribution of exchangeable sodium, potassium, chloride and calcium in the soil at five different places in the area of Numees (Richtersveld). Profile 1 was at the foot of a west-facing slope in a stand of *Brownanthus schlichtianus*; profile 2 was 100 m southwest of profile 1 between *Psilocaulon subnodosum* bushes; profile 3 was on an elevated plain (where *Cheiridopsis robusta* was common) aside a drainage creek 200 m north of profile 1; profile 4 was on an east-facing slope 300 m west of profile 1 between *Drosanthemum*, *Sphalmanthus trichotomus*, *Senecio longiflorus* and *Crassula grisea*; profile 5 was on a gently descending plain covered with the annual Mesembryanthemaceae *Opophytum aquosum* and *Mesembryanthemum pellitum*

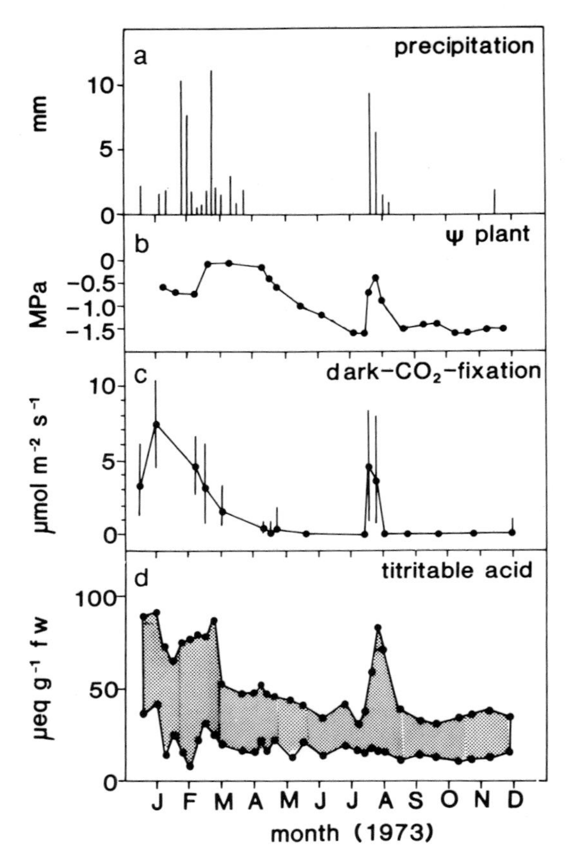

Fig. 4.44 (a) Rainfall in the course of 1973 at the measuring site in the Sonora desert. (b) Corresponding plant water potential in *Opuntia basilaris*. (c) Corresponding net CO_2 uptake of *O. basilaris* and (d) the titritable acid in the cell sap of *O. basilaris*; the dotted area between the two lines represents the acid accumulation overnight. (Data were taken from Szarek and Ting 1974.)

In the field in the Richtersveld in 1981 we used the pressure probe for direct measurements of turgor potential. Not only *Mesembryanthemum*, but also *Drosanthemum*, has large epidermal idioblasts. Species of *Drosanthemum* are dwarf shrubs with a woody stem and evergreen succulent leaves densely covered with epidermal idioblasts. In March 1981 we could not measure any turgor in these leaves because of the long dry period prior to the measurements. We once watered the plant with an equivalent of 20 mm of precipitation and obtained fully turgid leaves within three days. The Ψ_P of the idioblasts reached highest values of 0.39 MPa at predawn but they rapidly fell to 0.1 MPa. Only after sunset did the turgor rise again (Brinckmann *et al.* 1985). These results do not agree with the laboratory

Table 4.1. *Predawn water potentials of various succulents and non-succulents at Numees (Richtersveld). Water potentials were measured with a pressure bomb between 14 and 20 February 1987*

	Predawn water potential, Ψ_t (−MPa)
Succulents	
Crassula brevifolia	0.82
Drosanthemum sp.	1.80
Galenia dregeana	1.65
Psilocaulon subnodosum	3.20
Ruschia sp.	3.00
Salsola sp.	5.50
Tetragonia reduplicata	1.25
Trianthema triquetra	1.53
Zygophyllum prismatocarpum	5.48
Non-succulents	
Acacia karroo	1.58
Justicia orchioides	5.75
Monechma mollissimum	3.23
Rhus populifolia	2.03

measurements on *Mesembryanthemum crystallinum* (Rygol *et al.* 1986, 1989). This is not necessarily surprising. It may result from the different plant species as well as from other mechanisms that do not occur under the optimal conditions of the laboratory experiments.

4.2.2.3 Leaf water potential

The total water potential of succulents, which is the sum of the osmotic potential and the turgor potential, is largely determined by the osmotic potential. It ranges from −0.09 MPa (*Ferocactus acanthodes*, Nobel 1977a) to −4.0 MPa (*Psilocaulon subnodosum*, von Willert *et al.* 1985). It is obvious that the total water potential varies with equal strength to the osmotic potential depending on the water status of the plant. Szarek and Ting (1974) impressively showed this (Fig. 4.44).

Though the water potential of a cactus or a *Kalanchoe* well reflects the plant's water status, this is different for a richly branched dwarf shrub which, after a long period of drought, holds only a few leaves at the tips of its branches. In a shrub of *Psilocaulon subnodosum* with hundreds of twigs the predawn water potential varied from −0.7 to −4.0 MPa. This may reflect retranslocation of water in the plant. This mechanism is well developed in succulents and we will return to it in Section 4.2.4.

Table 4.1 gives a survey of some predawn water potentials in succulents

Table 4.2. *Comparison of parameters of water potentials obtained with different methods. Leaf water potential (Ψ_t) was estimated with a pressure bomb* (A) *and gives the first value of a* P–V *curve, or with the Wescor C-52 chamber* (B, C). Ψ_π *was taken from the* P–V *curve* (A), *by a dewpoint hygrometric measurement after freezing and thawing of leaf discs* (B) *or by freezing point depression of leaf cell sap* (C). Ψ_P *was calculated. All data are means with standard deviations and are given in* MPa. *(Data taken from Herppich 1989.)*

Method	$-\Psi_t$	$-\Psi_\pi$	Ψ_P	n
A	0.243 ± 0.027	0.446 ± 0.061	0.202 ± 0.059	17
B	0.260 ± 0.047	0.322 ± 0.038	0.058 ± 0.015	19
C	0.260 ± 0.047	0.446 ± 0.085	0.181 ± 0.058	4

and non-succulents at Numees. Measurements were carried out at the end of the dry season. They do not show any significant differences between succulents and non-succulents. However, we must admit that the selection of succulents is far from representative for the area. The reason is that measurement of the water variables of succulents often is impossible using a pressure bomb or a Wescor leaf chamber L-51. Consequently, simple field methods are lacking. Where possible Ψ_t is usually measured by pressure bomb, Ψ_π cryoscopically in the cell sap which is easy to obtain, and Ψ_P then can be calculated. Herppich (1989) carried out a comprehensive comparison of methods and came to the conclusion that, whenever possible, the *P–V* curve technique should be used. We complete this section on the equations describing the water budget of the plant with a summary of his results on *Plectranthus marrubioides* (Table 4.2).

4.2.3 Transpirational water loss

In Section 4.2.1 we learnt that transpiration is the product of two quantities, leaf conductance to water vapour (g_{wv}) and water vapour gradient between the substomatal cavity and the ambient air (ΔW).

$$J_{wv} = \Delta W \, g_{wv} \tag{4.37}$$

Due to the high temperatures and the low air humidity ΔW can reach rather high values in deserts. Prolonged periods with ΔW exceeding 70 mPa Pa^{-1} are not scarce. Such large differences in the molar fractions of water vapour in the substomatal cavities and ambient air would, on a physical basis, result in tremendous rates of transpiration if plants had not compensated for this physical effect by a number of inventions. They concern mainly morphological and anatomical features and are aimed to keep the leaf conductance, g_{wv}, low. We have seen that the leaf conductance consists of a series of fractional conductances. The most decisive quantities by far are the stomatal conductance and the resistance of the boundary layer to water vapour diffusion. Stomatal conductance is under the control of number and aperture of the stomata whereas boundary layer resistance is positively related to the thickness of the boundary layer.

According to Stalfeldt (1956), in 'normal species' the number of stomata ranges from 100 to 300 per square millimetre. More recent textbooks (Lüttge *et al.* 1988) give 100 to 860 mm^{-2} for herbaceous domesticated plants and trees. This source also gives values of 18 to 33 mm^{-2} for leaves of three CAM succulents. It is to be expected that all species of an arid habitat have comparably low numbers of stomata. This is not the case, however, as is shown in Table 4.3, for a selection of plants from the Richtersveld and the Central Namib. Nevertheless, the data allow some general conclusions. With 3 to 93 stomata per square millimetre succulents have a lower stomatal density than non-succulents (87 to 222 mm^{-2}). The low densities in the Crassulaceae of the Richtersveld are matched by similarly low values (30 to 50 mm^{-2}) for species of *Agave* in the Sonora desert (Nobel 1988). Although there are clear differences between succulents and non-succulents, the available data do not permit us to make a difference between C_3 and CAM succulents. We will see such a lack of difference repeatedly when we discuss the water and carbon budgets of these succulents.

The second important factor is the boundary layer and its influence on transpiration. The thickness of the boundary layer is determined by the wind speed and the surface features of the leaf, as we have seen in Section 4.1.6. Thus the development of windless spaces not only increases the thickness of the boundary layer but also decreases the diffusion of water vapour. In this respect both sunken stomata (Fig. 1.12) and epidermal idioblasts are important. They increase the length of the path of diffusion of water vapour and thus reduce the conductance. In numerous Mesembryanthemaceae (e.g. *Drosanthemum*, *Brownanthus*, *Mesembryanthemum*, *Monilaria*, *Dorotheanthus*) the edges of the idioblasts meet each other so that no stomata are visible at the surface. That is why Table 4.3 lists no figures for Mesembryanthemaceae with protruding idioblasts. When fully turgid the idioblasts often open up and allow sight of the proper epidermis and the stomata. When turgor decreases, under water shortage, the idioblasts deform and close the openings between them, greatly increasing the boundary layer resistance (Brinckmann *et al.* 1985). *Brownanthus schlich-*

Table 4.3. *Stomatal densities in numbers per square millimetre (means and standard deviations of at least ten replicates) of a selection of succulent and non-succulent species of the Richtersveld and the Central Namib*

C_3 species	
Codon royenii	182 ± 11
Cucumis meeusei	127 ± 15
Euclea pseudebenus	113 ± 3
Galenia dregeana	161 ± 18
Othonna opima	71 ± 7
Pachypodium namaquanum	70 ± 2
Pelargonium sp.	222 ± 12
Trichodesma africanum	182 ± 1
Welwitschia mirabilis	87 ± 11
Zygophyllum longicapsulare	58 ± 1
Zygophyllum stapfii	44 ± 2
CAM species	
Adromischus maculatus	30 ± 3
Aloe pearsonii	66 ± 3
Aloe ramosissima	10 ± 0.5
Bulbine sp.	33 ± 0.1
Ceraria fruticulosa	41 ± 1
Ceraria namaquensis	55 ± 3
Cheiridopsis robusta	12 ± 1
Conophytum minutum	7 ± 0.6
Crassula atropurpurea	25 ± 2
Crassula brevifolia	5 ± 0.4
Crassula clavata	12 ± 1
Crassula deceptor	26 ± 2
Crassula elegans	13 ± 2
Crassula erosula	12 ± 1
Crassula expansa	48 ± 4
Crassula grisea	9 ± 0.6
Crassula pseudohemisphaerica	23 ± 2
Crassula rupestris	37 ± 1
Crassula sericea	58 ± 4
Delosperma pergamentaceum	3
Haworthia tesselata	4
Leipoldtia constricta	67 ± 2
Psilocaulon subnodosum	93 ± 4
Ruschia subaphylla	81 ± 2
Senecio corymbiferus	38 ± 3
Stoeberia beetzii	59 ± 2
Tylecodon paniculatus	7 ± 1
Tylecodon pearsonii	54 ± 5
Tylecodon reticulatus	12 ± 1
Tylecodon wallichii	16 ± 2
Range for succulents	3–93
Range for non-succulents	87–222

tianus, an evergreen stem and leaf succulent that is drought deciduous (see Figs. 4.71 and 4.72), has epidermal idioblasts on both the leaves and the stem internodes. Wide open fissures remain between the idioblasts on the leaves, while those on the stem practically completely close off its epidermis. This represents an impressive example of correspondence between structure and function of organs.

Similarly wax bloom and hairiness can influence transpiration because they increase the diffusion path. However, Ehleringer and Mooney (1978) found that leaf pubescence in *Encelia farinosa* did not affect water vapour diffusion to the extent that it could be considered of ecophysiological importance. Wax bloom and hairiness can indirectly affect transpiration, however. In Section 4.1.4 we have already shown that such structures can reduce the global radiation load, resulting in a lower energy input and lower temperatures. Lower temperatures imply a lower ΔW and thus a lower transpiration rate at environmental conditions that are otherwise similar.

Our discussions on transpiration and the relevant quantities g_{wv} and ΔW have shown that no definite statements on the amount of water loss through transpiration can be made as long as no exact definition of conditions is given. For a valid comparison at least ΔW should be the same in order to evaluate rates of transpiration against different leaf conductances for water vapour. A constant ΔW and a different g_{wv} of the same species at different seasons also allows conclusions about the water status of the plant and the availability of water in the soil, even when Ψ_t or Ψ_s are not known.

In Fig. 4.45 we show the daily courses of the transpiration of C_3 succulents of the southern Namib. As these on their own are not informative we also provide the courses for g_{wv} and ΔW. At sunrise the stomata are opened up but ΔW is still very low and the rate of transpiration increases slowly with increasing ΔW. Soon the stomata close, however, as can be seen particularly clearly in *Trianthema*, and the rate of transpiration again largely follows ΔW. In Fig. 4.46 the relations between transpiration, leaf conductance and ΔW are clearly shown. The diagram shows that stomata are rapidly and widely opened at the beginning of irradiance, provoking a gulp of photosynthetic CO_2 uptake of which the maximum characteristically precedes the maximum in transpiration. In Section 4.3 we will return to this phenomenon.

Data on transpiration of succulents under field conditions are very scarce. This not only results from theoretical difficulties as discussed above, but also from the unfavourable type of plant material, the complicated instrumentation needed and the remoteness of sites with abundant succulents. From 1981 to 1987 we repeatedly carried out measurements on

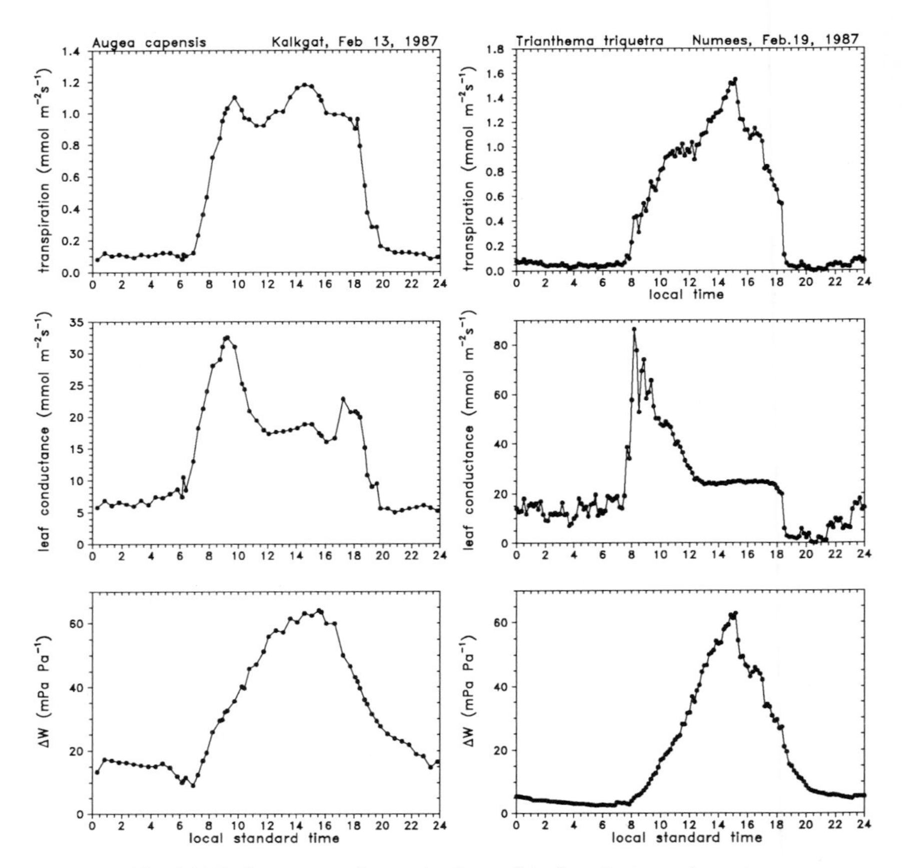

Fig. 4.45 Daily courses of transpiration and leaf conductance for water vapour for two succulents. The diurnal changes in $\triangle W$ are also given to show that climatic conditions during the period of measurement were very similar and thus allow comparison.

succulents and non-succulents in the Richtersveld and the Knersvlakte. In Table 4.4 we summarize some results for both plant groups. Aware of the restraint needed when comparing such data, we nevertheless dare to conclude that, apart from a few exceptions, the leaf conductance and transpiration rates of succulents are lower than those of non-succulents.

We now should pay attention to the transpiration of C_3 succulents as against CAM succulents. Do pronounced differences in the amounts of transpiration exist and is it justified to strictly distinguish between both types? The values in Table 4.4 might suggest a generally lower transpiration for CAM succulents. During the daytime CAM plants, at least under

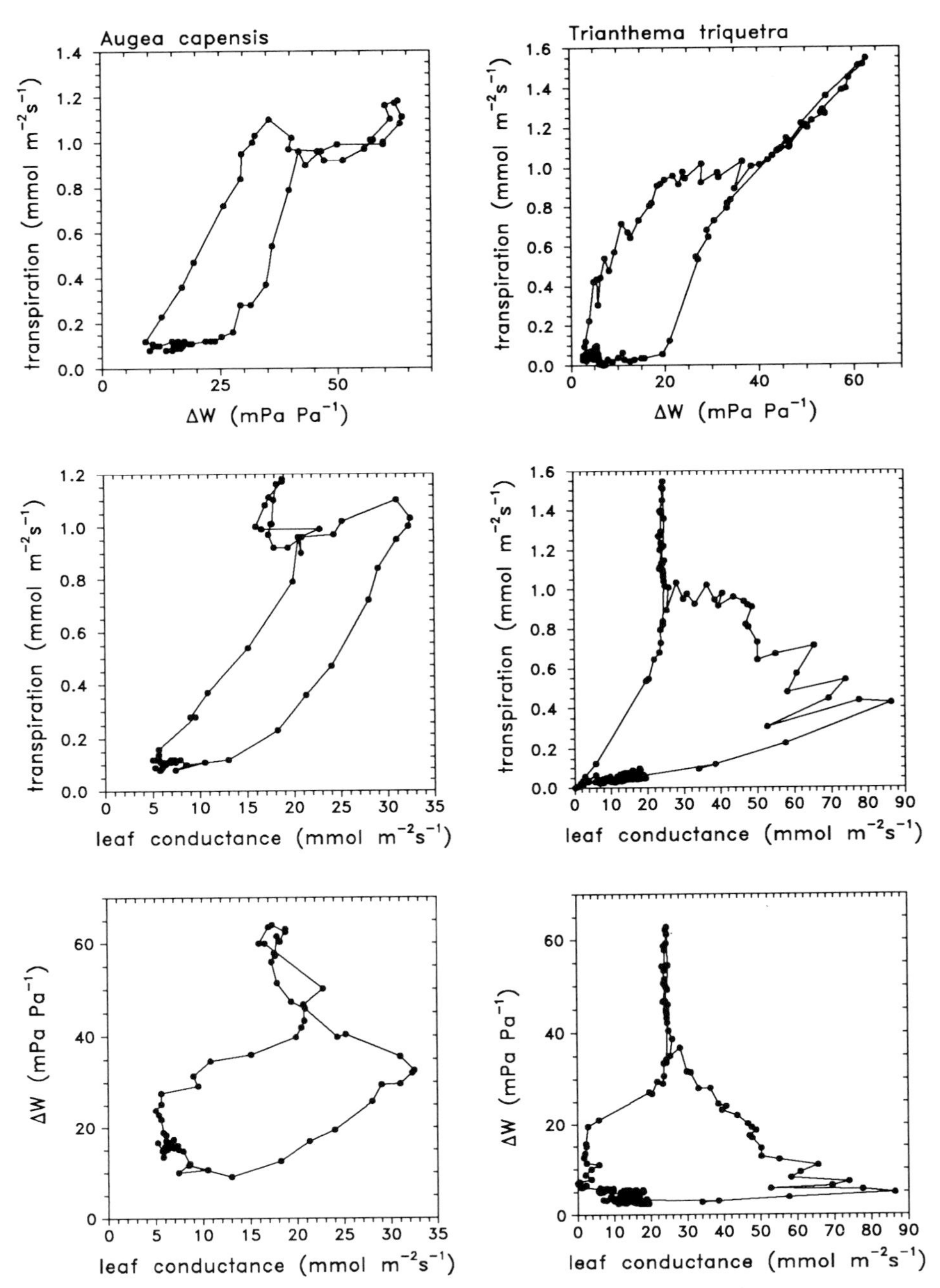

Fig. 4.46 Various relationships between transpiration, leaf conductance and $\triangle W$ for the same succulents as presented in Fig. 4.45. Graphs show that the succulents open their stomata rapidly and wide at dawn when $\triangle W$ is rather low but stomata are closed soon afterwards.

Table 4.4. *Transpiration, leaf conductance and mode of photosynthesis of some succulents in the dry parts of southern Africa*

Species	Transpiration mmol m^{-2} s^{-1}	Leaf conductance mmol m^{-2} s^{-1}	Mode of photosynthesis	Comments, season Max. ΔW (mPa Pa^{-1})
Succulents				
Aloe ramosissima	0.22	46	CAM	September during a period of
Brownanthus schlichtianus	0.49	52	CAM	long-lasting bergwind.
Cheiridopsis robusta	0.31	14	CAM	Max. ΔW: 51 mPa Pa^{-1}
Mesembryanthemum barklyi	0.38	19	CAM	
Mesembryanthemum pellitum	0.39	30	CAM	
Opophytum aquosum	0.32	33	CAM	
Tylecodon paniculatus	0.49	13	CAM	
Cotyledon orbiculata	1.34	46	CAM	March, four days after an
Psilocaulon subnodosum	0.42	26	CAM	abundant rainfall event.
Othonna opima	2.42	78	C_3	Max. ΔW: 43 mPa Pa^{-1}
Stoeberia beetzii	1.55	54	CAM	
Zygophyllum prismatocarpum	0.74	47	C_3	
Augea capensis	1.20	33	C_3	February, end of dry season.
Trianthema triquetra	1.55	83	C_3	Max. ΔW: > 60 mPa Pa^{-1}
Zygophyllum prismatocarpum	0.18	20	C_3	
Non-succulents				
Acacia karroo	4.2	191	C_3	October, four days after
Codon royenii	4.17	227	C_3	abundant rainfall.
Didelta carnosa	4.07	128	C_3	Max. ΔW: > 60 mPa Pa^{-1}
Fingerhuthia africana	4.02	80	C_4	
Forsskaolea candida	7.01	238	C_3	
Justicia orchioides	7.34	211	C_3	
Monechma mollissimum	0.96	49	C_3	
Rhus populifolia	2.34	76	C_3	
Solanum namaquanum	2.34	92	C_3	
Welwitschia mirabilis	2.61	123	C_3	October, Central Namib. Max. ΔW: > 60 mPa Pa^{-1}

drought stress, have their stomata opened up only very early in the morning (Fig. 4.47). This opening apparently is induced by the irradiance. During this early morning gulp of CO_2 all CAM succulents studied by us strongly increase their pool of malic acid formed during the night. Only after the stomata have been closed does the amount of malic acid decrease. In many species the pool of malic acid is exhausted by noon (e.g. *Aloe ramosissima*, Fig. 4.47), but in others (e.g. *Brownanthus schlichtianus*) it lasts till the late afternoon. In any case the stomata are never re-opened when the reserves of malic acid have been used.

Fig. 4.47 concerns only three CAM species, a leaf succulent (*Aloe ramosissima*), an annual succulent (*Mesembryanthemum pellitum*) and a stem succulent (*Brownanthus schlichtianus*). Leaf conductance to water vapour, transpiration, malic acid content and the most important climatic data are given. The differences with the C_3 succulents of Fig. 4.45 are striking. Fig. 4.47 is very illustrative in evaluating the maximum values for conductance (g_{wv}) and transpiration (J_{wv}) given in Table 4.4. A proper comparison should be made only for a 24-hour period of measurement, but reliable measurements of that length of time in the natural habitat are rarely possible. Therefore the curves given in Figs. 4.45 and 4.47 should be evaluated with some restraint. The reasons are as follows. Because of strong nightly emission of thermal radiation the succulents cool down to temperatures considerably below ambient air temperature. As a result, dewfall on the plants occurs frequently, often as soon as from about midnight. In chambers clamped to the plant dewfall does not occur, and as a result the measured values are not relevant for the dew conditions. In Section 4.2.4 we return to this question.

During moist nights, CAM succulents fix CO_2 which means that their stomata are open. In C_3 plants this is not the case. Because of the small differences in the molar fractions of water vapour in the substomatal cavity and the ambient air, water loss during the night is limited. When the leaf temperatures are below those of the ambient air and dewfall occurs, water loss does not occur at all and even an inward moisture gradient can be established around the leaf (von Willert *et al.* 1985).

As regards water economy, the CAM mode of photosynthesis seems advantageous for succulents compared to the C_3 mode. However, when we consider our data from the Richtersveld it seems doubtful whether this really is true (von Willert and Brinckmann 1986).

For a proper comparison it is absolutely necessary that the species only differ in their mode of photosynthesis and are otherwise similar in life form and other features. Among the species listed in Table 4.4 *Cotyledon*

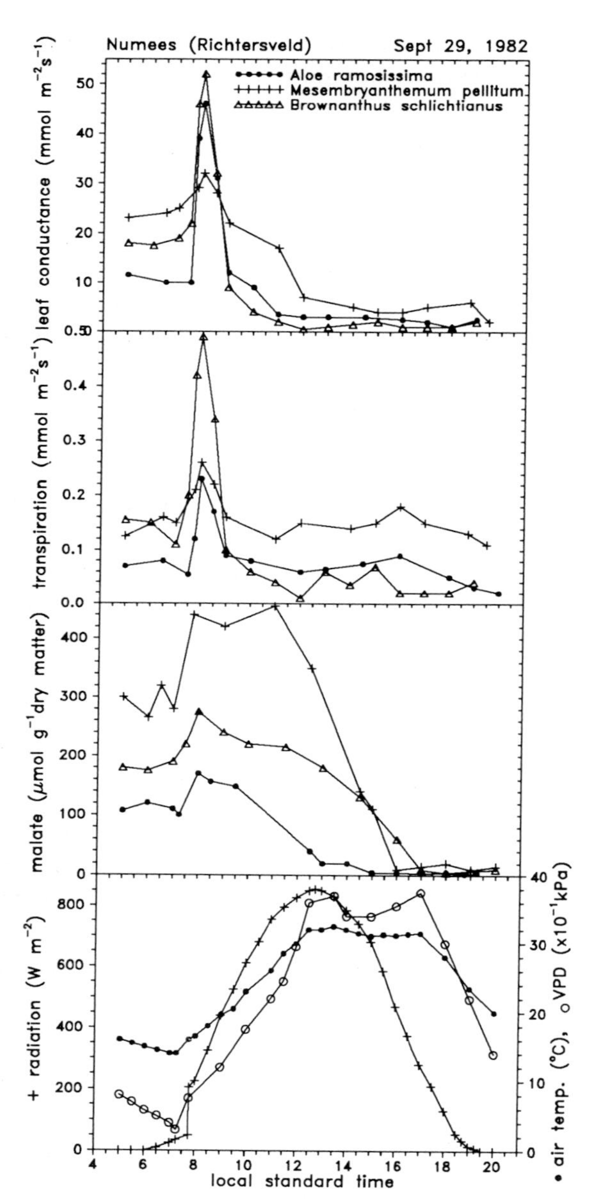

Fig. 4.47 Leaf conductance for water vapour, transpiration and malic acid concentration in the cell sap of three different CAM succulents from predawn to dusk determined in the natural habitat. Radiation, air temperature and water vapour pressure deficit of the air (VPD) are also given.

orbiculata and *Othonna opima* meet these requirements. Both are dwarf shrubs with evergreen succulent leaves and they differ only in their mode of photosynthesis, *Cotyledon* being a CAM, *Othonna* a C_3 succulent.

At abundant water supply *Othonna* loses 0.98 l water per day and square metre whereas *Cotyledon* transpires only 0.28 l at the same time. Both species significantly reduce transpiration when soil water becomes short and bergwinds cause additional drought stress. *Othonna* then only loses $0.046\,l\,m^{-2}\,day^{-1}$ compared to the $0.074\,l\,m^{-2}\,day^{-1}$ of *Cotyledon*. Under this extreme drought there was no sign of stomatal opening in *Othonna* while there was still a slight opening for *Cotyledon* during the night. At a VPD of about 40 $mPa\,Pa^{-1}$ this is responsible for the higher water loss of *Cotyledon* (von Willert and Brinckmann 1986). Consequently, under extreme short-term drought the water balance of a C_3 succulent might even be better than that of a CAM succulent. However, this single example is by no means enough evidence for a generalization. Further studies and more examples are needed to draw a firm conclusion.

In all the data we have presented so far there is no clear evidence that succulents, either CAM or C_3 really use their stored and utilizable water to keep their stomata open when non-succulents are forced to close them. Even a short-term drought is quickly responded to by a powerful stomatal closure. What really counts in water economy is saving; storing provides an additional advantage only when closely linked to saving. An example for that is given in Fig. 4.48. Branches of succulents and non-succulents were cut off from the plants and were exposed to the prevailing mesoclimate of the natural habitat. The actual water content of the excised plant parts was followed over the next two weeks. From this graph it becomes clear that saving, only in combination with storage of utilizable water, guarantees survival. Again it shows that C_3 succulents might be even better savers than CAM succulents. But we have to be cautious with this statement until more evidence becomes available.

Many succulents belonging to the most desired species in plant collections are more or less completely embedded in the soil (*Lithops*, *Fenestraria*, *Haworthia*, *Bulbine*, etc.). These succulents possess window leaves. One might speculate about the advantages and disadvantages of being buried in the substrate. We have already discussed the temperature regime of such succulents (see also Nobel 1989). We should now focus on their water relations. Experiments done under controlled environmental conditions revealed that embedding saves water (Eller and Ruess 1982). Transpiration of embedded *Lithops lesliei* is about 20% lower than that of completely exposed ones. Only for embedded plants is the water uptake by roots able to

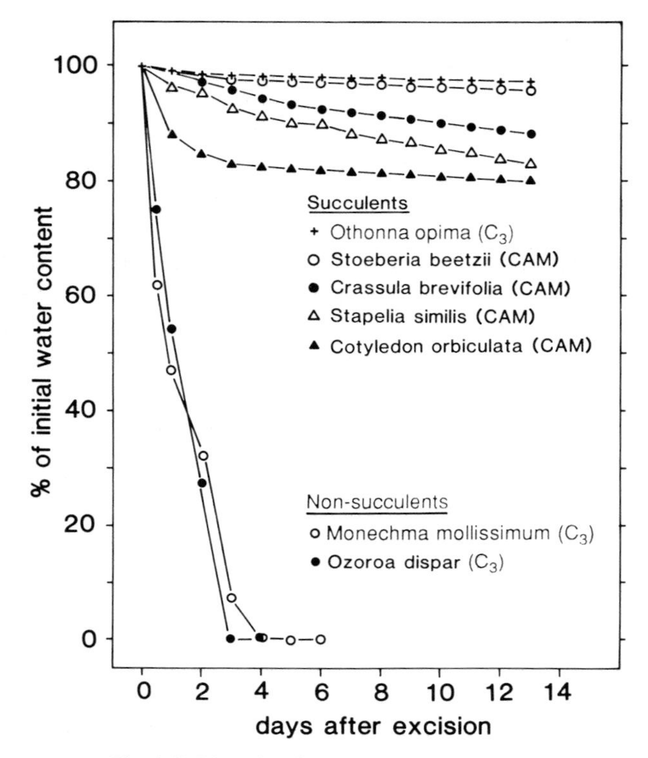

Fig. 4.48 Water loss in percentage of the initial water content of leaves from succulent and non-succulent desert plants. Plant parts were cut off and the decrease in actual water content was followed over a period of two weeks. During that time plant parts were exposed to the prevailing mesoclimate of the natural habitat.

meet the daily water demands. For exposed plants water uptake lags behind water loss; this results in a shrinkage of the plant and finally in its death. Here, too, much more effort is needed to fully elucidate the phenomenon of embedding and generalize on its adaptational values.

4.2.4 Water uptake

Sources for water uptake by the plant are soil water and atmospheric water in the form of rain, dew, fog or air humidity. Consequently water can be extracted from the soil or the atmosphere. Once water has entered the plant it can only travel down an energy gradient and always moves from source to sink.

4.2.4.1 Uptake from the soil

The main characteristic of desert soils is dryness and hence water uptake from the soil is possible only during very short periods after rainfall events. Water uptake from the soil is via the root system of the plant.

As far as is known, succulents have a shallow root system and the main rooting horizon is 50 to 150 mm below the soil surface. Euphorbias have deeper reaching roots. Often the roots have tubers or originate from bulbs and rhizomes. Some examples of root systems are presented in Fig. 4.49.

Features of the root system of a number of desert plants including succulents were described by Cannon (1911). Nobel (1988) reviewed the root system of cacti and agaves and Weisser (1967) reported the very shallow (only 10–20 mm below surface) side roots of the earth cactus *Neochilenia napina*. Despite the shallowness of the root system of most succulents their lateral extensions can cover long distances. We followed roots of succulents over 10 to 20 m away from the plant. Some of them finally ended underneath a big rock. Here they branched intensely and, together with roots coming in from other directions and other plants, formed a dense network of fine roots.

The rock reduces evaporation from the soil so that humid conditions persist longer underneath the rock than elsewhere. Additionally, dew and insufficient rainfall will run down and gather under the rock. Consequently, there is a higher probability of water available under the rock and a more reliable moisture supply. Such a source of water is well exploited by the plants. How many plants compete for the water underneath a rock can be impressively demonstrated by watering the rock. It resulted in the growth of plants in all directions away from the rock, sometimes of far distances. From this we have to conclude that water might be available to plants even if the bulk soil is definitely dry.

From our experience these water sources, though limited, are of outstanding importance for the succulents in the southern Namib. They permit survival at a very low metabolic level during extreme drought. In contrast with plants in pots in a greenhouse, desiccation in the natural habitat is very slow as water might be available at some secret places.

Water uptake by roots requires a downhill gradient of water potential from the soil to the root. The prevailing water potentials of succulents range between -0.8 and -3.5 MPa for the majority of species (Table 4.1) and can drop to -5.5 for some halophilic species. Water potential of the root will be somewhat higher. In case of sandy soil a biologically active rainfall will raise soil water potential quickly to zero or values just below zero. But the soil will not stay moist for a long time.

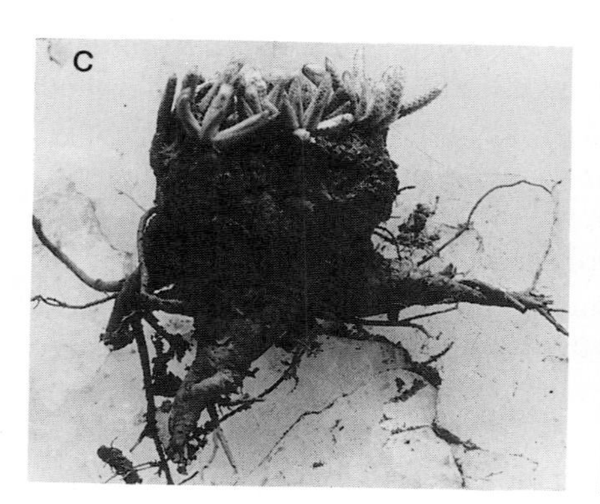

Fig. 4.49 Root systems of different succulents in the southern Namib desert. (a) *Euphorbia* sp. (Knersvlakte): stem and main root are succulent and supported by a large, strong root system. (b) *Pelargonium* sp. (Richtersveld): water storage above and below ground is almost equal. (c) *Anacampseros alstonii* (35 km east of Warmbad (Namibia) on a quartz hill): a very large water storage organ is developed below ground. (d) *Ruschia schneiderana* (Richtersveld): this leaf succulent collects water via an extended but very thin root system with apparently no storage capacity.

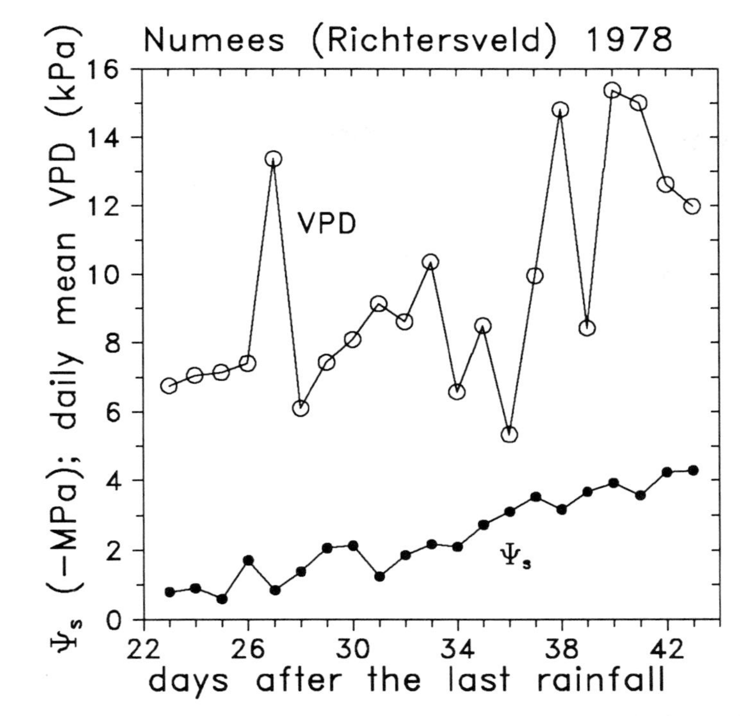

Fig. 4.50 Decrease of soil water potential in a desert soil after an abundant rainfall event. Soil water potential was measured in the main rooting horizon at 100 mm depth. Additionally, the VPD values as means over 24 hours are given. The peaks on days 27, 38, 40, and 41 are due to bergwinds blowing during the night.

At 10 cm depth, corresponding with the main rooting horizon of the succulents at Numees, water potential decreased slowly to −0.8 MPa during the first three weeks after a 8-mm rainfall event but then declined quickly to −4.2 MPa over the next three weeks (Fig. 4.50). Plants can follow this decrease only during a short period which means that periods of water uptake are short and very much depend on the amount of rainfall. The shallow root system of succulents and their pronounced lateral extension might reflect a valuable adaptation to the scarce rainfall events and might help exploit rainfalls which will not penetrate deeply into the soil. For *Agave deserti* and *Ferocactus acanthodes*, which have a main rooting horizon of only 80 mm, Nobel (1976, 1977a, b) gives 6–7 mm as a lower limit of biologically active precipitation.

A rapid exploitation of soil moisture requires either the permanent existence of living roots able to absorb water whenever it becomes available, or a quick generation of new water-absorbing roots. For cacti and agaves it has been shown that 'rain roots' develop upon rewetting the soil. Within 6 to 8 hours new exclusively water-absorbing fine roots develop on already established roots (Kausch 1965, Nobel and Sanderson 1984). Nevertheless, the first gulp of water uptake is by the established roots. It takes as long as four days before the 'rain roots' contribute around 50% to the total water uptake (Nobel and Sanderson 1984). Although 'rain roots' are usually shortlived, they sometimes become established roots. In any case, 'rain roots' represent a massive sink for carbon. Unfortunately nothing is known about similar structures in other families with succulents.

To compensate for the very short periods allowing water uptake, exploitation of soil water is rapid and starts almost instantaneously upon rewetting of the soil. There is no possibility of measuring water uptake in the field other than by indirect methods. Leaf water potential, osmotic potential or water content are good indicators of an improved water budget of the plant and they were used to demonstrate rapid water uptake after periods of desiccation. Fig. 4.38 summarizes some examples of rapid water uptake after watering succulents and also shows the effect of an abundant rainfall.

For *Opuntia basilaris* Szarek and Ting (1974) showed that soil water and plant water potential recover quickly after a rainfall event (see Fig. 4.44). Fig. 4.51a gives additional evidence with a higher resolution. After watering, the plant water potential first increases slightly but then, with a delay of 18 hours, increases rapidly to about one-third of the original value. Similar effects were obtained for a variety of succulents in the Richtersveld when plants were artificially watered twice with an equivalent of 20 mm of precipitation each time. Irrigation was done one square metre around the plant. Actual water content of the leaves was followed over a period of more than a month. At the end of this period 26 mm of rain fell (the first abundant rain after two years of severe drought). The responses to this event were also monitored. From Fig. 4.51b and c it becomes clear that an artificial watering, though supplying more water per square metre, is less effective than rainfall. This impressively demonstrates that only part of the rooting system of a succulent is within the watered area around the plant. For *Stoeberia beetzii* water uptake after rain is about twice as rapid as after watering. As a first approximation, and assuming that water uptake is directly correlated with root surface, this can be taken as an indication that about one-half of the roots of the dwarf shrub *Stoeberia beetzii* occur within the watered square metre around the plant. The same holds for *Drosanthemum* sp., a succulent with an epidermal water storing tissue. Fig. 4.5lb also

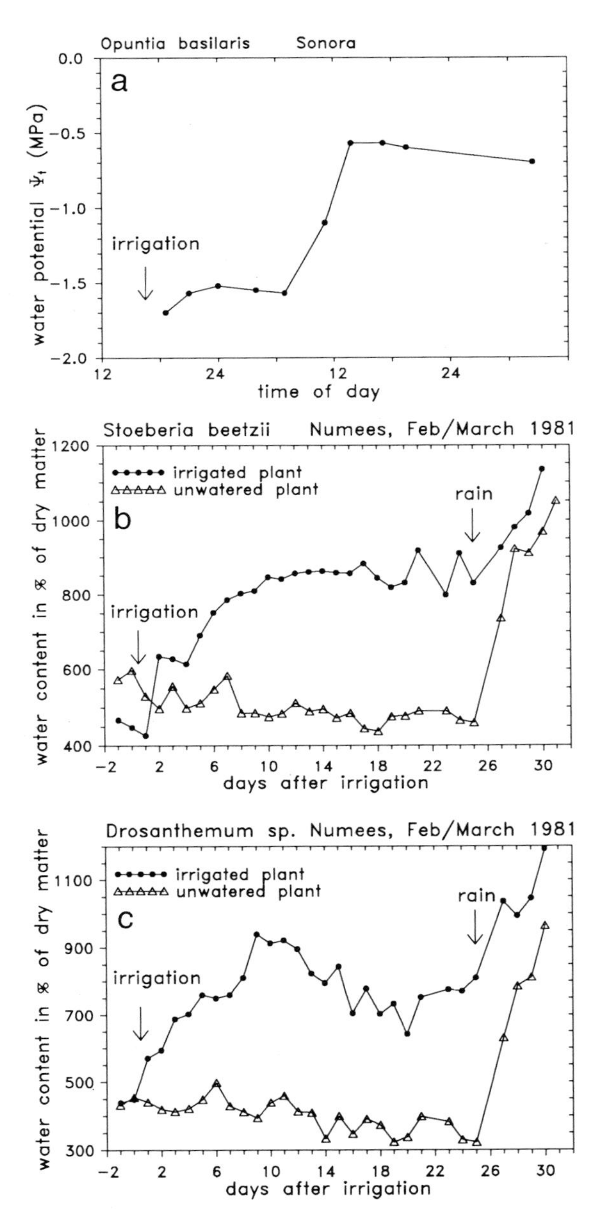

Fig. 4.51 (a) Rapid water uptake of a desiccated *Opuntia basilaris* shown by the increasing plant water potential after irrigation (data taken from Szarek and Ting 1974): (b) and (c) Increase of the leaf water content of extremely desiccated *Stoeberia beetzii* (b) and *Drosanthemum* sp. (c) Dwarf shrubs after irrigation and rainfall. (Data partially taken from von Willert and Brinckmann 1986.)

shows that 40 mm of water on one square metre around the plant, given on two successive days, did not establish water saturation in the plant, as a rain 4 weeks later further increased plant water content. Fig. 4.51 might also explain the deleterious effect of too much water supply which, as shown in Fig. 1.2, can result in a burst of the leaf. In both species irrigation and rain lead to a more than three-fold increase in leaf water volume and shows how desperately desiccated the leaves were at the end of the prolonged drought.

This strong uptake of water brings us to the phenomenon of refilling the water storage tissue. The sudden availability of water in the soil leads to a tremendous first gulp of water uptake (Achmed effect, von Willert and Brinckmann 1986) which then slows down until finally the daily increment of water content approximates zero. The capacitance of the water storage tissue is responsible for the process of refilling. We briefly introduce the meaning of capacitance and consider its consequences.

The capacitance, C, is defined as that quantity of water a cell can gain or lose if the water potential, Ψ, of that cell is changed by one unit (Dainty 1976). This concept of capacitance can also be applied to tissues or organs (Nobel 1983). The relation

$$C=\frac{dV}{d\Psi}\ (m^3\ MPa^{-1}) \tag{4.46}$$

then expresses the change in the volume of water of, for instance, a water storage tissue that results from a given change of the water potential of that tissue. We can also write equation (4.46) as

$$C=\frac{dV}{(d\Psi_\pi + d\Psi_P)} \tag{4.47}$$

Since Ψ_π and Ψ_P respond differently to changes in the water volume V we can make two statements:

(1) There must be a strong dependency of the capacitance on the actual water volume especially in the range between saturation and the point of turgor loss where Ψ_P becomes zero and consequently Ψ_t equals Ψ_π.

(2) Below the point of turgor loss only Ψ_π and changes in Ψ_π determine the capacitance.

These considerations show that the synthesis or consumption of osmotically active solutes play a key role in water uptake into or release from a water storage tissue. It is easy to understand that the number and size of the cells of a tissue determine the absolute water volume that can be withdrawn or gained, as this fraction only depends on the total volume. In order to

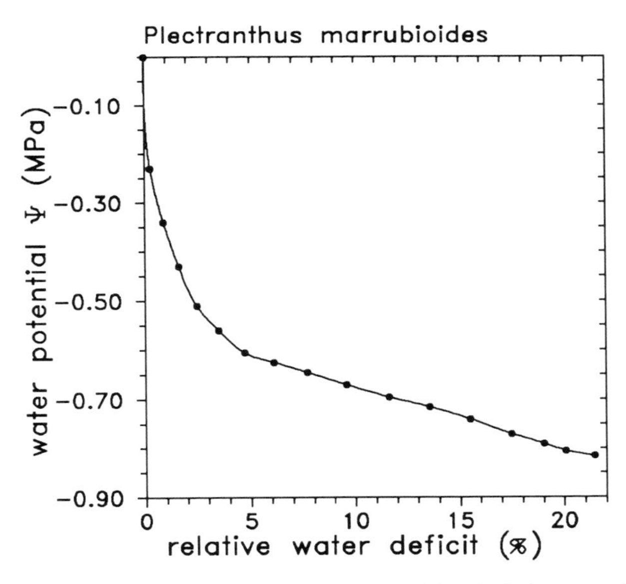

Fig. 4.52 Relationship between water potential and relative water deficit (100 − RWC) of a branch of the CAM succulent *Plectranthus marrubioides*.

eliminate this dependency and for the sake of a proper comparison of the capacitance of various tissues or organs one can base the capacitance on the total water volume of the tissue or organ in question. Since the actual water volume V is related to the maximum water volume at saturation V° by $V/V^{\circ} = \mathrm{RWC}$ (relative water content) we get for the capacitance which is based on the unit water volume and termed C'

$$C' = \frac{\mathrm{dRWC}}{\mathrm{d}\Psi} \tag{4.48}$$

Hence it is possible to determine the capacitance by means of a P–V analysis (for details we refer to Koide *et al.* 1989). A plot of the water potential Ψ_t over the RWC or the relative water deficit, as done here, results in a relationship shown in Fig. 4.52. The capacitance C' is then simply the reciprocal of the slope at each point of the curve of this relationship. Fig. 4.52 documents that the capacitance of a tissue or organ increases with decreasing RWC. For reasons of comparison it is better to standardize and compare the slope at saturation for an estimation of the capacitance (Nobel 1988). Nevertheless, it is necessary to emphasize that minimum and maximum capacitance of a tissue can easily differ by an order of magnitude as has been demonstrated for *Plectranthus marrubioides* by Herppich

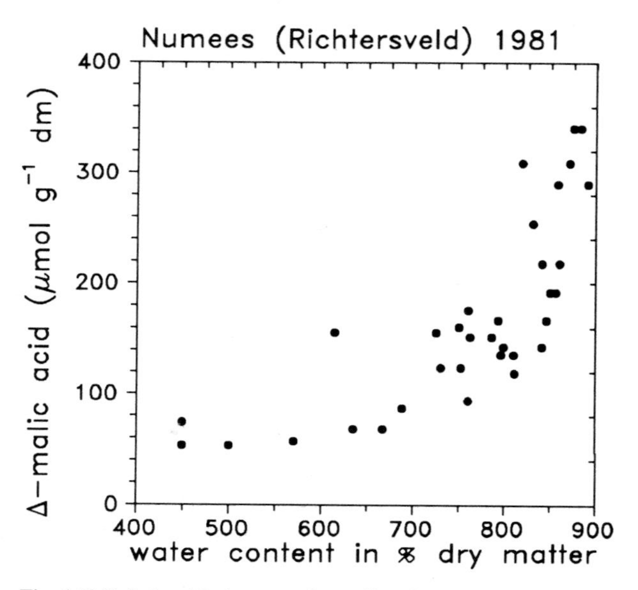

Fig. 4.53 Relationship between the malic acid accumulated overnight and the water content of the leaves in percentage of dry matter during the rehydration process of a severely desiccated *Stoeberia beetzii* dwarf shrub after irrigation.

(1989). For the sake of completion we should also mention that the capacitance cannot only be based on total water volume but also on surface area or weight of the organ.

The succulents *Stoeberia beetzii* and *Drosanthemum* sp. whose water uptakes are shown in Figs. 4.51b and 4.51c are CAM succulents. That means they synthesize and accumulate organic acids during the night. What role, if any at all, might this accumulation of solutes play in the context of water uptake and water gain? The important consequence of the relationship between water potential and RWC (Fig. 4.52) is that the capacitance of a tissue increases with decreasing RWC. This means that the same change in Ψ results in a higher amount of water gained by a slightly desiccated tissue as compared to a less desiccated tissue. Or, in other words, the change in Ψ necessary to maintain or improve a reached degree of water saturation increases with increasing water content. From this it follows that refilling of a partially emptied water storage tissue will be a biphasic process. The first phase will be a rapid water uptake and dilution of the cell sap with only little increase in solute concentration. This results in a gradual recovery of the osmotic water potential. To overcome this dilution and to sustain water gain demands the synthesis in or translocation of solutes to the water storage tissue. This means that the second phase of refilling will be a slow process that requires an increasing $\Delta\Psi$ when nearing water saturation. This second phase is characterized by a pronounced increase in solute concentration accompanied by only small gains of water. Fig. 4.53 shows the relationship between the amount of overnight malic acid accumulation and the refilling of the water storage tissue of heavily desiccated leaves of *Stoeberia beetzii* after irrigation of the desiccated plant.

We cannot finish the section on water uptake from the soil without dealing briefly with the question of what happens to the direction of the water flux if the soil water potential is much lower than the osmotic potential of the succulent. From a physical point of view this should result in a water loss from the plant (root) to the soil as water always runs down an energy gradient. Recently Caldwell and Richards (1989) demonstrated this. Nothing is known about similar reactions in the shallow root system of succulents. On the other hand, Nobel and Sanderson (1984) could show that the roots of agaves and cacti act like rectifiers. Responsible for this physically strange behaviour is the hydraulic conductivity of the root which changes during desiccation of the soil allowing a water flow only from the soil to the plant and not in the reverse direction.

4.2.4.2 The abundance of CAM succulents in the Richtersveld

Any visitor to the southern Namib desert will be overwhelmed by the abundance of CAM succulents in this area (for details see Section 4.3). This instantaneously raises the question why are they so successful in that area? What are the advantages of CAM and which environmental stresses and selective pressures may have favoured CAM succulents in this area?

Several years of experience in the southern Namib and suffering, together with the succulents, from drought gave us the idea that one, if not the main, function of CAM with its nocturnal accumulation of osmotically active solutes is water harvesting in its widest sense (von Willert *et al.* 1983). Such a water gain promoted by the accumulation of malic acid or other acids concerns either the whole plant or only parts of it and includes water uptake from the soil and/or water uptake from the atmosphere and/or translocation of utilizable water within the plant.

Our assumption that malic acid accumulation in CAM succulents in the southern Namib desert has something to do with the water status of the plant was based on a series of observations (von Willert *et al.* 1979a, b, 1983).

(1) For rooted plants nocturnal acid synthesis and accumulation is directly correlated with water availability in the soil and water status of the plant.
(2) Excised plants limit acid accumulation to a very low level which increases only in nights with dewfall or very high air humidity.
(3) Stomatal opening and the onset of a marked malic acid synthesis of rooted CAM succulents coincides with a soil water potential exceeding a critical threshold. In a desiccating soil this threshold is reached progressively later in the night which results in a gradual decrease in the amount of overnight malic acid accumulation.
(4) The onset of a bergwind during the night imposes an atmospheric drought stress and instantaneously diminishes or ceases nocturnal malic acid formation.
(5) The evaporative demand during bergwind conditions is so high that even CAM succulents with sufficient water supply (i.e. shortly after an abundant rainfall event) run into a water deficit as the enormous water loss cannot be compensated by water uptake via the roots.

All these observations hint that CAM succulents are able to monitor their water status (water content) or water budget. We shall give evidence for this statement below and also in Section 4.3.

In dry habitats the availability of water in the soil is restricted to short periods after a rainfall event. In a desiccating soil the soil water potential undergoes substantial diurnal fluctuations. That is why a high predawn soil water potential tells little about the availability of water during the rest of the day. Fig. 4.54 gives an example for the diurnal changes in soil water potential for a typical stand of succulents in the southern Namib. In the main rooting horizon at 100 mm depth the soil water potential guarantees water uptake by the vegetation only during the second part of the night and in the early morning when values are higher than -2 MPa. During the rest of the day soil water potential is lower and thus, for the majority of the succulents growing in this soil, too low for water uptake.

One argument for the connection of malic acid formation and water gain is that the withdrawal of water from the transpiration stream requires a sink for water within the plant. This sink can be produced by decreasing the osmotic potential via a malic acid accumulation. From Fig. 4.54 it follows that a water uptake from a desiccating soil cannot start earlier than about 02:00 in the night. If we overlay the soil water potential with the sap concentration of malic acid of a CAM succulent measured at the same time

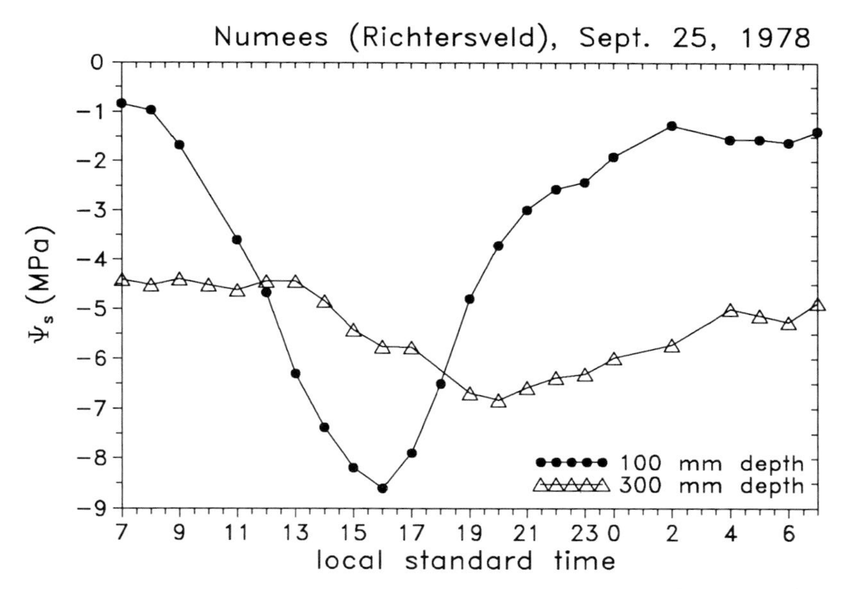

Fig. 4.54 Diurnal variation of water potentials in a desert soil in the area of Numees (Richtersveld). Soil water potential was determined at 100 and 300 mm depth on a slightly descending plain at the foot of an east-facing slope.

and at the the same site we realize that a significant malic acid synthesis coincides with both stomatal opening and the possibility of water uptake via the roots (Fig. 4.55).

The idea to link nocturnal acid formation to the water status of a CAM succulent rather than to its carbon budget (von Willert *et al.* 1979a, b, 1983) received further support in the following years. In a series of growth chamber and greenhouse experiments either using the water displacement method (Chen and Black 1983), or a potometer technique (Ruess and Eller 1985), or the pressure probe and pressure bomb (Lüttge and Nobel 1984), it could be shown by simultaneous measurement of water uptake, transpiration and acid formation that a water (=volume) gain was always positively correlated with malic acid accumulation. Lüttge (1986) reviewed some of the available information related to this subject and concluded that osmotic changes caused by nocturnal malic acid storage in CAM succulents evidently drive water uptake by and water storage in the plant.

Despite all this evidence and reasonable argumentation further studies are necessary to fully elucidate this aspect of the ecological importance of CAM and its possible role in CAM succulent abundance in the southern Namib desert.

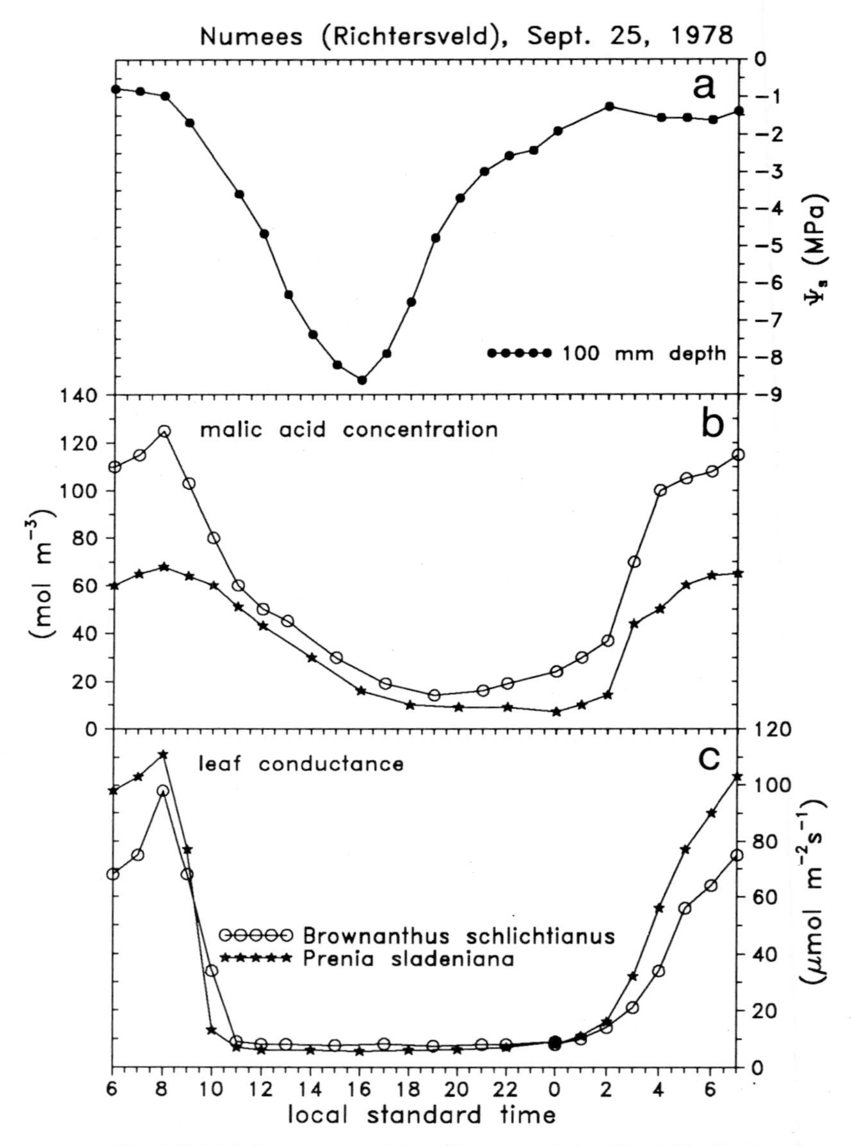

Fig. 4.55 (a) Soil water potential at 100 mm depth (see Fig. 4.54), (b) diurnal changes of malic acid concentration in the cell sap of two CAM succulents growing in the same soil, and (c) leaf conductance to water vapour for the same succulents. Sunrise was at 06:45 local standard time.

4.2.4.3 Uptake from the atmosphere

There is no doubt that for all higher plants rooting in a substrate water uptake via the roots is the main form of water gain, but the question may be asked whether this is the only form (Rundel 1982). This question becomes even more meaningful if one considers that in many arid areas water deposition on leaves and stems frequently occurs during the night. This water deposition results from dewfall or from advective fog blowing inland off the ocean. Since these atmospheric water sources are more reliable than rainfall, it suggests that plants growing in such areas might have developed structures and strategies to exploit or utilize this water at least partially.

Lower plants like lichens and poikilohydric mosses are able to live exclusively from dewfall, fog and even moist air. The Namib, in particular, gives an excellent example for this. Resurrection plants like some grasses, ferns and herbs are able to take up water via the whole surface as can easily be demonstrated if dry parts of these plants are sprayed with water. Within hours they unroll their leaves and gain turgidity.

Further examples among higher plants for water harvesting from the atmosphere are all rootless species which cover their water demand by uptake via other organs and surfaces. Prominent examples are found in the family of the Bromeliaceae where special water scales serve the uptake of liquid water via leaves and stems.

Can succulents also do this and are they especially equipped for it in areas where succulents are abundant and rich in numbers? As has already been outlined and shown in Fig. 3.4 and 3.6, the Namib is characterized by a high number of fog days. It is also shown in Fig. 3.10 that especially during the dry summer months the relative air humidity in the southern Namib reaches very high values in the night which, together with the undertemperature of the succulent leaves and stems, during calm and clear nights results in regular dew deposition on these organs. The time of dewfall can vary considerably. We experienced dewfall prior to midnight but also just before dawn. Consequently the amount of dew will also vary, but no data are available besides those given in Section 2.1.1 which show that 0.2 mm is a long-term mean per dewfall event. Assuming dew and fog on about 200 days per year would result in an equivalent of about 40 mm of rainfall. For the Richtersveld the long-term annual mean of rainfall is 64 mm which means that dew and fog add a considerable amount to the total precipitation. We must stress, however, that (1) even if we assume an uptake of liquid water via the plant's surface, only a fraction of the deposited water will be utilized, and (2) the total amount of dew and fog precipitation is never enough to wet the soil down to the rooting horizon except at places where it gathers in rock crevices or underneath stones and then is available to roots. Consequently, the absolute quantity of water that might be gained by uptake of liquid water through leaves and stems is rather small. Whether a plant might benefit from such a water gain depends on the amount of water the same plant loses unavoidably by transpiration.

We have seen in the previous section that in response to the climatic prehistory and the prevailing environmental conditions succulents are excellent at saving water. The C_3 succulent *Othonna opima* loses less than 50 ml water per square metre a day which is equivalent to 0.05 mm water deposition per square metre. That is only one-quarter of the mean dew precipitation per dewfall event and might indicate that even small amounts of dew that can be utilized by succulents could considerably improve the diurnal water budget.

The formation of dew requires a surface temperature that is lower than the dewpoint temperature of the adjacent air. It is known from spines of cacti that they serve this purpose and that dew condenses at the cooler tips of the spines. The leaf surface of many succulents is not smooth but characterized by numerous protruding papillae, bladder-cells and trichomes. All these structures might get cooler than the rest of the leaf and the ambient air by radiational heat loss and might therefore initiate dew formation. Since little is known about this function of the leaf epidermis, research on this subject is urgently needed.

Dew or fog deposition on the surface is essential if liquid water is to be taken up by leaves or stems, but it still requires an uptake mechanism. It is a well-known physical rule that where water enters easily it will leave easily as well. For water fluxes to, from and within the root we have seen exceptions from this rule. If we accept this concept for the roots why should we not apply a similar concept to the epidermal tissue of leaves and stems?

One possible way for water to enter into the leaf is via the stomata. Since the aperture of the stomata can be controlled by the plant it would meet the requirement of controlled fluxes to and from the leaf and would by-pass the physical rule mentioned above. A second way would be across the cuticle and cell wall of the epidermal cells. This path, equal to a reversed cuticular transpiration, would somehow require that the cuticle and cell wall alter their permeability properties when they come into contact with water droplets and that they quickly return to the former status as soon as the surface is dry again. At a first glance this might sound rather peculiar and speculative but we will present some data later in this chapter indicating

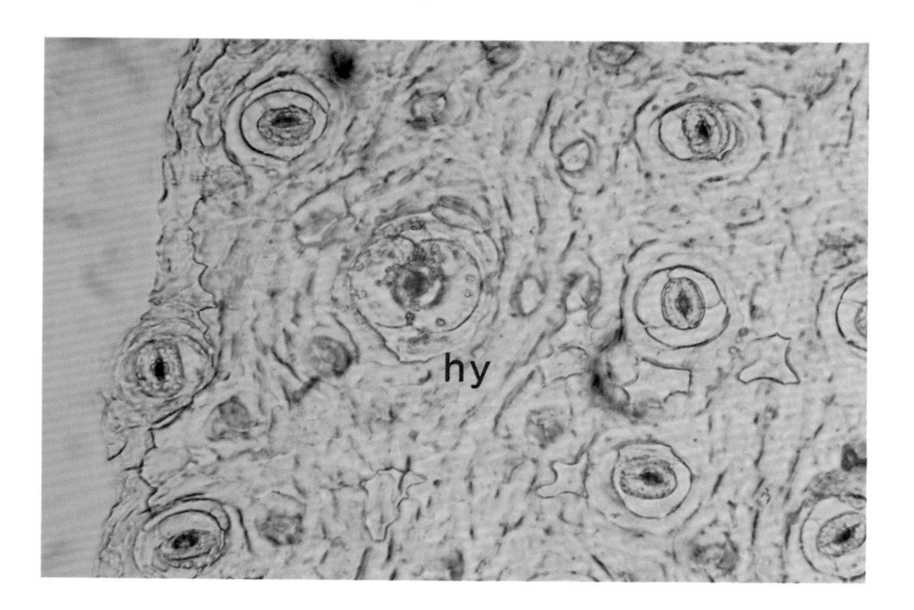

Fig. 4.56 Silicone print of the upper leaf epidermis of *Crassula erosula* showing a typical hydathode (hy) surrounded by six stomata. (By courtesy of B. Hendricks.)

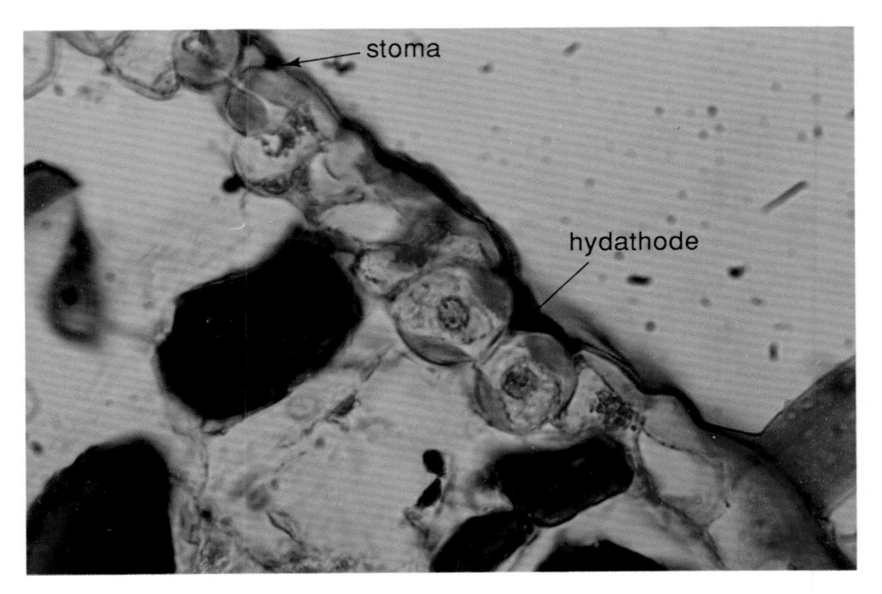

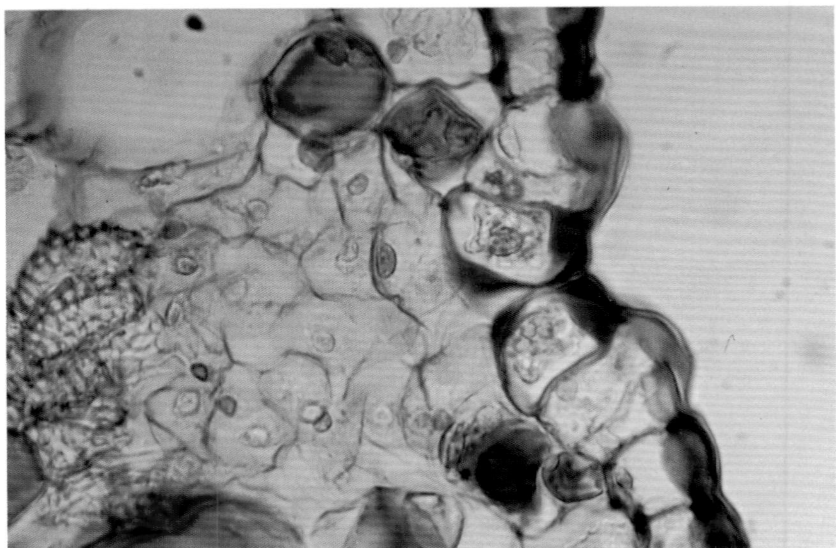

Fig. 4.59 Cross-section of a leaf of *Crassula clavata* showing a normal stoma and a hydathode. Note the much bigger guard cells of the water pore of the hydathode and the sheath of tannin-rich cells around the hydathode. (By courtesy of B. Hendricks.)

Fig. 4.60 Cross-section through a leaf of *Crassula* sp. showing the different parts of a hydathode: the guard cells of the water pore, the protruding large neighbouring cells, the epithem and the tracheids of the xylem. The sheath of tannin-rich cells surrounding the hydathode is not completely developed. (By courtesy of B. Hendricks.)

that something like water or water vapour uptake via the surface of leaves should be considered to happen.

To our knowledge there has not been any convincing report about a water absorbing tissue in the epidermis of succulents with the exception of the so-called hydathodes found in nearly all species of the genus *Crassula*. Generally the term hydathode is used for a special structure that is responsible for guttation, a form of active water excretion. Since nobody would assume that succulents of the genus *Crassula* suffering from drought in the southern Namib desert would excrete what they are short of, water, the function of their hydathodes must differ from that which we normally summarize under this term.

4.2.4.4 The hydathodes of crassulas

Since members of the genus *Crassula* survive severe droughts in the Richtersveld better than any other group of succulents, and since all but one *Crassula* species (*Crassula brevifolia*) growing in this area possess hydathodes, it seems likely that hydathodes play a key role in this context.

The life forms of crassulas vary considerably from rosettes pressed on the soil surface via columnar forms as shown in Fig. 1.9 to forms with a 'normal' shoot carrying distinct leaves. We examined several specimens of at least 43 different species from the southern Namib and found hydathodes in all but one species. There is a clear distribution pattern of the hydathodes. It can be generalized that hydathodes occur at all exposed surfaces, i.e. predominantly on the upper surface of the leaves, at the edges of a leaf and in columnar life forms only on the outer exposed leaf surfaces. The lower leaf surface might be free of hydathodes or have considerably fewer hydathodes than the upper surface. Further features of the hydathodes are:

(1) The number of hydathodes per square millimetre ranges from less than one to more than 20 and is always lower than the number of stomata.
(2) The hydathodes can easily be detected on surface images (Fig. 4.56). They are about twice to three times larger than stomata.
(3) Hydathodes often occur in clusters and are sunken into the epidermis (Fig. 4.57).
(4) In many species hydathodes are surrounded by trichomes (Fig. 4.58) which might trigger and facilitate condensation of water.
(5) On cross-sections hydathodes can easily be distinguished from stomata by their size as well as size and shape of the neighbouring cells (Fig. 4.59).

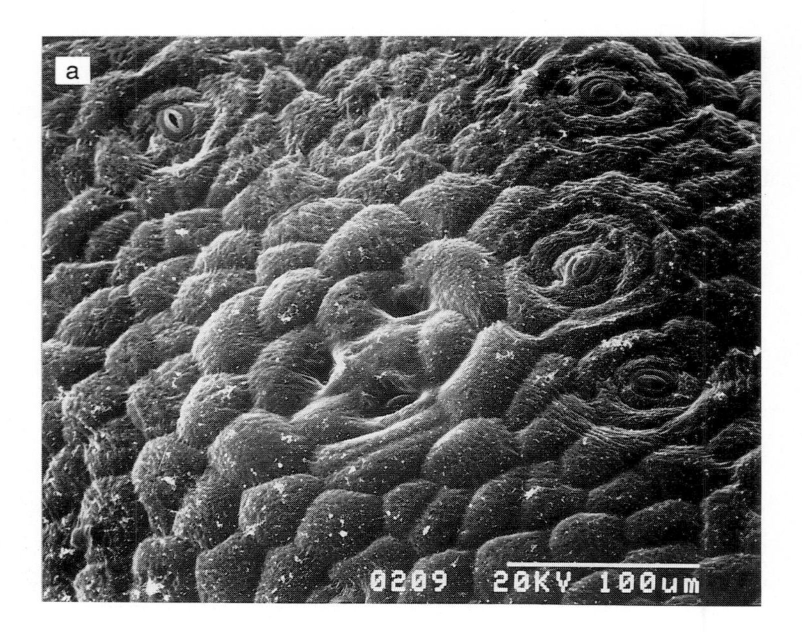

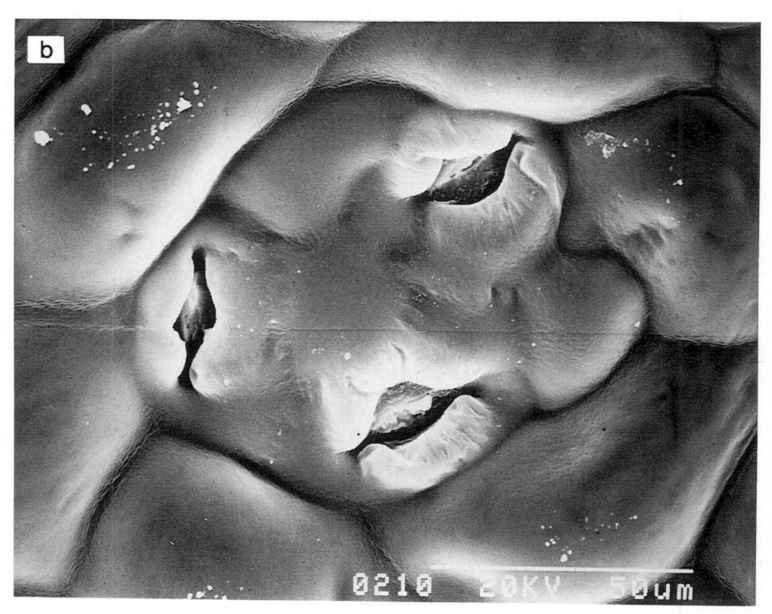

Fig. 4.57 (a) SEM picture of the epidermis of *Crassula tetragona* showing a cluster of three hydathodes in the centre of the picture and normal stomata. Hydathodes are sunken into the epidermis. (b) The hydathode cluster in more detail. (By courtesy of B. Hendricks.)

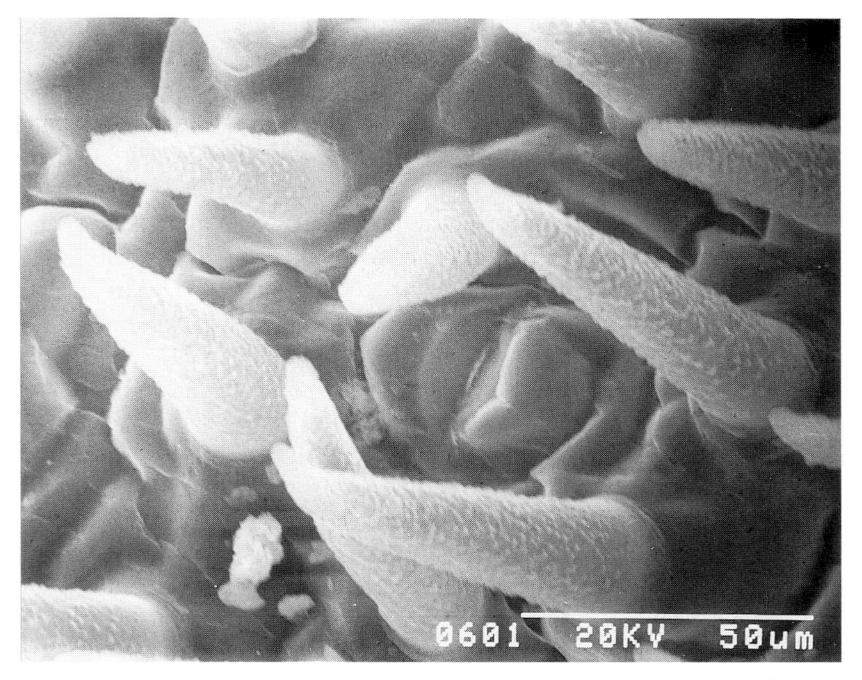

Fig. 4.58 SEM picture of a hydathode of *Crassula nudicaulis* surrounded by several trichomes. (By courtesy of B. Hendricks.)

(6) A complete hydathode consists of at least three different parts: the xylem tracheids, the epithem and the guard cells. Many hydathodes are additionally surrounded by a sheath of tannin-rich cells which always accompany the vascular bundles of the leaves (Fig. 4.60).

There is a report that crassulas are able to take up liquid water via the hydathodes (Tölken 1977). Water droplets stained with a water soluble dye were applied to the surface of leaves and after a while coloration of the xylem in the leaves was observed. However, no quantification was made. Although we never examined this feature of the crassulas in the southern Namib we strongly support the idea of water uptake via hydathodes as many of the given features of these structures imply such a function. Furthermore, crassulas often settle exposed habitats with a higher probability of dewfall and fog. They have a very low transpiration which means that even small quantities of water taken up via the leaf surface will substantially improve the diurnal water budget and help the plants survive longer rainless periods.

Any research on this subject is highly recommended as it will help our understanding of the biology and distribution of crassulas in arid areas.

Uptake of liquid water is one possible mode of water uptake through epidermal tissues; the other possibility is uptake of water vapour by reversed transpiration. Since this process requires open stomata and only occurs at night it should be restricted to CAM succulents.

A basic requirement for reversed transpiration is a water vapour gradient from the atmosphere to the substomatal cavities of the leaf. It has been shown that such gradients of up to 4 mPa Pa^{-1} occur in the Richtersveld (von Willert *et al.* 1985, von Willert and Brinckmann 1986). The steepness of this 'reversed' water vapour gradient depends on air and leaf temperatures, on relative air humidity and on the degree of saturation of water vapour in the substomatal cavities.

If recent considerations that water vapour partial pressure within the intercellular airspaces is significantly different from saturation (Ward and Bunce 1986) also prove valid for non-fully turgid succulents then the process of reversed transpiration would become even more likely. What is the evidence so far available for this process?

If branches from CAM succulents were cut off at the end of the dry summer period when no water was available in the soil these excised branches behaved absolutely like the rooted plant. CO_2-gas exchange was reduced to a minimum and overnight malic acid formation was negligible except for very humid nights. The cut-off branches were then exposed on nets at the same height above ground as they had been on the bush. Their fresh weight was determined every morning and evening over a period of 15 days. During this time a bergwind blew twice during the night, dewfall occurred on three nights and finally rain fell. As can be seen from Fig. 4.61, fresh weight of all succulents decreased during the light period. But there was also a considerable weight gain overnight except for the very dry nights with bergwind. Fig. 4.61 also shows that weight loss (in a first approximation equal to water loss) for the different species was very different and that species losing much water during the day were able to gain greater quantities overnight. A plot of the sum of all losses in per cent of the initial weight over the sum of all gains in per cent of the initial weight during the period of 15 days demonstrates this physically to be expected relationship impressively (Fig. 4.62). From 17 to 49% of the daytime water loss could be compensated for by overnight uptake of either liquid water or water vapour. Apart from the rainfall the period of 15 days was rather representative for the climate of the Richtersveld.

Earlier we saw that water gain of CAM succulents was linked to the

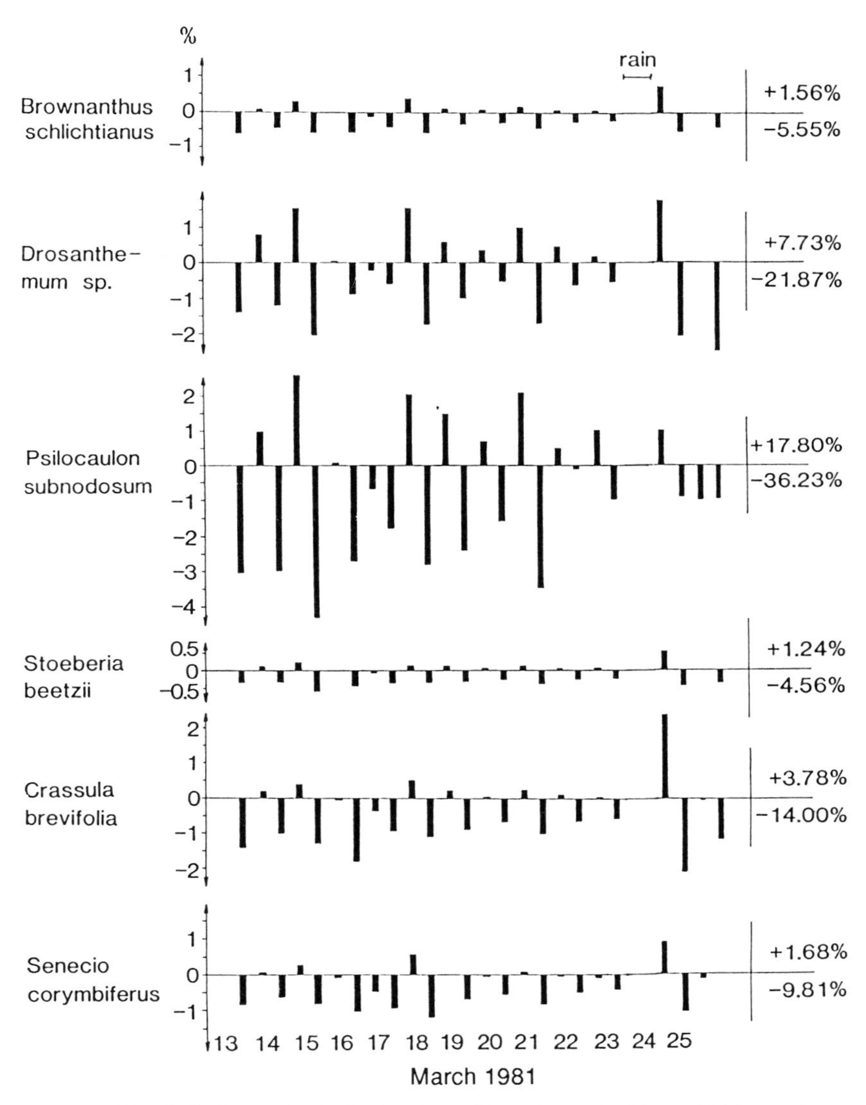

Fig. 4.61 Changes in the fresh weight of branches excised from CAM succulents of the Numees area (Richtersveld). The excised branches were exposed on nets 100 mm above ground in the same microclimate as the mother plant. Fresh weight was determined every day at sunrise and sunset. Bars represent weight losses and weight gains in per cent of the initial fresh weight. The experiment started one week after excision and proceeded for two further weeks. Bergwind, dew and even rain occurred during the course of the experiment. The numbers on the right side represent the sum of all losses and gains in per cent of the initial weight.

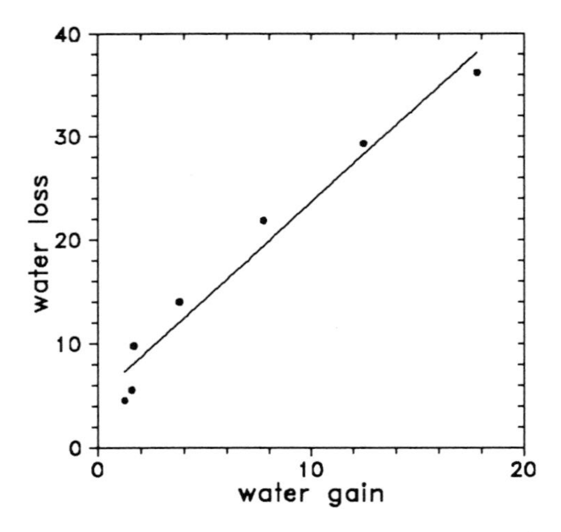

Fig. 4.62 A plot of the weight losses during the daytime over the weight gains during the night (in a first approximation reflecting water loss and water gain) of the data presented in Fig. 4.61 reveals that plants with a high transpirational water loss also gain higher quantities of water overnight. The solid line is the linear regression line for all data points.

formation and accumulation of malic acid. This is also valid for the uptake of atmospheric water by excised branches. A representative example of the direct proportionality of the amount of overnight accumulated malic acid and the water gained over the same period is presented in Fig. 4.63.

As outlined above, the steepness of the water vapour gradient between the intercellular air spaces of the leaf and the air adjacent to the leaf also depends on the deviation from vapour saturation in the intercellular airspaces. Since this deviation increases the lower the water potential of the leaf is, water gain should increase with desiccation of the leaf and should be highest for succulents with lowest water potentials. It has been shown for *Psilocaulon subnodosum* that with lowered osmotic potential of the cell sap water gain of excised branches increased (von Willert *et al.* 1985). The fact that in areas with frequent humid nights in the southern Namib and along the coast of the northern and Central Namib, the salt-accumulating Mesembryanthemaceae predominate may reflect their better ability to take up water from the atmosphere. This consequently will lead to a better diurnal water budget in this group of succulents. It then turns out that salt accumulation from a non- or weakly saline soil is one important element of the life strategy of these plants. But benefit is altered to death if prolonged atmospheric drought periods occur and endogenous salt concentrations rise to a toxic level.

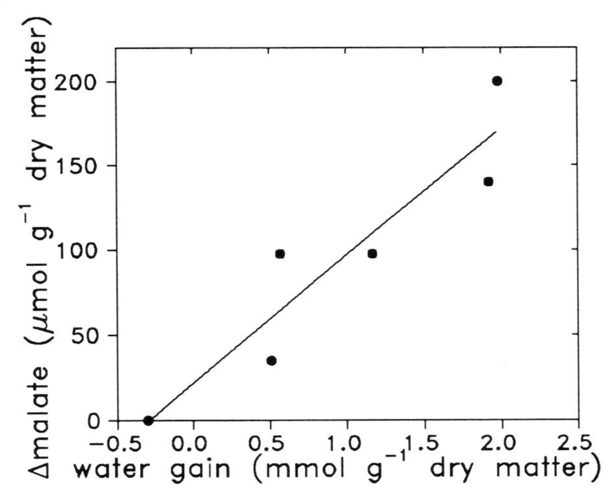

Fig. 4.63 Relationship between the amount of overnight accumulated malic acid (△malate) and the corresponding water gain of excised branches of the CAM succulent *Psilocaulon subnodosum*. The solid line is the linear regression line for all data points.

A very important consequence of dew or fog deposition on leaves and stems must be mentioned in this context. There can be no doubt that the water droplets on the surface will diminish the transpirational water loss in the morning, will keep leaf temperatures low for a longer period of time and will act as a short-term relief from the harsh environmental conditions of a desert. This implies that plants growing on west-facing slopes should benefit more from dew and fog than plants growing on east-facing slopes.

Measurements and observations in the southern Namib led us to assume that water uptake from the atmosphere is an important feature of CAM succulents for bridging drought gaps. It might explain the abundance of CAM succulents in arid areas where humid nights prevail. However, much more study has to be undertaken and further evidence is necessary to fully elucidate this subject.

4.2.5 The significance of succulence for the water balance

So far we have dealt with the driving forces of water fluxes and with the fluxes to or from the plant. Much of what has been outlined can be generalized for all plants and is not special for succulents. We will now focus on the typical character of succulents which is the storage of water or, in morphological terms, their succulence. Water storage necessarily implies water fluxes within the plant: fluxes into the water storing organ or from the storage tissue to the various sinks where water is needed.

The integral of the inward and outward water fluxes over a period of time is defined as the water budget of a plant. Evaluation of the water fluxes within the plant defines its water balance. The important quantity of the bulk water which can be translocated from the water storage site to target organs is the utilizable water. We shall now try to characterize the term utilizable water more accurately and to highlight possibilities of evaluating this quantity before we finally deal with water translocation in more detail.

4.2.5.1 Utilizable water

In Section 1.1 we gave a definition of a succulent and concluded that the most important criterion in this respect was utilizable water. In Section 1.2 we discussed the anatomical and histological structures which serve the storage of water. We addressed the principles of the driving forces for water fluxes early in Section 4.2 and now we must investigate whether we can determine a variable of the water status of the plant that allows us to estimate the amount of utilizable water.

Using a *P–V* curve it is easy to determine the maximum symplastic (or osmotic) and apoplastic water volume of a tissue. This has been done for only a few succulents that allow the application of this technique. Both the apoplastic and symplastic water volume can be considered as reversible water reserves of the plant, i.e. water can be withdrawn from the stream of transpirational water into these volumes and water can also be added from these volumes to the stream of transpirational water and into other sinks.

Central or peripheral water storage tissues in a leaf of a succulent are usually symplastic reserves, whereas apoplastic reserves are found in the stems of, for example, *Aloe*, *Tylecodon* and *Cyphostemma* spp. Similar apoplastic water reserves are typical for, for example, the wood of central and north European trees which can use these reserves to keep up their transpiration for several hours (Waring and Running 1978, Milne 1989). In most succulents apoplastic water reserves are of minor importance only. Here symplastic water reserves are more important, both in all-cell succulents and in partial succulents with a distinct water storage tissue. However, neither of the water volumes so far discussed are identical with the utilizable water.

One could be inclined to regard the difference in water volume between full saturation and the zero turgor level as utilizable water – or with respect to the actual water status the volume difference between the actual water level of the plant and the zero turgor level. This volume can be quantified using the *P–V* curve technique. It is only part of the utilizable water, however, it is the minimum amount. Herppich (1989) found that *Plectranthus marrubioides*, a succulent Lamiaceae from the Yemenitic mountains with leaves of the all-cell succulent type, can lose 30% of its leaf water content at saturation (that corresponds to about 38% of its maximum of symplastic water) before an irreversible damage of the photosynthetic apparatus could be measured. Much more limited saturation deficits were tolerable, however, when rerooting of cuttings was taken as a criterion. Attempts to substitute the normally slow desiccation by pressing out water volumes with a pressure bomb narrowed the tolerable limits.

Measurement of the variable fluorescence of chlorophyll a or testing whether shoots can become rooted again after reversible withdrawal of water from them using a pressure bomb are certainly suitable methods to approximate the amount of utilizable water. But at present we cannot precisely measure this amount with such methods. The reasons are obvious. With the pressure bomb the water necessarily is withdrawn very fast and the plant has no possibility to osmoregulate or to form protective substances as observed during gradual desiccation. Using the pressure bomb also makes it impossible to consider the retranslocation of water from the water storage tissue to the chlorenchyma, which is a common process in leaf and stem succulents. (We will shortly return to this question.) Finally, the pressure bomb method does not consider the possibility of an irreversible withdrawal of water with a simultaneous abandonment of entire organs or parts of organs. This is common in many succulents and is impressively shown by *Prenia sladeniana*, an evergreen leaf succulent in the Mesembryanthemaceae occurring in the Richtersveld.

Prenia forms large prostrate shrubs with circular, opposite leaves of the all-cell succulent type. The water content of the leaves is high; in September 1978 at its natural habitat in Numees it was 1220% of its dry weight. When a *Prenia* shrub is cut at soil surface and consequently all water uptake from the soil is prevented, the leaf water content decreases drastically by 120%. At the same time there is a clear drop in the osmotic potential of the leaf's water from −2.0 MPa to −2.4 MPa (Fig. 4.64). Measurements show that over the next few days the water content and the osmotic potential in the leaf fully recuperate and they reach the same values as before the plant was cut off. This is possible in the first place because of an osmotic adaptation which results within three days in a rise of the osmotic potential of the leaf's water. Secondly, the plant is able to withdraw water from the oldest leaves and make this available to the rest of the plant. It is remarkable that this process of water withdrawal does not occur evenly over the leaf. Starting from the leaf tip and progressing to the leaf's base, cell by cell is fully drained of its water. As a consequence, a leaf of *Prenia* can be fully turgid up to its middle while the remainder is just a parchmenty piece of skin.

Not all leaf succulents withdraw water from the oldest leaves until they

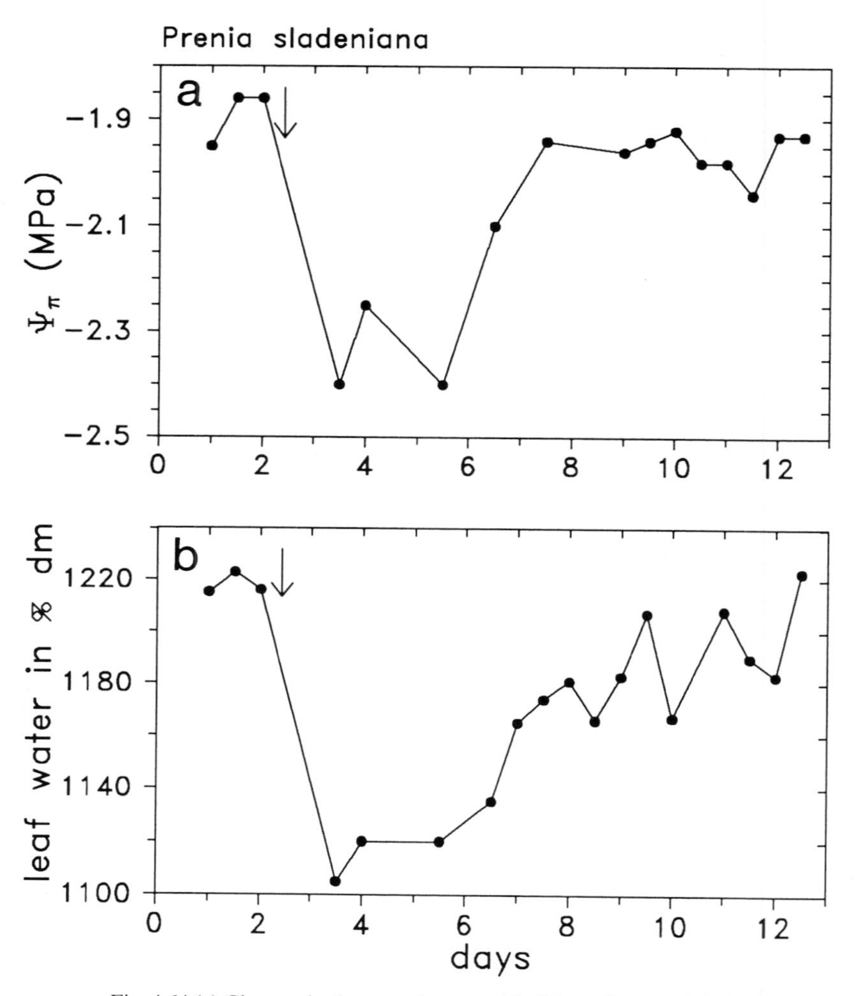

Fig. 4.64 (a) Changes in the osmotic potential of the cell sap and (b) in the actual leaf water content in percentage of the dry matter of leaves of the CAM succulent *Prenia sladeniana* at its natural habitat in the Richtersveld. At the arrow the shoot was separated from the root. Subsequently both parameters were followed over a period of ten days.

are fully dry and re-use this water in the rest of the plant, as we have described for *Prenia*. *Senecio medley-woodii*, for example, shows quite a different mechanism. Here water is withdrawn equally from all over the leaf, and the leaf does not dry out from its tip. In Fig. 1.5 we showed that a desiccating succulent leaf decreases in thickness (shrinks). This can be used in a non-destructive manner to obtain a series of relative measures of the reduction of water availability inside a single leaf. Such an analysis on

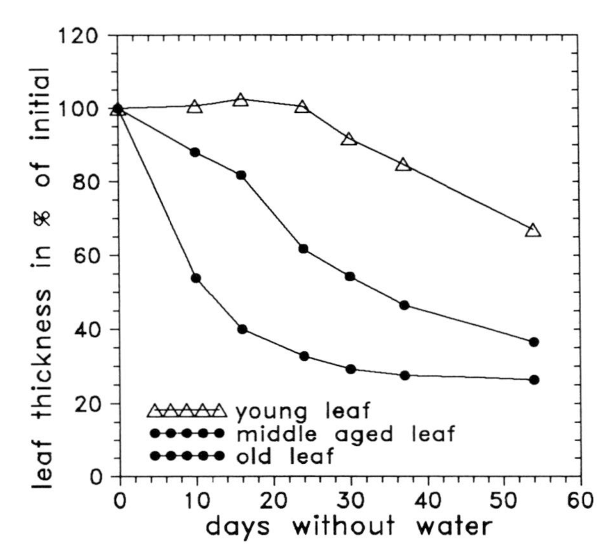

Fig. 4.65 Changes in the thickness of leaves of three different age classes after stopping the water supply of the CAM succulent *Senecio medley-woodii*.

Senecio medley-woodii showed (Donatz, pers. comm.) that at the beginning of a period of desiccation, first the oldest leaves make their water available and then the middle leaves, while only after 25 days of desiccation do the youngest leaves start to lose more water than they receive from the older ones (Fig. 4.65). Watering the plant at this stage shows that, in contrast to *Prenia*, the leaves of all ages, as far as they have not been dropped, keep their vitality and take up water again.

A shoot of *Prenia sladeniana* cut off from the plant can stay alive for a considerable period of time and even grow. The length of this period of survival depends on the number of leaves on the shoot. The leaves at the apex of a shoot of *Prenia* initially remain sealed to each other by their margins and form a bud that gradually increases in size. When this bud reaches a certain size it bursts and releases the next bud that was formed inside the old one; this grows again then bursts, etc. This process continues on shoots that have been cut off. The photographs in Fig. 4.66 show this. The pictures clearly show that the basal leaves dry up but remain on the plant while the bud keeps growing and then bursts. Finally water is also withdrawn from the bud leaves. But at that moment water is retranslocated from the apex. A basal axillary bud draws the water from the shoot's apical bud. This basal bud can be clearly seen in the picture (see arrow) taken one year after the shoot had been cut off. Apparently the plant tries to keep one bud that is situated more closely to the spot where normally the roots would

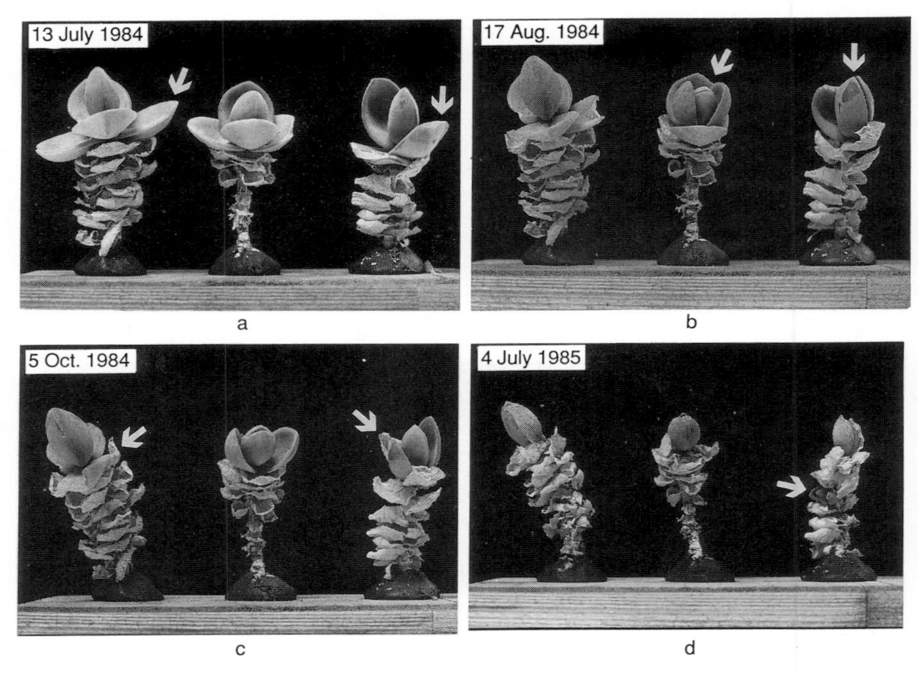

Fig. 4.66 Water translocation, die back and growth of three branches cut off from the CAM succulent *Prenia sladeniana*. Excised branches were sealed with lute and held in a growth chamber with a photoperiod of 12:12 hours. Daytime temperature was 25 °C, night temperature 15 °C. The relative air humidity was 50% during the day and 97% during the night. Photograph (a) was taken directly after cutting. Note the shrinkage of the turgid leaves (arrows) in the following photograph. Photograph (b) was taken 35 days after cutting. Note that the buds of two branches have opened (arrows). Photograph (c) was taken 85 days after cutting. Note that leaves died back completely but the buds still remain turgid. Photograph (d) was taken 355(!) days after cutting. Note that the main buds start drying while a turgid side bud appears (arrow).

be alive so that water taken up by the roots would have a shorter distance to cover. We will return to the question of translocation of water and then give some more examples.

This example from *Prenia sladeniana* not only demonstrates the extraordinary importance of utilizable water for the survival of a succulent but also shows that the quantity of utilizable water cannot be determined by pressure bomb techniques. While pressure bomb techniques allow us to measure some variables that are of importance for the water balance of a plant, even on some succulents, we cannot use them to measure the quantity of utilizable water which is so crucially important for survival through rainless periods.

4.2.5.2 *Water translocation*

In the discussion on utilizable water we saw an example of water translocation in succulents. Obviously, in succulents the process of water translocation is of utmost importance. We would like to give another example.

In the Mesembryanthemaceae there are many species in which the green parts consist of a pair of succulent leaves that can be connate to different degrees. Prominent examples are offered by the genera *Cheiridopsis*, *Conophytum*, *Argyroderma*, *Lithops*, etc. Apparently these plants are evergreen, as they always have their leaves. They renew their leaves annually, however, and also when rains fail to come, and the new pair of leaves is always formed at the cost of the water and biomass reserves in the old pair. Consequently, the new plant is smaller than the old one until it can take up sufficient water from the environment again and fully expand. We followed the development of the leaves of the same individual of *Cheiridopsis robusta* in its natural habitat in the Richtersveld from 1977 to 1983. During this period the annual precipitation values reached their minimum (Fig. 3.8) and this was reflected in the leaf sizes of *Cheiridopsis* (Fig. 4.67). From March 1977 to February 1981 leaf lengths decreased from *c*. 100 mm to 50 mm, and in the subsequent years when rainfall was above average leaf lengths increased again till they reached their former values.

The leaves of *Cheiridopsis robusta* are of the partially succulent type. They contain a central water storage tissue. We have seen that *Prenia sladeniana* possesses an excellent system of desiccation avoidance by water translocation and utilization keeping the water content of the leaves constant, even in periods of drought. The water content in the leaves of *Cheiridopsis robusta* continuously decreases during a period of drought. However, we always measured the water content of the total leaf and not of the individual tissues inside the leaf. Therefore we cannot rule out that also in *Cheiridopsis* the water content of the chlorenchyma was kept constant by translocation of water from the central storage tissue to the chlorenchyma. It is probable that we here have an analogous case to the cacti and agaves (Nobel 1988) and that it is precisely the function of the separate water storage tissue to keep the water status of the chlorenchyma at an appropriate level.

Reduction of the water volume of a partially succulent leaf leads to characteristic folds at predisposed sites. As a result the leaf becomes shorter and flatter. In extreme cases this process can turn a cylindrical leaf, as for example in *Othonna opima*, into a flat leaf. We observed this during the drought period of 1979/1980 in the Richtersveld (Fig. 4.68).

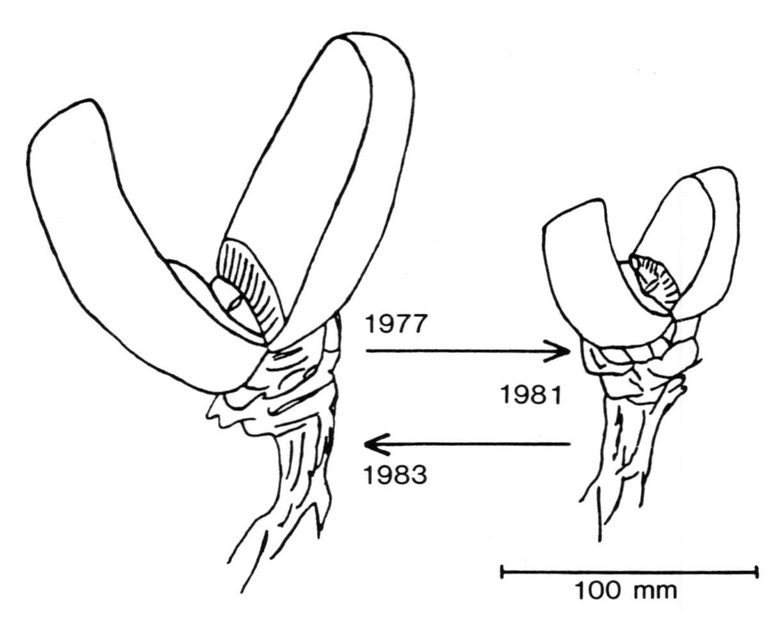

Fig. 4.67 Insufficient rainfall caused a steady decrease in the size of the leaves of *Cheiridopsis robusta*, a CAM succulent consisting of only one pair of leaves which are replaced every year. The renewal process is conducted by water translocated from the existing leaves to the newly formed leaves. As rainfall initially was minimal, these leaves subsequently could not grow out to their potential size and they became gradually smaller from year to year. Sufficient rainfall in the following years restored the original size of the leaves.

The shrinkage of the leaf as shown in Fig. 4.67 is always accompanied by a reduction of the actual leaf water content and is directly related to the amount of rain received during the rainy season. In Fig. 4.69 we show this for *Cheiridopsis robusta* and *Aloe pearsonii*. This, together with the leaf cross-sections showing the dehydration–rehydration cycle of *Aloe pearsonii* (Fig. 1.5), demonstrates how large a fraction the quantity of utilizable water can be. During prolonged periods of drought cacti and agaves also can reversibly lose up to 50% of their water reserves (MacDougal *et al.* 1915, Nobel 1976, 1977a, b).

We have seen that we can distinguish at least two radically different forms of water storage tissue among the succulents: (1) irreversible water storage tissue; this occurs in many succulents with all-cell succulent leaves and we have exemplified it for *Prenia sladeniana*; (2) reversible water storage tissue; this is exceptionally well developed in succulents with leaves of the partially succulent type, e.g. in *Aloe*. The broad spectrum of variation

Fig. 4.68 *Othonna opima* (Richtersveld): (a) The succulent leaves are fully turgid and round after sufficient winter rainfall but (b) appear highly desiccated and nearly flat after a prolonged drought.

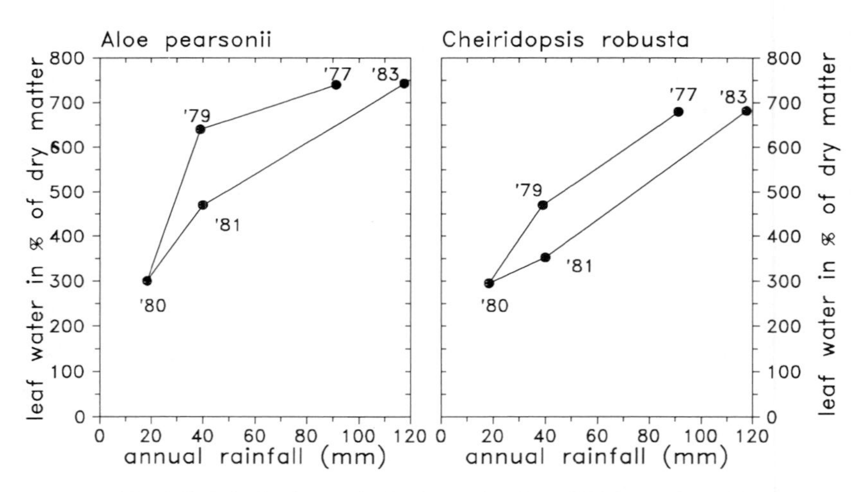

Fig. 4.69 Dehydration – rehydration cycle of two CAM succulents. The actual leaf water content in percentage of dry matter of the two succulents with partially succulent leaves as measured at the end of the dry season is plotted over the annual rainfall. Figures at the dots indicate the respective years of the determination.

on the basic anatomic–histological scheme is mirrored in the variation on this basic functional pattern: there are all-cell succulent leaves that dry back like an *Aloe* leaf as well as leaves with a central water storage tissue that dry back like a *Prenia* leaf.

Having described the phenomena, we will now try to understand the mechanism of water translocation. In cacti, which have a central water storage tissue, the pathway of the water from the storage tissue to the xylem and then in the peripheral chlorenchyma could be completely explained (Nobel 1988). We can apply a similar mechanism for the water translocation in succulent leaves.

The van't Hoff law given in equation (4.28) ($\Psi_\pi = -\mathrm{R}T\Sigma c_i$) describes the relation between the osmotic pressure π of a solution and the concentration of the dissolved substances in that solution. This can also be written as

$$\Psi_\pi = -\frac{\Sigma n_i}{V}\,\mathrm{R}\,T \tag{4.28a}$$

If we apply this relation to the water storage tissue or the chlorenchyma cells then V stands for the respective water volumes V_{ws} and V_{ch}. From the equations

$$\Psi_\pi = -\frac{\Sigma n_{ws}}{V_{ws}}\,\mathrm{R}\,T \tag{4.49}$$

for the water storage tissue and

$$\Psi_{\pi} = -\frac{\Sigma n_{ch}}{V_{ch}} R\,T \tag{4.50}$$

for the chlorenchyma and from the fact that the cells of the water storage tissue are usually considerably larger than those of the chlorenchyma, it follows that the water storage tissue experiences a larger change in volume than the chlorenchyma when the change in $d\Psi_{\pi}$ in both tissues is similar and Σn_i stays constant ($n_{ws} = n_{ch}$). This relation also implies that a change in Σn_i, for instance an increase in the chlorenchyma and a decrease in the water storage tissue or a translocation of solutes from the water storage tissue into the chlorenchyma, results in a directed symplastic water flow from the water storage tissue into the chlorenchyma.

Consequently, under drought stress the water storage tissue will lose its turgor relatively rapidly as a result of water loss and its cells will fold as shown in Fig. 1.6. The chlorenchyma, on the other hand, can still maintain its turgor at the cost of the water storage tissue and thus remain active for a considerably longer period of time during the period of drought stress.

Essential preconditions for the proper functioning of the water storage tissue are that it must have a low content of inorganic ions with limited mobility and that osmotically active substances can rapidly be synthesized and broken down. This makes it understandable why the water storage tissues contain high concentrations of organic acids, particularly malic acid, but these acids do not fluctuate diurnally. Denius and Homann (1972) showed this for *Aloe arborescens* and we demonstrated it for many succulents in the southern Namib.

In the Mesembryanthemaceae there are many succulents with central storage tissues and others with all-cell succulence. Unfortunately we did not separately analyse the different types of tissue. However, our data on ion concentrations in the total leaf water (Fig. 4.70) clearly confirm the point made above. It is even possible that the chlorenchyma accumulates salt. This would again increase the translocation of water towards the chlorenchyma in periods of drought.

Transport of inorganic ions and other substances play an important role in the translocation of water. We demonstrate this in an example.

Fig. 4.71 shows *Brownanthus schlichtianus*, a Mesembryanthemaceae from the Richtersveld. The plant forms shrubs up to about 40 cm high and has erect, articulated shoots. With sufficient water supply every segment bears a pair of succulent leaves at its base. Leaves and shoots are of the all-cell succulent type. At the start of a period of drought the leaves start drying back progressively from their tips to their bases. The drying back process of

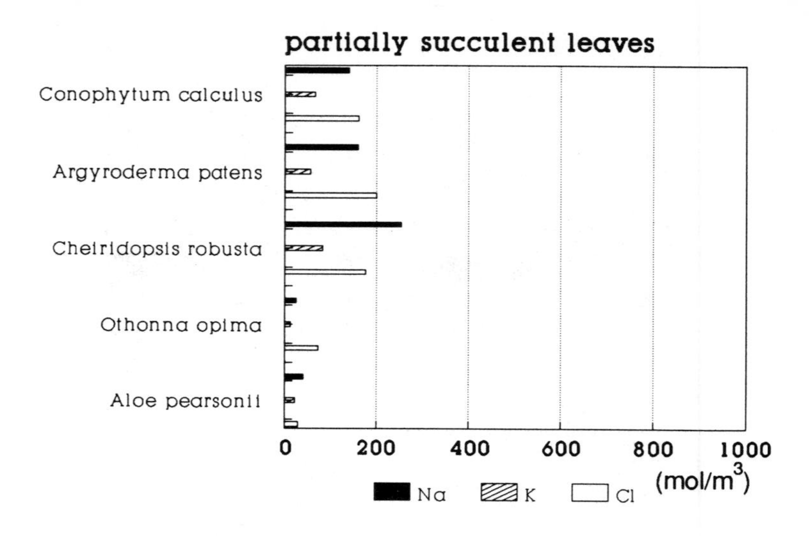

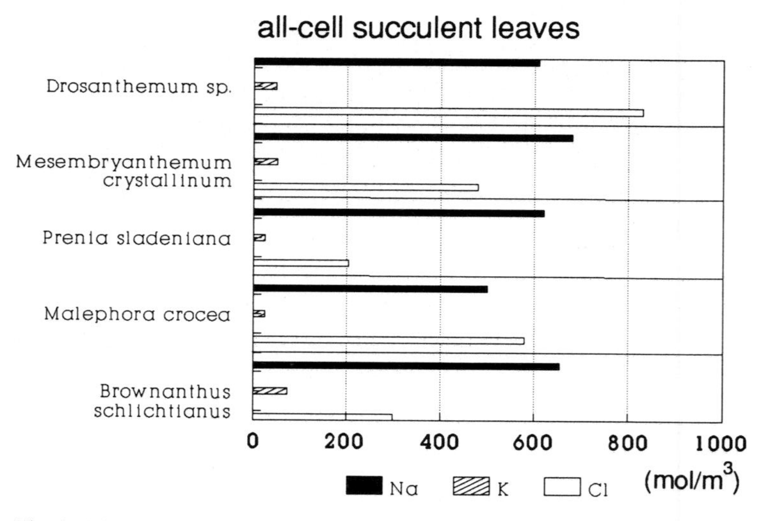

Fig. 4.70 Sodium, potassium and chloride concentration in the cell sap of (a) succulents of the partially succulent type and (b) succulents of the all-cell succulent type.

Fig. 4.71 The CAM succulent *Brownanthus schlichtianus.*

the leaves starts at the shoot's base and rapidly progresses to the shoot's apex. When all leaves have died the potassium gradient along the shoot is similar to that in all other plants: potassium concentrations are highest in youngest parts of the shoot and decrease towards the shoot's base. Chloride shows a similar pattern along the length of the shoot, whereas sodium increases with age of the plant part. With continuing drought the segments start to die back, first the oldest and then progressively toward the shoot's apex. With persisting drought the direction of this process of die-back is reversed: the shoot starts to die back from its apex and maintains the oldest segments that are still vital. An analysis of the ionic gradient in this stage shows that it has been reversed. Potassium is being translocated again from the shoot's apex towards its base and its absolute amount in the oldest living segments greatly increases. If rains come now, or if the plant is watered in this stage, the plant sprouts with new axillary shoots at the base of the oldest vital segment and in the new shoot the normal ionic gradients are restored. The various phases of this process, and the potassium contents in μmol g^{-1} of dry matter are presented in Fig. 4.72.

From a comprehensive screening programme we conclude that proline does not play an important role as a protective substance during drought or heat stress in the succulents of the Richtersveld (Treichel *et al.* 1984). However, *Brownanthus schlichtianus* is one of the few species that increase their proline content under such stress conditions. After the leaves have died back the apex of the shoot contains about 17 μmol and the base of the shoot about 4 μmol of proline per gram of dry matter. When the drought stress is at its peak the proline gradient is reversed as well and the basal vital segments then show values of 50 μmol. Towards the apex the proline content decreases to about 10 μmol g^{-1} of dry matter. The highest value we ever measured is, with 52 μmol $^{-1}$ of dry matter, rather modest compared to 590 μmol g^{-1} of dry matter measured in *Welwitschia mirabilis* (von Willert 1985). However, the segments of *Brownanthus* completely contain the woody part of the central cylinder and it is quite possible that the proline content in the cortical parenchyma is higher than the values measured by us. It seems clear that *Brownanthus schlichtianus* – just as *Prenia sladeniana* – tries to save a basal axillary bud. At least it is possible, and we say this with some restraint, to interpret the proline gradient in this sense.

During the many years we visited the Richtersveld we often wondered why the *Brownanthus* shrubs, which clearly have an apical growth pattern, always all had the same height. The processes described above explain this phenomenon.

We stipulated that in the succulents with partially succulent leaves a change in the concentration of solutes in the cell sap opened the possibility for a directed symplastic translocation of water from the storage parenchyma into the chlorenchyma. We also pointed out that a low concentration of inorganic ions in the storage parenchyma is an essential precondition and our measurements confirmed this. But the succulent Mesembryanthemaceae of the all-cell succulent type typically contain high concentrations of inorganic ions (see Fig. 4.42). The examples of *Prenia sladeniana* and *Brownanthus schlichtianus* discussed earlier also demon-

Fig. 4.72 The die-back process of the stem succulent *Brownanthus schlichtianus* during increasing drought and the accompanying alterations in potassium content in the stem parts showing a water and solute translocation from the bottom to the top of the shoot and also in the reverse direction.

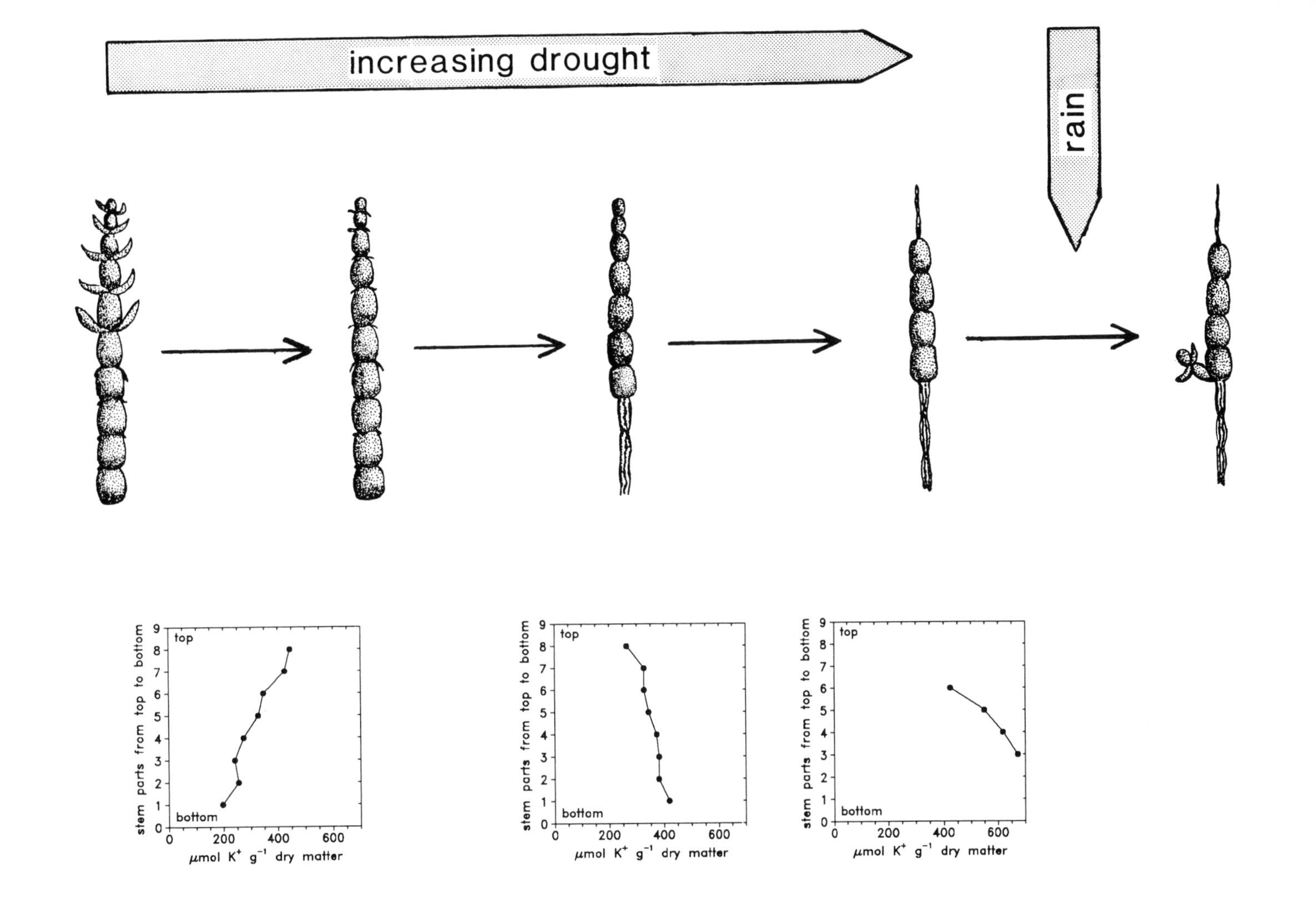
increasing drought
rain
top
bottom
stem parts from top to bottom
μmol K+ g−1 dry matter
top
bottom
stem parts from top to bottom
μmol K+ g−1 dry matter
top
bottom
stem parts from top to bottom
μmol K+ g−1 dry matter

strate that in these all-cell succulents the translocation of water functions perfectly. It is difficult to understand how, in a leaf that is dying back, the symplastic transport of water is carried out from cell to cell despite their high ionic contents. Perhaps the transport of inorganic ions as measured in *Brownanthus* may explain this problem.

The overriding significance of succulence for water status and water transport is undoubtedly that the fraction of utilizable water is made available. Precisely how this quantity of water is made available has not yet been clarified.

4.3 Carbon fluxes

We may define the carbon budget of a plant in a similar way to the water budget, simply as the difference between carbon gain and carbon loss expressed as the respective carbon fluxes. We can write

$$J_c(\text{net}) = J_c(\text{gain}) - J_c(\text{loss}) \qquad (4.51)$$

This definition includes all forms of carbon gain and carbon loss. One must, however, consider that carbon fluxes to or from the plant are mainly by CO_2 fluxes and that fluxes of organic compounds are rather small and perhaps mainly restricted to some specialists, although the release of organic matter by root systems and secretory tissues might be considerable. Nevertheless, at a first approximation the carbon budget of a plant is given by the CO_2 fluxes into and out of a plant. This is termed the CO_2 exchange with the environment. Consequently we will first deal with such CO_2 exchange reactions and mechanisms before we focus on the question of the costs for succulence of a special tissue for water storage. Besides, we have seen that the water balance of succulents markedly benefits from succulence and the fraction of utilizable water. Similar considerations should be made for the carbon balance. Here also only part of the biomass can be regarded as utilizable biomass. Its quantity is very different in evergreen and deciduous species, perennials and/or annuals. These facts will be outlined in more detail in Chapter 5.

When dealing with the significance of succulence for the CO_2 balance we will see the overriding importance of the quantity of utilizable water but also its limits, and the total water economy of succulents will become obvious again.

Since the term succulent is very often used synonymously with CAM plant we should give special attention to CAM and its role in the context of the carbon balance of succulents.

4.3.1 The CO_2 budget

The CO_2 budget of a plant depends on CO_2-fixing and CO_2-releasing reactions. For a given period of time we can define:

$$\Delta(CO_2) = \int J_{ph}\, dt - \int J_{mi}(\text{dark})\, dt - \int J_{mi}(\text{light})\, dt - \int J_{phr}\, dt \quad (4.52)$$

where $\Delta(CO_2)$ is the CO_2 budget, J_{ph} is the gross photosynthesis, J_{mi} is the mitochondrial respiration in the dark and the light respectively, and J_{phr} is the photorespiration.

This equation is valid for the entire plant and includes green and non-green parts.

For the following considerations it is necessary to define standard conditions and to make some simplifications. Photon irradiance must saturate photosynthesis, temperatures should be optimal for the day (e.g. 25 °C) and also for the night (e.g. 15 °C), air humidity should be 40% during the day (that means ΔW will not exceed 20 mPa Pa^{-1}) and near saturation during the night with a ΔW not exceeding 1 mPa Pa^{-1}. The plant must be sufficiently supplied with water to prevent stomatal closure due to water deficit. The allocation of carbon from non-green tissue or vice versa should not be considered.

From a theoretical point of view the CO_2 budget can be improved by either increasing gross photosynthesis or by decreasing one of the CO_2-releasing processes.

4.3.1.1 Photorespiration

Photorespiration originates from the activity of the CO_2-fixing enzyme, rubisco, which besides its carboxylating activity also has an oxygenase activity, i.e. the rubisco reacts with O_2 as well as with CO_2. With CO_2 it processes the carboxylation of RuBP which leads to the formation of two molecules of PGA, while with O_2 as substrate it splits RuBP to give a C_2 and C_3 compound. The C_3 compound is PGA which can directly enter the CALVIN-cycle. The C_2 compound is P-glycolate. From two of these C_2 compounds one PGA is formed in a complex series of reactions which take place in three different compartments of the cell. During these processes O_2 is consumed and CO_2 is released and the initial splitting only occurs upon illumination. Therefore this series of processes is termed photorespiration.

Detailed studies of the carboxylating and oxygenating properties of rubisco have shown that the ratio of the partial pressures of CO_2 and O_2 at the site of the rubisco determine its properties. CO_2 and O_2 are not only substrates but also competitive inhibitors of the respective opposite reaction. High partial pressures of O_2 inhibit the carboxylation while high partial pressures of CO_2 inhibit the oxygenation. With the naturally prevailing composition of the air of about 0.03% CO_2 and 21% O_2, which corresponds with a CO_2 and O_2 concentration in the chloroplast of about 10 μmolar and 250 μmolar respectively, the ratio of carboxylation to oxygenation ranges between 3:1 and 4:1. Consequently, photorespiration and photosynthetic CO_2 fixation are competitive reactions and photorespiration can significantly reduce the efficiency of photosynthesis.

The partial pressure of CO_2 is substantially reduced in the chloroplasts when photosynthesis proceeds and stomata close partially or completely. Under such conditions the ratio of the partial pressure of CO_2 and O_2 (pCO_2:pO_2) at the site of the rubisco shifts towards O_2, favouring photorespiration. As the solubility of CO_2 decreases more rapidly with temperature than the solubility of O_2, increasing leaf temperature will also favour photorespiration via decreasing pCO_2:pO_2 ratios. It is obvious that photorespiration lowers the efficiency of CO_2 fixation, and consequently productivity. This decrease very much depends on the prevailing environmental conditions. It has been reported that up to 50% of the photosynthetic carbon gain can be chewed up by photorespiration.

Later we shall come back to the possible ecological significance of this, at first glance, wasteful reaction but we have to emphasize here that the CO_2 budget of a plant could be effectively improved by minimizing photorespiration. A possible tool to achieve this might be found in a regulation of the partial pressures of CO_2 and O_2 at the site of the rubisco.

4.3.1.2 Mitochondrial respiration in the light

The main question is: does the so-called 'dark respiration' proceed under illumination or not? For years it has been a dogma that it does. As it is not possible to distinguish precisely with the available methods between CO_2 production and consumption of the simultaneously running processes of photorespiration and mitochondrial respiration together with the O_2 production and CO_2 consumption of photosynthesis, a definite answer cannot be given.

A crucial question is whether the photosynthetic ATP production is sufficient to provide ATP for all other ATP-requiring reactions inside and outside the chloroplast or whether the chloroplast ATP is completely consumed for CO_2 reduction. ATP originating from photophosphorylation in the chloroplast can, in case of necessary transportation to the cytosol, pass the chloroplast membranes to the cytosol via the PGA/dihydroxyacetonephosphate shuttle. An increasing ATP level in the cytosol would slow down or even switch off glycolysis so that mitochondria run short of or even out of substrate. Hence, export of ATP from the chloroplast would diminish mitochondrial respiration in the light. This influence would be less pronounced if the substrate for mitochondrial respiration in the light comes from other reactions by-passing the ATP regulation of phosphofructokinase, e.g. pyruvate that results from decarboxylation of malic acid. There is still the interaction of citratesynthetase with ATP. Since cytosolic ATP cannot enter the mitochondria, inhibition of this mitochondrial enzyme can only derive from ATP generated in the mitochondria. The formation of mitochondrial ATP can be overcome by partial uncoupling or by electron flow through the alternative CN-resistant pathway.

Summarizing, we must conclude that it is unlikely that dark respiration proceeds in the light unaffected by photosynthesis. However, the degree of interaction and perturbance is unknown. Furthermore, one must mention that ^{14}C-labelled intermediates when fed to plants in the light run through the TCA cycle in the same way as in the dark. This is a strong argument for the operational activity of the TCA cycle and for mitochondrial respiration in the light.

4.3.1.3 *Mitochondrial respiration in the dark*

We will not give an outline of mitochondrial respiration here but will focus on the possibility of minimizing CO_2 losses during respiration in the dark.

An excellent tool for improving the CO_2 budget would be the refixation of respiratory CO_2 that is generated during the night. This refixation cannot run via the rubisco because this enzyme has a very low activity in the dark. The required reducing equivalents and ATP are not available. Yet, another enzyme, PEPC, is capable of fixing CO_2 in the dark. The CO_2 acceptor of the PEPC-mediated reaction is PEP which is taken from glucan degradation. The first-formed product is OAA.

$$\mathrm{PEP} + \mathrm{HCO_3}^- \xrightarrow{\mathrm{PEPC}} \mathrm{OAA} + \mathrm{Pi} \tag{4.53}$$

Although the phenomenon of refixation of respiratory CO_2 has been well established for several decades, very little is known about its mechanism, pathway and regulation. It is only clear that the metabolically very active α-oxoacid, OAA, is instantaneously metabolized to more stable compounds. Usually a reduction to malic acid by the action of MDH

$$\mathrm{OAA} + \mathrm{NADH} + \mathrm{H}^+ \underset{}{\overset{\mathrm{MDH}}{\rightleftharpoons}} \text{malic acid} + \mathrm{NAD}^+ \tag{4.54}$$

or a transamination by AAT

$$\text{OAA} + \text{glutamic acid} \underset{}{\overset{\text{AAT}}{\rightleftharpoons}} \text{aspartic acid} + \alpha\text{-oxoglutarate} \quad (4.55)$$

are assumed, but as citric acid is often found as a fixation product the sequences indicated in Fig. 4.73a would also be possible. The more stable compounds are usually not further metabolized in the dark but accumulate in the cell.

Refixation of respiratory CO_2 in the dark prevents CO_2 release, but the synthesized and accumulated compounds are of only limited value for the plant in this form. Therefore they must be activated on the following day and must flow back to the metabolism. When speaking of malic acid – which is commonly the major product accumulated overnight – two reactions are possible. The first is a complete oxidation to CO_2 and water in the TCA cycle. The second is a single decarboxylation resulting in CO_2 and one C_3 compound. All released CO_2 can again be refixed via rubisco, and the C_3 compound may enter neoglucogenesis. The different possible pathways of the degradation of malic acid are given in Fig. 4.73b. Depending on how much of the CO_2 released during the day is refixed by photosynthesis, the CO_2 budget of the plant can increase slightly or substantially.

It is necessary to mention further advantages of this mechanism. Due to the decarboxylation of malic acid in the light endogenous CO_2 will increase in concentration which in turn affects photorespiration by inhibiting the oxygenase activity of the rubisco. When malic acid is completely oxidized to CO_2 and water, mitochondria are involved. This means that with the mechanism of refixation of respiratory CO_2 in the night, dark respiration (or at least part of it) is shifted into the light period. This has the consequence that if mitochondrial respiration proceeds in the light it does not consume simultaneously generated assimilates but finishes its respiration from the preceding night which stopped at the level of PEP.

4.3.1.4 Gross photosynthesis

The assumption that gross photosynthesis can be augmented via an increase of rubisco is not valid since all photosynthetically active cells possess rubisco in excess (rubisco accounts for up to 50% of the soluble leaf protein). Where light is a minimum factor an increase in number of chloroplasts and chlorophyll per unit leaf area can be observed. This leads to better light harvesting. For desert plants light is not a limiting factor in photosynthesis. A higher absorption of incident radiation is not required as this would only increase leaf temperature (the chloroplasts account for

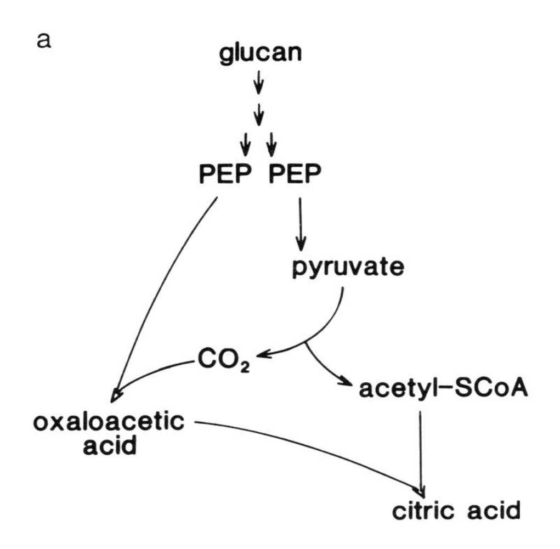

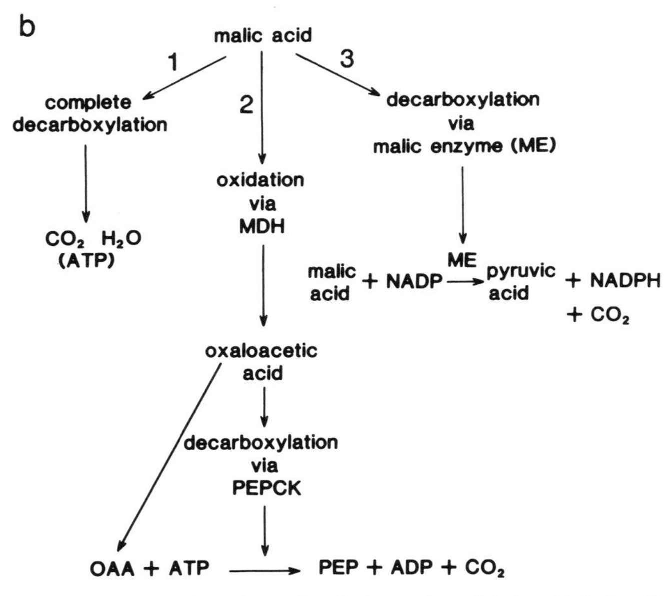

Fig. 4.73 (a) Simplified scheme for the formation of citric acid during the night without additional uptake and fixation of external CO_2. (b) The three possibilities for malic acid degradation during the day: (1) Involves a complete degradation with the participation of mitochondria, (2) is a two-step reaction involving the oxidation of malic acid to OAA by MDH and the subsequent decarboxylation of OAA by PEPCK, and (3) is decarboxylation by the cytoplasmic malic enzyme. In (2) and (3) the resulting C_3 compound might enter neoglucogenesis or path (1).

about 70% of the absorbed radiation from which only 1–2% is utilized in photosynthesis). In general, constant environmental conditions leave no valuable possibilities for a plant to increase primary CO_2 fixation. Hence, a substantial improvement of the CO_2 budget can be achieved only by minimizing CO_2 loss either by refixation of respiratory CO_2 or by reducing photorespiration.

4.3.2 Modes of CO_2 fixation

For the fixation of CO_2 two enzymes are available, PEPC and rubisco. Primary CO_2 fixation can take place exclusively in the light or in the dark or in both light and dark. It can occur in all cells that have chloroplasts or in a special tissue, and it can utilize external CO_2 or endogenously generated CO_2. According to this variation we distinguish C_4 photosynthesis, C_3 photosynthesis and CAM.

All three modes of photosynthesis occur in desert succulents. That is why we have to deal with the features of the different modes of carbon pathways in photosynthesis. We shall do this rather briefly because there are numerous textbooks and reviews available. We shall focus on the question in what way the different modes allow the plant to make use of the different possibilities to improve its CO_2 budget.

4.3.2.1 C_3 plants

C_3 plants fix CO_2 via rubisco and regenerate the CO_2 acceptor RuBP in the CALVIN-cycle. As the first stable fixation product is a C_3 compound (PGA) their mode of photosynthesis is called the C_3 mode and plants operating in this way are called C_3 plants. C_3 plants 'suffer' extremely from photorespiration especially when high radiation, high temperatures and insufficient water supply cause the stomata to close. Despite the high affinity of the rubisco for CO_2 (K_m 5–15 μmol) a proper photosynthesis requires widely open stomata. The reason for this lies in the fact that the rubisco utilizes CO_2 as substrate. Due to the prevailing pH the concentration of CO_2 in the chloroplasts is rather low (about 7 μmol). This results in a low diffusion of CO_2 within the cell as the CO_2 concentration difference between sink and source cannot be greater than the source concentration. For a detailed description we refer to Edwards and Walker (1983). The necessity of widely open stomata unavoidably favours water loss by transpiration. As a result of this, C_3 plants consume 500 to 1000 g of water to build up 1 g of dry matter.

As there are excellent descriptions of the path of carbon in C_3 plants in many textbooks, we present here a very short and reduced schedule only (Fig. 4.74).

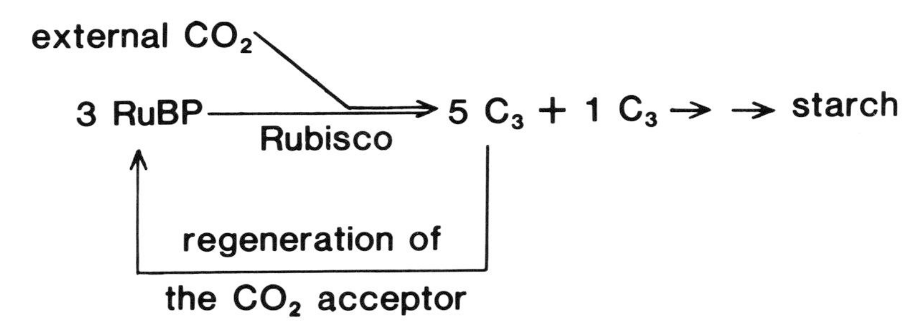

Fig. 4.74 Simplified scheme of the path of carbon in the CALVIN-cycle of C_3 plants.

Edwards and Walker (1983) emphasized the uniqueness of this cycle, being the only one in nature that produces more than it consumes. Hence, the CALVIN-cycle is the basis for all life on our planet. We will see in the following sections that though other mechanisms of CO_2 fixation have evolved, the final step in CO_2 fixation is always the CALVIN-cycle.

4.3.2.2 C_4 plants

C_4 plants fix atmospheric CO_2 twice. However, these two processes do not run in the same cell nor in the same tissue.

Characteristic for leaves of C_4 plants is that they have two distinct and separated types of assimilation tissue, the mesophyll and the bundle sheath. The latter consists of relatively large cells surrounding the vascular bundle. They have rather large chloroplasts, which are often characteristically located at the centripetal cell walls. The mesophyll cells are responsible for the primary fixation of CO_2. Since they do not have the enzyme rubisco their CO_2 fixation runs via PEPC. With the substrate HCO_3^-, OAA is formed and instantaneously converted to either malic acid or aspartic acid. These more stable compounds are then transported to the bundle sheath where they are either directly (in case of malic acid) or after a re-transamination (in case of aspartic acid) decarboxylated. The released CO_2 is subsequently refixed by the rubisco of the bundle sheath chloroplasts.

When malic acid is transported to the bundle sheath NADP malic enzyme converts it to CO_2 and pyruvate. The latter goes back to the mesophyll cell. When aspartic acid is transported it is re-transaminated to OAA in the mitochondria of the bundle sheath cells. Subsequently OAA is reduced to malic acid which is then decarboxylated to pyruvate via the NAD malic enzyme. In order to keep the nitrogen of the bundle sheath and mesophyll balanced pyruvate is first transaminated to alanine before it flows back to the mesophyll.

From the back-flown pyruvate or alanine the primary CO_2 acceptor is regenerated in the mesophyll. Alanine is first transaminated to pyruvate before the key enzyme of the C_4 photosynthesis the pyruvate-P_i-dikinase converts pyruvate to PEP. This reaction

$$\text{Pyr} + \text{ATP} + \text{P}_i \longrightarrow \text{PEP} + \text{AMP} + \text{PP}_i \tag{4.56}$$

is an energy consuming process. In comparison to the C_3 path, the C_4 path requires two additional ATPs for the fixation and reduction of CO_2. One of these ATPs is used to form PEP and the other is used in the reaction mediated by the adenylate kinase which converts the AMP produced in the above reaction to ADP

$$\text{AMP} + \text{ATP} \rightleftharpoons 2\ \text{ADP} \tag{4.57}$$

The PEPC is not sensitive to O_2 and the mesophyll cells do not possess rubisco. Consequently, there is no photorespiration in the mesophyll. Photorespiration can only occur in the bundle sheath.

As long as these processes (primary CO_2 fixation in the mesophyll, transport of the C_4 compound to and CO_2 liberation in the bundle sheath cells, retransport of the C_3 compound to and regeneration of the primary CO_2 acceptor PEP in the mesophyll) run at a higher rate than the secondary CO_2 fixation via rubisco, CO_2 accumulates in the bundle sheath. The elevated CO_2 concentration increases the ratio CO_2:O_2 in the bundle sheath. This in turn favours the carboxylating activity of the rubisco and inhibits photorespiration. For a number of C_4 species and in detail for *Zea mays* it has been shown that only rubisco and pyruvate-P_i-dikinase are rate-limiting enzymes for the CO_2 fixation. The activities of all the other enzymes involved in C_4 photosynthesis are far in excess of that required for the observed rate of CO_2 fixation. On the other hand only pyruvate-P_i-dikinase, responsible for the regeneration of the primary CO_2 acceptor, and NADP-dependent MDH, responsible for the reduction of the primary CO_2 fixation product OAA in the mesophyll, show an increase in activity with increasing light intensity. This activation proceeds more rapidly than photosynthesis (Usada *et al.* 1984, 1985). Consequently, the postulated requirements for an efficient CO_2 fixation in the bundle sheath are fulfilled.

C_4 photosynthesis is characterized by a potent primary CO_2 fixation via PEPC in the mesophyll. Subsequent transport and decarboxylation in the bundle sheath cells provide a concentrating mechanism at the site of the rubisco, thereby inhibiting photorespiration. Under atmospheric levels of CO_2 this concentrating mechanism leads to an approximately ten-fold higher CO_2 concentration in the bundle sheath cells than in photosynthesizing cells of C_3 plants where a CO_2 concentration of about 7 μmol prevails.

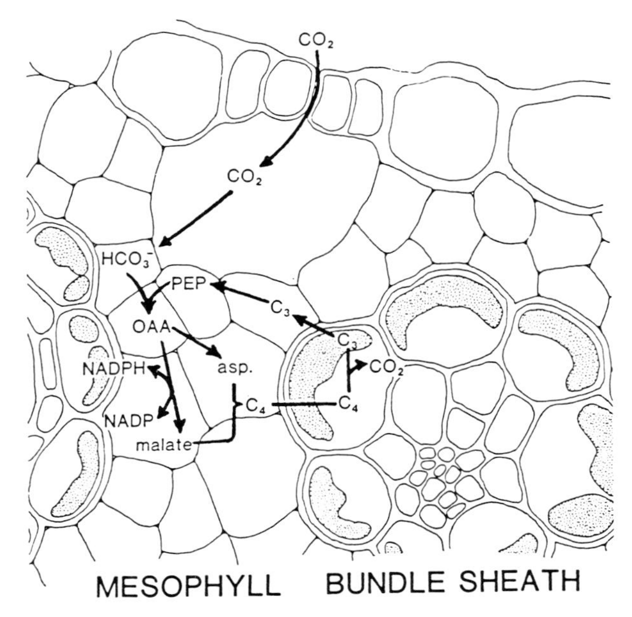

Fig. 4.75 Cross-section through a leaf of a C_4 plant showing the mesophyll, the bundle sheath and the epidermis with a stoma. The route of CO_2 from the atmosphere via the mesophyll to the bundle sheath and all reactions involved in this path are also given.

Under conditions of strong light and high temperatures this results in a higher primary production from the C_4 path than from C_3 photosynthesis.

The PEPC has only a slightly higher affinity for CO_2 than the rubisco but its reaction velocity is much higher especially at low CO_2 concentrations. The higher carboxylation capacity of the PEPC means that even at limited CO_2, when the stomata are relatively closed, the mesophyll cells fix enough CO_2 to maintain photosynthesis in the bundle sheath. Due to the large carboxylation potential the CO_2-compensation point of C_4 plants lies between 0 and 5 ppm while that of C_3 plants ranges from 25 to 50 ppm. The ability of the C_4 plants to fix CO_2 sufficiently even with only slightly opened stomata and the possibility of overcoming photorespiration result in a high water use efficiency. For the formation of 1 g dry matter C_4 plants utilize 200 to 400 g of water. With regard to the CO_2 budget C_4 plants improve it by reducing photorespiration via a CO_2-concentrating mechanism at the site of the rubisco. A scheme of the path of carbon in C_4 photosynthesis when malate is the transported C_4 compound is shown in Fig. 4.75.

4.3.2.3 CAM plants

The path of carbon in CAM plants is not as easy to describe as it is with C_3 and C_4 plants, which fix CO_2 during the light period and release CO_2 during the night. Although the ideal CAM plant does not exist, a general feature is that under the conditions defined in Section 4.3.1 CAM plants shift part of their CO_2 uptake from the light period into the dark. This results in a completely different CO_2 exchange curve. A comparison of the CO_2 gas exchange of a CAM plant with that of a C_3 or C_4 plant is given in Fig. 4.76.

At the beginning of the night there is usually a clear net CO_2 release indicating mitochondrial respiration. This CO_2 release is then progressively diminished and converted into a net CO_2 uptake which increases, runs through an optimum and finally decreases slightly at the end of the dark period. The onset of light causes a gulp of net CO_2 uptake which is followed by a rapid decline. Around noon CAM plants operate at the compensation point or slightly above or below it. Towards the end of the light period a net CO_2 uptake may occur again.

Measurements of the O_2 uptake of CAM plants during the night revealed that there is a fairly constant O_2 consumption throughout the night indicating that the TCA cycle and mitochondrial electron transport chain function at a constant rate throughout the night (Brunnhöfer *et al.* 1968a, b, Andre *et al.* 1979). Thus the true CO_2 fixation in the dark is the sum of net CO_2 uptake and respiratory CO_2 production. The latter is equivalent to O_2 consumption.

Dark fixation of CO_2 is mediated by the PEPC and results in the formation of OAA as a first product which is then (via MDH) reduced to malic acid, but citric acid, isocitric acid, aspartic acid and glutamic acid can also be stable end-products which are accumulated in the cell. Since the free acids are synthesized the pH value of the cell sap drops considerably overnight. With the onset of light a second CO_2-fixing system, the rubisco, is switched on, and since the PEPC continues to operate both systems work on CO_2 fixation. In a malic acid and light mediated reaction the activity of the PEPC is continuously diminished (von Willert and von Willert 1979) presumably by changing the enzyme configuration. The K_m for PEP and the inhibition of the PEPC caused by malic acid increase substantially (von Willert *et al.* 1979a).

By feeding $^{14}CO_2$ during the early morning CO_2 gulp it was demonstrated that the CO_2 fixed via the PEPC path declines steadily while concomitantly the CO_2 running through the CALVIN-cycle increases until

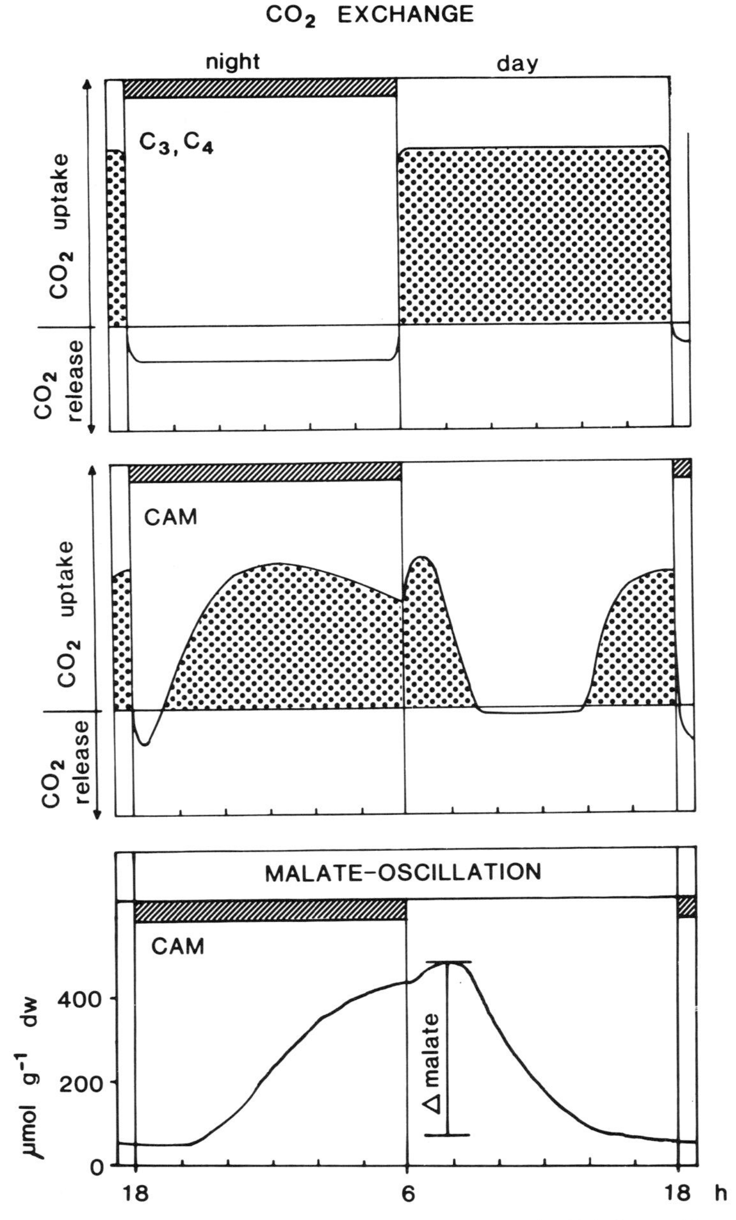

Fig. 4.76 Idealistic CO_2 gas exchange curves of C_3, C_4 and CAM plants with a photoperiod of 12 to 12 hours. Other conditions should be: night temperature at 15 °C, day temperature at 25 °C; VPD at night below 1 and during the day at about 20 mPa Pa^{-1}; photon flux density at a value allowing for optimum photosynthesis; plants sufficiently watered. The lowest graph shows the diurnal change in malic acid content in the leaves of a CAM succulent.

it reaches 100%. Consequently, the malic acid formation does not stop upon the onset of light but continues during the morning peak of CO_2 uptake as shown in Figs. 4.55 and 4.76. Only when the net CO_2 uptake in the morning declines substantially and the rubisco is the main or the only CO_2-fixing enzyme does the decarboxylation of malic acid start. The degradation of malic acid ends when all available malic acid is consumed. It is only then that net CO_2 uptake in the light occurs again. Fig. 4.76 shows this relation between CO_2 gas exchange and leaf malic acid content.

Malic acid can be decarboxylated directly by NADP malic enzyme in the following reaction

$$\text{malic acid} + NADP^+ \longrightarrow \text{pyruvic acid} + CO_2 + NADPH + H^+ \quad (4.58)$$

or is alternatively first oxidized to OAA and subsequently decarboxylated by the PEP carboxykinase reaction

$$OAA + ATP \longrightarrow PEP + CO_2 + ADP \quad (4.59)$$

The released CO_2 can be refixed by the rubisco. When the malic acid consumption has stopped, the formerly closed stomata open again and atmospheric CO_2 is taken up. While in the C_4 photosynthesis the C_3 skeleton resulting from the decarboxylation of malic acid in the bundle sheath is transported back to the mesophyll for regeneration of the CO_2 acceptor, the fate of the C_3 skeleton after decarboxylation of malic acid in CAM plants is not completely resolved. One assumption is that pyruvate is phosphorylated to PEP which then enters neoglucogenesis. Another assumption is that pyruvate is completely oxidized to CO_2 and water. As there is evidence for both assumptions we suggest that especially in the natural habitat environmental factors decide which path is followed.

From a theoretical point of view CAM plants realize all three mentioned possibilities to improve their CO_2 budget.

(1) They refix their respiratory CO_2 during at least parts of the night.
(2) Although photorespiration potentially exists, it is unlikely to run during the phase of malic acid degradation. Decarboxylation provides a CO_2:O_2 ratio at the site of the rubisco which favours the carboxylation activity. During decarboxylation endogenous CO_2 concentrations of up to 3% have been measured, which is a hundred-fold higher CO_2 concentration than in the ambient air. Thus photorespiration should only occur when decarboxylation of malic acid has ceased.
(3) If we assume a complete oxidation of malic acid to CO_2 and water in the mitochondria, pyruvate would provide a substrate for the

mitochondrial respiration in the light that is a leftover from the preceding night and that does not derive from simultaneously running photosynthesis. In that case CAM plants would not only shift one part of their CO_2 uptake from the light period into the night but would simultaneously shift one part of the glucan degradation which stopped in the night at the level of PEP, into the following light period.

The description of CAM given so far represents the prevailing textbook opinion and can be summarized in at least five characteristics of CAM plants:

(1) Net CO_2 uptake in the night (we will come back to this point later when we try to define CAM in terms of our experience);
(2) diurnal oscillation in acids, usually malic acid but citric acid or isocitric acid can also be main oscillators;
(3) diurnal oscillations in reserve carbohydrates in antiphase to the oscillations of the acids;
(4) an inverse rhythm of stomatal opening;
(5) a large vacuole to store the acids.

Stomata are always open when CO_2 is taken up, e.g. in parts of the night, in the early morning and late afternoon. Around noon stomata are closed. This means that stomata are only open when the water vapour gradient from the leaf to the atmosphere (ΔW) is low. Consequently, CO_2 uptake of CAM plants is connected with a low water loss. This results in a water use efficiency which is even better than that of C_4 plants. For the formation of 1 g dry matter CAM plants utilize only 50 to 100 g of water.

4.3.2.4 ***Intermediates***

The classical subdivision of the plants into C_3, C_4 and CAM plants is an artificial though well established and confirmed concept. Nevertheless, there exist numerous exceptions, or should we better say intermediates, which exhibit features of more than one strict pathway. The intermediates known so far belong either to C_3–C_4 intermediates or – and this group is of greater interest for us – to succulent species which shift from C_3 to CAM. A recent review lists 23 species from seven genera which belong to the C_3–C_4 intermediates (Monson 1989). They all exhibit reduced levels of photorespiration which result from a refixation of photorespired CO_2 rather than from a CO_2-concentrating mechanism. Members of the C_3–C_4 intermediates are mostly weedy species and grow predominantly in disturbed habitats. Since succulents have not been reported in this group we will not go into further details. Of much greater importance for us are those plants that mediate between C_3 and CAM.

The first report that a succulent undergoes transition from C_3 to CAM concerned the halophilic ephemeral *Mesembryanthemum crystallinum* (Winter and von Willert 1972). Among even-aged plants salt-treated specimens showed all features of CAM whereas untreated plants continued to perform a 'normal' C_3 photosynthesis. Since then much attention has been devoted to characterize further this reaction which was interpreted at first as induction of CAM. It could be shown that even untreated *Mesembryanthemum crystallinum* plants slowly shift from C_3 to CAM and that salt treatment is not necessarily required for this transition (von Willert and Kramer 1972). It further became evident that this shift also occurred in a natural habitat on the coast of Israel where seedlings start with C_3 and CAM becomes more and more established with plant age (Winter *et al.* 1978). Since plant age, thermoperiod, photoperiod, increasing sodium chloride and decreasing soil water availability all changed together, no discrimination between the various environmental factors and their contribution to the shift from C_3 to CAM was possible in this investigation.

The transition from C_3 to CAM is accompanied by a substantial increase in extractable PEPC activity which has been shown to result from an augmentation of the enzyme protein (von Willert *et al.* 1976, 1977). This was recently confirmed by Ostrem *et al.* (1987) who demonstrated an increase of the level of translatable mRNA for PEPC upon salt treatment of *Mesembryanthemum crystallinum*. The same authors (Vernon *et al.* 1988) also confirmed earlier findings, that retransfer of salt-stressed *Mesembryanthemum crystallinum* plants to non-saline conditions leads to a rapid decline of PEPC activity and nocturnal acid accumulation (von Willert *et al.* 1977) by showing that the mRNA level also decreased by 77% within 2.5 hours after salt removal.

The results suggest a causal relationship between PEPC activity and overnight acid formation. In fact, a plot of the overnight accumulated malic acid over the actual activity of the PEPC reveals a good correlation (Fig. 4.77). Nevertheless, two facts must be mentioned which are not in full agreement with this assumption. Firstly, the increase in PEPC activity precedes the onset of malic acid accumulation by days. This gap in responses to salt is greatest when seedlings are subjected to salt stress and gradually decreases the older the plants are which are treated with salt. Secondly, stress relief results in a sudden and dramatic decline in PEPC activity without markedly influencing nocturnal acid formation (Winter 1973, Winter and Lüttge 1979). This suggests that the elevated PEPC level is not really necessary for a better CAM performance.

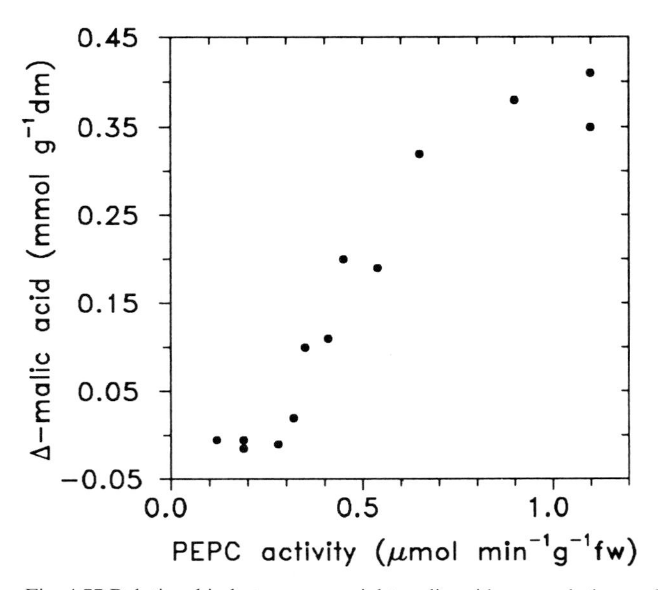

Fig. 4.77 Relationship between overnight malic acid accumulation and the corresponding activity of the CO_2-fixing enzyme PEPC during the shift from C_3 to CAM in NaCl-treated plants of *Mesembryanthemum crystallinum*. (Data taken from M. Herppich, unpublished.)

Though many details are known about the phenomenology of the transition from C_3 to CAM exhibited by *Mesembryanthemum crystallinum*, the regulation and sequence of steps of this reaction is still not completely understood. The original idea was that the *Mesembryanthemum crystallinum* plants have to reach a special ontogenetic stage before they become susceptible to the stress-triggered transition from C_3 to CAM. A very carefully performed survey with many replications clearly demonstrated that this concept is wrong (Herppich, unpublished). In a series of experiments the appearance of CAM was followed from seed germination to plant death by subjecting the plants to salt at different stages of their life cycle. A comparison of the development of CAM in non-stressed *Mesembryanthemum crystallinum* plants with that in plants which were treated with NaCl solutions after germination showed that CAM appears earlier in the treated plants but revealed also that even the cotyledons are able to shift from C_3 to CAM (Fig. 4.78). The figure also makes clear that untreated plants grow faster and that they exhibit CAM rather late in their life cycle. It further becomes evident that even in the salt-treated plants the respective youngest leaf always starts with a normal C_3 photosynthesis. Consequently, the transition from C_3 to CAM is correlated with maturation but is

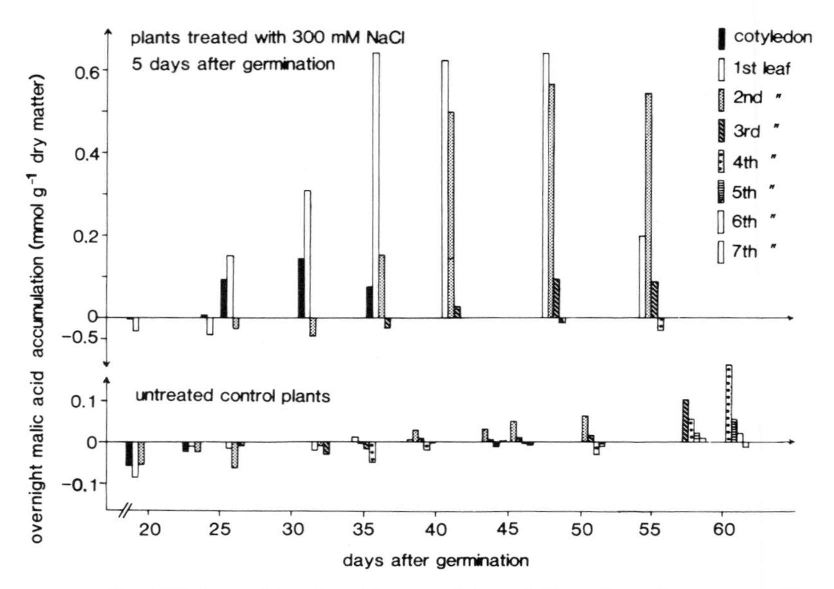

Fig. 4.78 Overnight malic acid accumulation of *Mesembryanthemum crystallinum* plants that were either irrigated with a 300 mM NaCl solution five days after germination or left untreated. Plants were harvested at different intervals and overnight malic acid accumulation in the different leaves was separately determined. Note that already in the cotyledons a considerable overnight malic acid accumulation occurs and that even in the treated plants emerging leaves do not perform a CAM. (Data taken from M. Herppich, unpublished).

independent of the ontogenetic stage of the plant. These findings force us to reconsider and re-investigate some allegedly solved questions in connection with the metabolic transition in *Mesembryanthemum crystallinum*.

The fact that young leaves of rapidly growing CAM plants do not exhibit CAM but perform C_3 photosynthesis is a well-established feature of many CAM succulents. At present, we will only give one representative example of this feature. Since overnight acid accumulation is the easiest measure for CAM, we will show how this parameter varies with leaf age in a CAM succulent of the genus *Plectranthus* (Fig. 4.79).

Another CAM induction is shown in *Kalanchoe blossfeldiana* where the photoperiod controls CAM (Brulfert *et al.* 1973). This reaction draws a link to a further prominent succulent for which a C_3–CAM shift is well established, *Portulacaria afra*. For this endemic succulent of southern Africa photoperiod, drought and leaf age trigger the expression of CAM. When sufficiently supplied with water *Portulacaria afra* behaves like a C_3 plant but then has already high though non-fluctuating levels of malic acid.

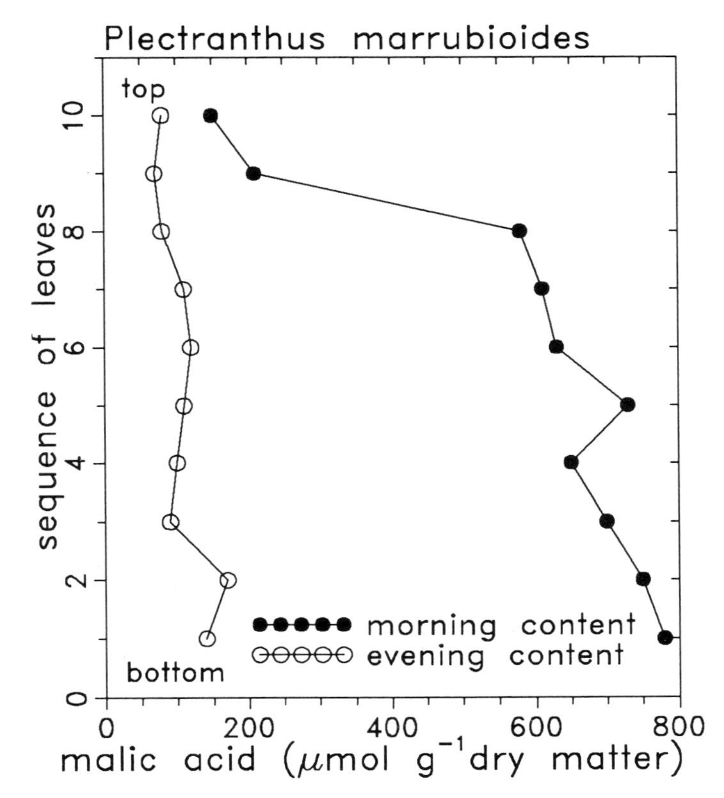

Fig. 4.79 Malic acid content in leaves of *Plectranthus marrubioides* in the evening and in the morning. Leaves were numbered from bottom to top. Overnight acid accumulation increases with leaf maturation. The youngest leaves show only a slight accumulation of malic acid.

With increasing drought, daylength and leaf age CAM gradually develops (Ting and Hanscom 1977, Guralnick *et al.* 1984a, b).

We assume that the *Portulacaria* type of reaction is widely distributed among succulents in southern Africa. We will give some further examples in Section 4.3.4.3. The possible ecological significance might be seen in a limitation of carbon and water losses which contribute to the longevity of the leaf.

Summarizing this Section we have to conclude that under certain environmental conditions which have to be determined from case to case even C_3 plants are capable of a nocturnal acid production and reversible accumulation. To our present knowledge, however, this ability is restricted to C_3 succulents.

4.3.2.5 Comparative aspects

According to their mode of CO_2 fixation, C_3, C_4 and CAM plants can be precisely defined and classified into distinct groups. The differences and similarities of the three modes of CO_2 fixation are summarized in Fig. 4.80.

An often used means to distinguish between these groups is by the so-called $\delta^{13}C$ value. The basis for this $\delta^{13}C$ value is the isotopic composition of the CO_2 in the atmosphere. Of the total carbon atoms of the atmospheric CO_2 98.9% contain the ^{12}C but 1.1% the ^{13}C isotope.

We have seen that there exist two enzymes for the primary fixation of CO_2 in plants, rubisco and PEPC. Of these two enzymes only rubisco discriminates against the heavier isotope ^{13}C. Consequently, C_4 plants and CAM plants contain more ^{13}C in their dry matter than C_3 plants. By comparing the ratio of ^{13}C:^{12}C in the plant material with the same ratio in a standard one gets the so-called $\delta^{13}C$ value which is given by the relation

$$\delta^{13}C\ (‰) = \left(\frac{^{13}C/^{12}C \text{ in the plant}}{^{13}C/^{12}C \text{ in a standard}} - 1\right) \times 10^{-3} \tag{4.60}$$

In an *in vitro* test of the two enzymes one gets

Enzyme	$\delta^{13}C$ value	Test condition
PEPC	−3.0‰	at 25 °C and pH 8.5
rubisco	−33.7‰	at 24 °C and pH 8.2
rubisco	−18.3‰	at 37 °C and pH 8.2

The different discrimination of the two enzymes leads to the following range of $\delta^{13}C$ values for the different plant groups:

C_3 plants	−23 to −34‰
C_4 plants	−10 to −18‰
CAM plants	−14 to −33‰

Though the method of determining the $\delta^{13}C$ values is rather expensive and time consuming it is a fantastic tool for a screening of the vegetation. One must, however, be cautious with final statements based only on the $\delta^{13}C$ values of field collected samples, as they can be misleading. Judging from such $\delta^{13}C$ values *Welwitschia mirabilis* was classified a CAM plant (Schulze *et al.* 1976), but detailed investigations in the Namib desert demonstrated the complete absence of any CAM feature in *Welwitschia mirabilis* despite a $\delta^{13}C$ value of −19.6‰ (von Willert *et al.* 1982).

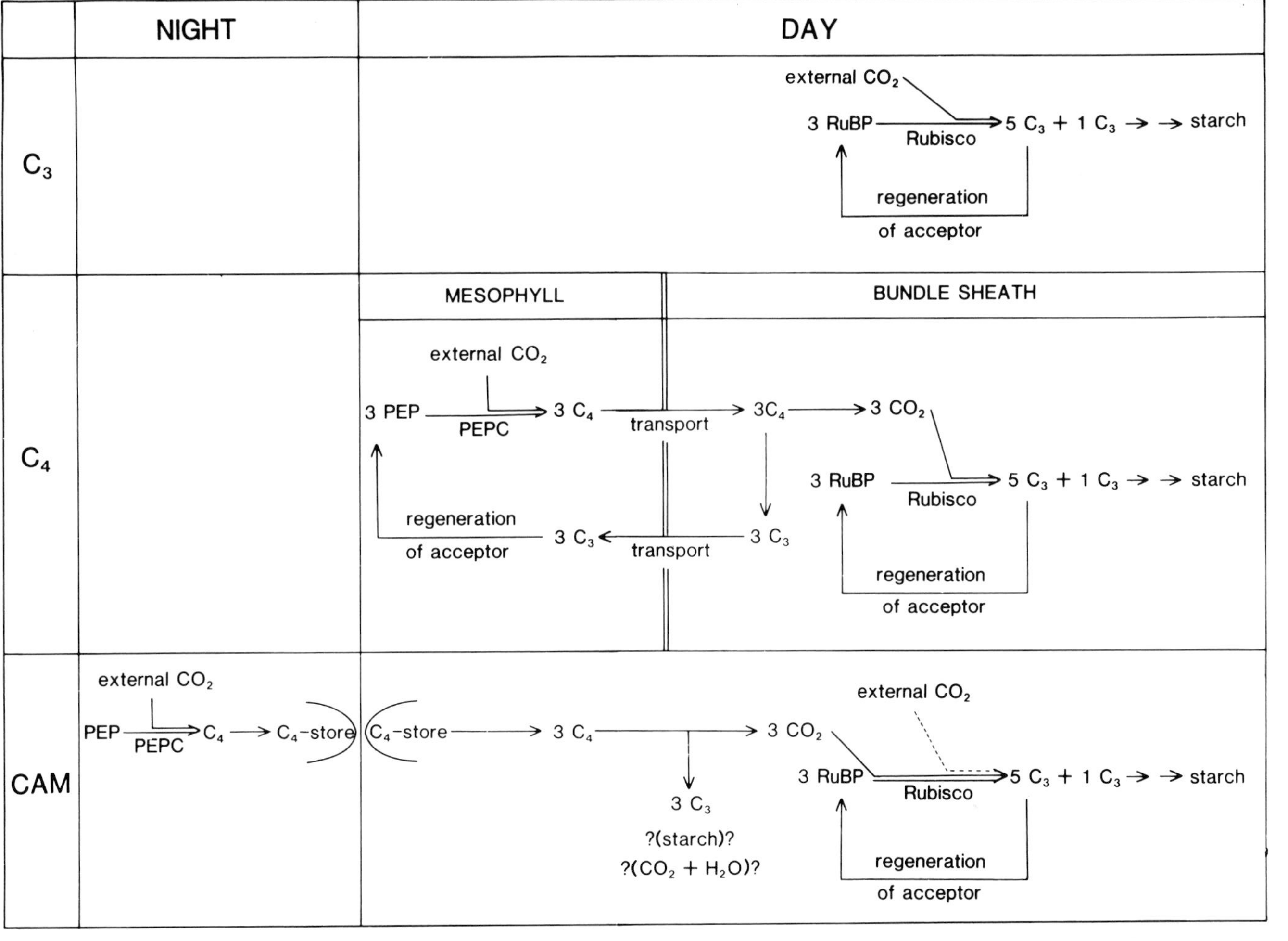

Fig. 4.80 A scheme of the path of carbon in C_3, C_4 and CAM plants.

Despite these clear-cut differences between the three plant groups their responses to environmental parameters do not reflect such a strict classification. This was not to be expected since members of all three groups grow in areas with a wide ecological amplitude. From what is known to date, C_3 plants occur in all climate and vegetation zones and show the highest plasticity and adaptability, whereas C_4 and CAM plants occur in a more restricted fashion and never inhabit an area exclusively. C_4 plants predominate in the subtropics with summer rainfall but there are prominent examples even in the cooler temperate zones (*Spartina*). CAM plants occur in tropical rain forests as well as in subtropical deserts or even submerged in lakes of the temperate zones. Even the response of photosynthesis to photon flux densities is no longer a criterion allowing us to distinguish between C_3 and C_4 plants since it has been shown that light acclimatization and not the mode of photosynthesis determines light saturation of photosynthesis (Osmond *et al.* 1980, Pearcy and Ehleringer 1984). The same is ruled out for the relation between photosynthesis and temperature.

We will not go into more detail, as it was our aim only to illustrate that the path of carbon in photosynthesis is not strictly linked with adaptational values and consequently does not allow any prediction about the ecological significance *per se* even if the arguments sound reasonable.

If we consider only the photosynthetic efficiency and productivity of a plant, then photorespiration appears to be a waste. Of course, even in arid areas with their unpredictable rainfall events productivity is crucial but of much greater importance is the ability to survive rainless periods in a way that allows maximum yield as soon as water becomes available again. As long as this is connected with maintaining the assimilating organs (evergreen) care must be taken of the photosynthetic apparatus which should not be damaged. Is photorespiration a useful tool to achieve this?
Some aspects might throw light on this question.

(1) Photorespiration is in fact a mechanism preventing photo-oxidation of the photosynthetic apparatus as it counteracts an overflow of the electron transport chain when CO_2 becomes rate limiting. The presence of low oxygen (2%) stops photorespiration of C_3 plants and damages them within a short time. This was also found for CAM succulents (Thomas, unpublished).
(2) Photorespiration should be most active at limited CO_2 supply. Such conditions prevail in drought periods during the day when stomata are closed and the plant operates at the CO_2 compensation point.
(3) During such periods a net growth is impossible and consequently a

photosynthesis with an efficiency of zero or worse can be tolerated as long as the photosynthetic apparatus remains functioning.

(4) Consequently, plants naturally suppressing photorespiration should suffer more from drought stress than plants with a free-running photorespiration. Perennial evergreen plants in extreme deserts are predominantly or exclusively C_3 plants. They can but do not have to be succulents. If they are CAM plants they generally cease any net CO_2 exchange and sometimes also stop refixation of respiratory CO_2. Since the photosynthetic apparatus of such plants (cf. Fig. 4.49) is not damaged, as has been shown by von Willert and Brinckmann (1986), photorespiration must have helped in keeping them alive. Perennial C_4 plants with evergreen succulent leaves, if they exist at all, seem to avoid extreme deserts. We assume that the two strategies, increase of plant productivity by using the C_4 path and survival in extreme arid areas with evergreen succulent leaves, exclude each other. This would mean that diminishing photorespiration cuts down the ecological plasticity and the ability to establish successfully and survive in extreme dry habitats.

Among succulents the most common photosynthetic pathways are the CAM and C_3 type whereas the C_4 type is restricted to some halophilic succulents and a few annuals such as *Zygophyllum simplex* which occur in dry areas with summer rainfall.

4.3.3 Carbon investments for succulence

Succulence, though obviously beneficial for plants under certain circumstances, has its costs and consequently disadvantages. We have seen in Chapter 1 that heavy leaves of succulents require a strong shoot. That means a high proportion of the carbon gain is used to build up structural elements serving mechanical strength. Furthermore, the often unfavourable ratio of assimilating tissue (chlorenchyma) to non-green tissue exhibited by many succulent organs means a high proportion of maintenance respiration which also diminishes carbon gain. This explains the characteristically slow growth of most succulents even under favourable conditions.

From an anatomical point of view we were able to distinguish between succulents of the all-cell succulent type and those of the partially succulent type. We have also documented that with regard to the fraction of utilizable water the groups differ significantly from each other. The all-cell water storage is more or less a one-way storage while the separate central storage can be emptied and refilled several times. The different functions imply different expenditures to build up the necessary structures. All-cell succulent leaves are normally characterized by thin cell walls, especially the outer cell walls of the epidermal tissue. Consequently, the investment of carbon for storing a unit of water is much less in this group compared to the partially succulent type. Water exchange with the atmosphere is normally higher for all-cell succulent leaves and the whole water capacity of the plant depends on the number of leaves. A vigorous growth during favourable conditions is a valuable element of the life strategy of these succulents.

From detailed measurements of fresh weight, dry weight, ash content, total leaf surface and caloric values of the dry matter we were able to calculate the energy in kJ that had been expended to envelop one millilitre of water either by leaf or by stem tissue. Results for the energy input for a number of succulent and non-succulent species growing in the Numees area of the Richtersveld are presented in Table 4.5. The respective ash content of the dry matter is also given in this table. From the data several features can be deduced.

The most interesting feature is that succulents with a high ash content expend less energy and hence carbon to envelop one unit of water than succulents with a low salt content. This impressively supports the statement made in a different context in Section 4.2 where we briefly discussed the ecological advantages of the high salt content and salt accumulation of succulent Mesembryanthemaceae. Salt facilitates uptake and storage of water. Due to the fact that less carbon is used to build up the storage tissue, growth of these succulents is rather vigorous compared with that of the low salt succulents. This feature, however, is not restricted to the succulents, either leaf or stem succulents, but is also exhibited by the non-succulent species as can be seen from Table 4.5.

The correlation is a bit disturbed by mixing plants from different families and with different growth forms, but becomes clearer if we, for instance, concentrate on the leaf succulent Mesembryanthemaceae. Table 4.6 gives the energy investment and the percentage ash for eight leaf succulent Mesembryanthemaceae listed according to increasing energy expenditure. The ash content of the dry matter exactly follows this sequence but in the reverse direction. Table 4.6 also shows a sequence of the same eight Mesembryanthemaceae but this time the listing follows the decreasing values in the last column. This column represents a newly defined succulence quotient (von Willert, unpublished) which was obtained by dividing the degree of succulence by the degree of sclerophylly and correcting the latter for the percentage of ash (= degree of sclerophylly°). Hence this quotient, which we term the succulence quotient, gives the amount of water that can be stored by the expenditure of one gram of organic matter.

Table 4.5. *Energy investment based on ash-free dry matter necessary to envelop one millilitre of water in leaves or stems of 39 different plant species of the Numees area (Richtersveld). The figures in parentheses give the percentage ash of the dry matter. Plant species are listed according to increasing energy investments and are grouped into succulents and non-succulent species*

	Energy investment in kJ for storing 1 ml water in tissues of		
	Succulents		Non-succulents
Plant species	Leaves	Stems	Leaves
Opophytum aquosum	0.12 (59)		
Mesembryanthemum barklyi	0.18 (57)		
Mesembryanthemum pellitum	0.24 (54)		
Prenia sladeniana	0.38 (44)		
Osteospermum microcarpum	0.45 (21)		
Mitrophyllum clivorum	0.58 (31)		
Zygophyllum longicapsulare	0.83 (38)		
Trianthema triquetra	0.89 (13)		
Othonna opima	0.98 (23)		
Psilocaulon subnodosum	1.05 (25)		
Brownanthus schlichtianus		1.08 (35)	
Zygophyllum prismatocarpum	1.17 (24)		
Didelta carnosa	1.18 (31)		
Senecio longiflorus		1.34 (14)	
Senecio corymbiferus	1.37 (13)		
Cotyledon orbiculata	1.52 (7)		
Aloe ramossissima	1.57 (12)		
Ceraria namaquensis	1.57 (12)		
Crassula deceptor	1.60 (8)		
Delosperma pergamentaceum	1.77 (26)		
Aloe pearsonii	1.82 (8)		
Acanthopsis disperma			1.83 (21)
Conophytum aequale	1.91 (14)		
Crassula macowaniana	2.07 (4)		
Ceraria fruticulosa	2.20 (6)		
Nicotiana glauca			2.38 (14)
Ceraria pygmaea	2.53 (11)		
Monechma mollissimum			2.64 (31)
Trichodesma africanum			2.99 (27)
Codon royenii			3.08 (21)
Euphorbia peltigera		3.64 (10)	
Forsskaolea candida			4.01 (30)
Pachypodium namaquanum	4.39 (13)		
Euphorbia dregeana		4.55 (5)	
Ozoroa dispar			4.99 (7)
Euphorbia gummifera		7.42 (4)	
Hermbstaedtia glauca			7.99 (6)
Maytenus cf. *tenuifolia*			13.43 (4)
Euclea pseudebenus			20.11 (4)

$$\text{succulence quotient} = \frac{\text{degree of succulence}}{\text{degree of sclerophylly}^{\circ}}$$

$$= \frac{\text{water in g at full hydration/leaf area in dm}^2}{\text{(dry matter} - \text{ash) in g/leaf area in dm}^2} \qquad (4.61)$$

With the same costs *Opophytum aquosum* can envelop ten times more water than *Delosperma pergamentaceum*. That means, *Opophytum aquosum* is a fast-growing succulent when compared with *Delosperma pergamentaceum*. This exactly reflects the field situation and the life forms of the two species. *Opophytum aquosum* is an annual CAM succulent while *Delosperma pergamentaceum* is a perennial CAM succulent. The sequence of plants when listed according to this new quotient is as expected nearly the same as when listed according to the energy expenditure to envelop 1 ml of water. This means that it is not necessary to determine the caloric values of the dry matter in order to calculate the investment of carbon for the storage of one unit of water. The succulence quotient is much easier to obtain and is highly informative. Neither the degree of succulence nor the degree of sclerophylly *per se*, which are also given in Table 4.6, proved to be valuable figures which can be used in ecophysiology. This is illustrated on a larger scale for 34 different succulent and non-succulent species from the Numees area in the Richtersveld including the Mesembryanthemaceae in Table 4.7. Species are ranked according to the new succulence quotient expressing how much water can be stored at the expense of a unit of organic matter. A comparison with Table 4.5 reveals that the sequence of species in both tables is fairly similar.

A second generalization can be made from Table 4.5. Expenditure increases with longevity of the water storing organ. The first eight species are either true annuals or seasonal leaf shedders or have at least rather short-living leaves. The same holds true for the stem parts of *Brownanthus schlichtianus*, as has been shown in Fig. 4.72. From all stem succulent forms *Brownanthus* shows the lowest investment for water storage, whereas the rather xeromorphic long-living *Euphorbia gummifera* has a seven-fold higher expenditure. This is even more pronounced if we compare *Opophytum aquosum* with the proper succulent leaf of *Aloe pearsonii*. Here the factor is as high as 15. The same statement as for the succulents is also valid for the non-succulent species. Here, too, the expenditure increases with longevity of the leaf but starts off at values significantly higher than for succulents.

The energy expenditure or organic matter input to envelop water is the basis for further calculations that throw light on the growth rates of succulents in comparison with non-succulent species.

Table 4.6. *(a) Energy investment necessary to envelop one millilitre of water in leaves of eight species of leaf succulent Mesembryanthemaceae growing in the Numees area (Richtersveld). Figures in parentheses represent the ash content in percentage of dry matter*

Plant species	kJ ml^{-1} H_2O	% ash
Opophytum aquosum	0.12	(59)
Mesembryanthemum barklyi	0.18	(57)
Mesembryanthemum pellitum	0.24	(54)
Prenia sladeniana	0.38	(44)
Mitrophyllum clivorum	0.58	(31)
Psilocaulon subnodosum	1.05	(25)
Delosperma pergamentaceum	1.77	(26)
Conophytum aequale	1.91	(14)

Table 4.6. *(b) Water content in percentage of fresh weight, degree of succulence, degree of sclerophylly and the amount of water that can be enveloped at the expense of one gram of biomass (organic dry matter) for the same eight species as in (a). Plants are listed according to decreasing values in the last column. Note that the order of species in both listings is nearly the same*

Plant species	% water	Degree of succulence	Degree of sclerophylly	g water per g org. mat.
Opophytum aquosum	96.6	13.149	0.456	69.02
Mesembryanthemum pellitum	94.9	19.269	1.032	43.49
Mesembryanthemum barklyi	93.6	20.931	1.428	31.96
Prenia sladeniana	92.1	13.645	0.798	30.54
Mitrophyllum clivorum	93.0	12.841	0.967	19.28
Psilocaulon subnodosum	90.7	7.296	0.752	12.93
Conophytum aequale	85.6	15.619	2.627	6.86
Delosperma pergamentaceum	82.0	13.685	3.004	6.22

We want to compare the C_3 succulent *Othonna opima* with the C_3 sclerophyllous tree *Ozoroa dispar*. A mature leaf of *Othonna opima* weighs about 16 g and contains about 14.7 ml water. According to Table 4.7 this is exactly the water volume which is enveloped at the expense of 1 g of organic dry matter. A mature leaf of *Ozoroa dispar* weighs only about 0.3 g and contains about 0.23 ml of water. That means, with 1 g of organic dry matter *Ozoroa dispar* can grow about 16 leaves whereas *Othonna opima* is only able to grow one leaf. Table 4.8 summarizes these results. The surface of one round leaf of *Othonna opima* is about 0.56×10^{-2} m^2 while the 16 leaves of

Table 4.7. *Water content in percentage fresh weight, degree of succulence, degree of sclerophylly, and the amount of water that can be enveloped at the expense of one gram biomass (organic dry matter) for 34 succulent and non-succulent plant species in the Numees area (Richtersveld). Plants are listed according to decreasing values in the last column. Note that neither the degree of succulence nor the degree of sclerophylly correspond with the sequence of plants, whereas the water content in percentage of fresh weight agrees rather well*

Plant species	% water	Degree of succulence	Degree of sclerophylly	g water per g org. mat.
Opophytum aquosum	96.6	13.149	0.456	69.02
Mesembryanthemum pellitum	94.9	19.269	1.032	43.49
Mesembryanthemum barklyi	93.6	20.931	1.428	31.96
Prenia sladeniana	92.1	13.645	0.798	30.54
Mitrophyllum clivorum	93.0	12.841	0.967	19.28
Trianthema triquetra	93.3	10.160	0.726	16.87
Othonna opima	91.9	9.832	0.869	14.68
Senecio corymbiferus	92.0	10.928	0.942	13.33
Zygophyllum longicapsulare	89.2	8.438	1.032	13.23
Psilocaulon subnodosum	90.7	7.296	0.752	12.93
Brownanthus schlichtianus	87.9	7.967	1.099	11.08
Ceraria namaquensis	90.2	4.149	0.453	10.34
Osteospermum microcarpum	89.1	3.270	0.402	10.25
Aloe pearsonii	90.2	16.899	1.849	10.00
Didelta carnosa	86.0	2.782	0.453	9.28
Aloe ramosissima	88.5	18.371	2.378	8.75
Ceraria fruticulosa	88.0	3.240	0.442	7.77
Zygophyllum prismatocarpum	85.4	7.497	1.280	7.76
Conophytum aequale	85.6	15.619	2.627	6.86
Acanthopsis disperma	83.9	4.983	0.958	6.67
Nicotiana glauca	85.0	3.303	0.584	6.63
Ceraria pygmea	85.4	14.507	2.493	6.52
Delosperma pergamentaceum	82.0	13.685	3.004	6.22
Crassula macowaniana	85.5	6.459	1.098	6.04
Crassula deceptor	82.9	5.342	1.103	5.22
Monechma mollissimum	77.0	3.120	0.932	4.78
Codon royenii	77.2	6.608	1.950	4.30
Trichodesma africanum	75.3	3.296	1.077	4.27
Pachypodium namaquanum	77.6	3.087	0.893	4.02
Ozoroa dispar	77.6	1.791	0.517	3.76
Forsskaolea candida	67.5	2.014	0.971	3.00
Hermbstaedtia glauca	70.4	2.045	0.858	2.55
Maytenus cf. *tenuifolia*	57.2	1.871	1.401	1.35
Euclea pseudebenus	51.0	1.212	1.167	1.04

Ozoroa dispar have a projected leaf surface area of about 2.55×10^{-2} m^2. That is, *Ozoroa dispar* has an assimilating leaf surface area 4.5 times greater than the corresponding leaf area of *Othonna opima*. Assuming a similar rate of carbon assimilation for both species would result in a 4.5 times higher carbon gain for *Ozoroa dispar*. Consequently, the acceleration of growth and hence the growth rate itself is much faster in *Ozoroa dispar* than in *Othonna opima*.

The same considerations can be made for two succulents with identical growth form, for *Prenia sladeniana* belonging to the all-cell succulent type and *Aloe pearsonii* belonging to the partially succulent type. Both species have more or less flat leaves and perform a CAM. Table 4.8 compiles all data necessary to calculate the number of leaves each plant can grow at the expense of 1 g of organic dry matter. This number is 12 for *Prenia sladeniana* but only three for *Aloe pearsonii*. Thus *Prenia sladeniana* is a fast-growing succulent when compared with *Aloe pearsonii* but it still grows slower than *Ozoroa dispar*.

One reason for the slow growth of the succulents lies in the enormous quantities of water which must be collected and transported through the plant body. The calculation presented above also makes clear that the occurrence and growth of succulents is mainly controlled by the availability of water.

It would be of great interest to have more data on more different plant species from one area as well as from different areas concerning leaf duration and investment of organic dry matter to build up water storage tissues. On the basis of such data combined with measurements of photosynthesis of these species and climatic data a rough cost–benefit balance could be established.

Also, very little information is available on root/shoot ratios of succulents. Under stable environmental conditions the root/shoot ratio of a particular plant is in equilibrium. When conditions change to a new but stable situation the root/shoot ratio shifts towards a new equilibrium value attuned to the new situation. Since age and phenological stage of the plant interfere and determine the root/shoot ratio, its proper analysis is rather difficult. Generally, environmental conditions causing stress to plant growth result in relatively high root/shoot ratios.

It is clear that the root/shoot ratio provides an important comparative measure of plant performance, though in desert species it needs a careful interpretation since species use their roots to different degrees as a storage organ for reserves. In the Numees area of the Richtersveld we were able to measure root/shoot ratios for some 20 species only and in most cases for

Table 4.8. *Energy expenditure to envelop one millilitre of water, amount of water that can be stored at the expense of one gram of biomass, percentage water based on fresh weight, the amount of water per leaf and the number of leaves that can be grown with one gram of organic dry matter. Further details are given in the text*

Plant species	kJ ml^{-1} water	g water per g org. mat.	% water	Water per leaf (ml)	Leaves per g org. mat.
Othonna opima	0.98	14.68	91.9	14.7	1
Ozoroa dispar	4.99	3.76	77.23	0.23	16
Prenia sladeniana	0.38	30.54	92.1	2.5	12
Aloe pearsonii	1.82	10.00	90.5	3.0	3

just one or two individuals. Consequently, the data presented here (Table 4.9) have no more than illustrative character but can be compared with values presented for desert plants in the literature. Our data agree with values from other desert areas in showing a wide range.

The root/shoot ratio measured for *Opophytum aquosum* (three replicates) is the lowest reported so far. *Opophytum* is an extremely succulent annual that uses its root very temporarily following a rainstorm to take up water that is stored in all parts of the above-ground plant body. Generally, succulents have very low root/shoot ratios and the succulent annuals *Trianthema triquetra* and *Mesembryanthemum pellitum* often still score lower than most of the perennial succulents like *Tylecodon* species, *Othonna opima*, *Senecio corymbiferus*, *Conophytum aequale*, *Crassula elegans* and *Euphorbia gummifera*. The non-succulent annuals, *Acanthopsis disperma* and *Trichodesma africanum*, also show values within the same range. The root/shoot ratio for the perennial *Ceraria fruticulosa* is the only one listed that is rather high. It is succulent in its leaves and stems as well as in its roots and presumably stores reserves in its root corm. Nevertheless, its value is far lower than those of some xerophytic perennials for which values of up to 7.3 (dry weight) have been reported (see Barbour 1981).

4.3.4 The significance of succulence for the carbon budget

In analogy to the water fluxes we can define that a carbon budget is the sum of all carbon fluxes out of and into a plant whereas the carbon balance also takes the carbon fluxes within the plant into account. For plants with high rates of CO_2 exchange, carbon fluxes will be more or less CO_2 fluxes and the carbon budget is then simply the net CO_2 flux. During

Table 4.9. *Root/shoot ratios on the basis of plant fresh weight and dry matter of some plant species of the area around Numees (Richtersveld). (After Werger 1983b.)*

Plant species	Root/shoot ratio	
	Fresh weight	Dry matter
Opophytum aquosum	0.002	0.007
Mesembryanthemum pellitum	0.01	0.03
Trianthema triquetra	0.01	0.06
Stoeberia beetzii	0.01	0.16
Acanthopsis disperma	0.02	0.04
Tylecodon paniculatus	0.03	0.11
Othonna opima	0.03	—
Crassula deceptor	0.03	—
Osteospermum microcarpum	0.05	0.11
Cheiridopsis robusta	0.06	0.19
Senecio corymbiferus	0.08	0.2
Conophytum aequale	0.08	0.29
Forsskaolea candida	0.11	0.13
Crassula elegans	0.12	0.23
Tylecodon reticulatus	0.14	0.29
Tylecodon pearsonii	0.15	0.43
Trichodesma africanum	0.34	0.42
Tylecodon sp.	0.48	0.67
Euphorbia gummifera	0.88	—
Ceraria fruticulosa	2.43	1.35

times of reduced CO_2 exchange, for instance when deciduous leaves translocate their reserves, when leaf growth starts, or when seed pods are filled, carbon fluxes other than CO_2 fluxes predominate.

As for the plant's water, the plant's biomass consists of two fractions, the utilizable biomass and the structural biomass. Only the utilizable biomass participates in carbon fluxes. Other than with water fluxes, carbon but not CO_2 fluxes are energy-dependent processes and require water as a solute. Consequently, not only CO_2 and water vapour diffusion out of and into the plant are linked to each other, but carbon and water movements within the plant also depend on each other.

Unequivocally, a leaf will die when its utilizable water is exhausted, regardless of whether or not its utilizable biomass is also exhausted. The fact that the reverse situation (utilizable biomass exhausted but water still available) is most unlikely to happen shows again that water availability is the overruling limiting factor also within the plant. The heading of this section is a reference to this fact. Since we have seen that the significance of succulence lies in the fraction of utilizable water, this section will deal with the significance of utilizable water for the carbon budget and carbon balance. The main question to answer will be: is utilizable water used to maintain a positive CO_2 budget despite a negative water balance to bridge drought periods?

There are two different forms of drought: firstly, no soil moisture and, secondly, a very reduced air humidity. For a plant both forms of drought result in a water deficit and a water demand especially if they act together which is generally the case in deserts.

With respect to the Namib desert we have seen that succulents dominate in areas where atmospheric drought occurs only temporarily either around midday or in winter during bergwind conditions. Thus, atmospheric drought which is characterized by a high Δw is normally a short-term drought. One might expect that because of their utilizable water succulents might be insensitive to such short-term droughts. We will try to answer this question.

In Fig. 4.50 we have seen how quickly soil moisture evaporates after an abundant rainfall event. Yet, in contrast to the sudden drop of air humidity accompanying the onset of a bergwind, desiccation of a soil is a slow process leaving enough time for the plant to adjust to decreasing water availability. However, a completely dry soil can last for a long period. Thus droughts caused by very low water potentials of the soil are long-term droughts and they are a real challenge for plant life.

How succulents handle their utilizable water to bridge such periods has been outlined in previous sections. We now have to deal with the consequences for the carbon fluxes and how utilizable water mediates such response.

4.3.4.1 The short-term drought

Following our definition, a short-term drought is an atmospheric drought which happens unpredictably and suddenly, forcing the plant to react quickly. Bergwinds can arise at any time of the day or night and may last for hours or days. The first report of the effect of a bergwind event on succulents of the Richtersveld dates from 1977 (von Willert *et al.* 1979a), when the overnight malic acid accumulation of Mesembryanthemaceae was recorded over a period of several days.

Table 4.10 summarizes the results, while Fig. 4.81 provides data on the air temperature and VPD conditions. Leaf succulent Mesembryanthemaceae from both types of succulents reduce their overnight accumulation of

Table 4.10. *Overnight accumulation of malic acid on a normal night, a night with a bergwind starting at 23:00 local standard time, and a night with a bergwind throughout. Measurements were done on three consecutive days from 9 to 11 March 1977 at Numees (Richtersveld). For temperature and VPD on the three days see Fig. 4.81. Data were taken from von Willert* et al. *(1979b)*

	Overnight malate accumulation (mol m^{-3})		
Plant species	Normal day	Bergwind starting at 23:00	Bergwind throughout 24 hours
Psilocaulon subnodosum	189	62	42
Sphalmanthus scintillans	90	28	0
Leipoldtia constricta	28	12	0

malic acid first by about 60% and on the following bergwind day by 80% or completely cease it.

Four years later at the climax of a drought period CO_2 gas exchange of *Psilocaulon subnodosum* was measured during a cool and humid night and during a night with a bergwind event starting late in the night at 4:30 local standard time. With a delay of 45 minutes nocturnal CO_2 uptake was converted within minutes to a CO_2 release (Fig. 4.82). Both measurements were done at the end of the dry season when no water was available in the soil. The decline of overnight malic acid accumulation is due to a complete stop of net CO_2 uptake and obviously also to a refixation of respiratory CO_2. This reaction can be the answer to atmospheric drought and/or to increasing night temperatures. Since nocturnal CO_2 uptake of CAM succulents has a broad temperature optimum around 20 °C (see Section 4.3.4), the elevated temperature of a bergwind at night should not have this deleterious effect on CO_2 uptake. Therefore, it seems to indicate stomatal closure.

Except for the delay of 45 minutes before stomata close nothing indicates that succulence because of its utilizable water diminishes the sensitivity of the succulent to short-term drought. One must, however, consider that these results were obtained with rather desiccated plants at the end of the dry season. At that stage utilizable water was at a very low level. Neither irrigated nor non-irrigated plants, nor CAM succulents after an abundant rainfall event change their responses to nocturnally arising bergwinds. Fig. 4.83 and 4.84 give two examples. In all cases stomatal closure and a

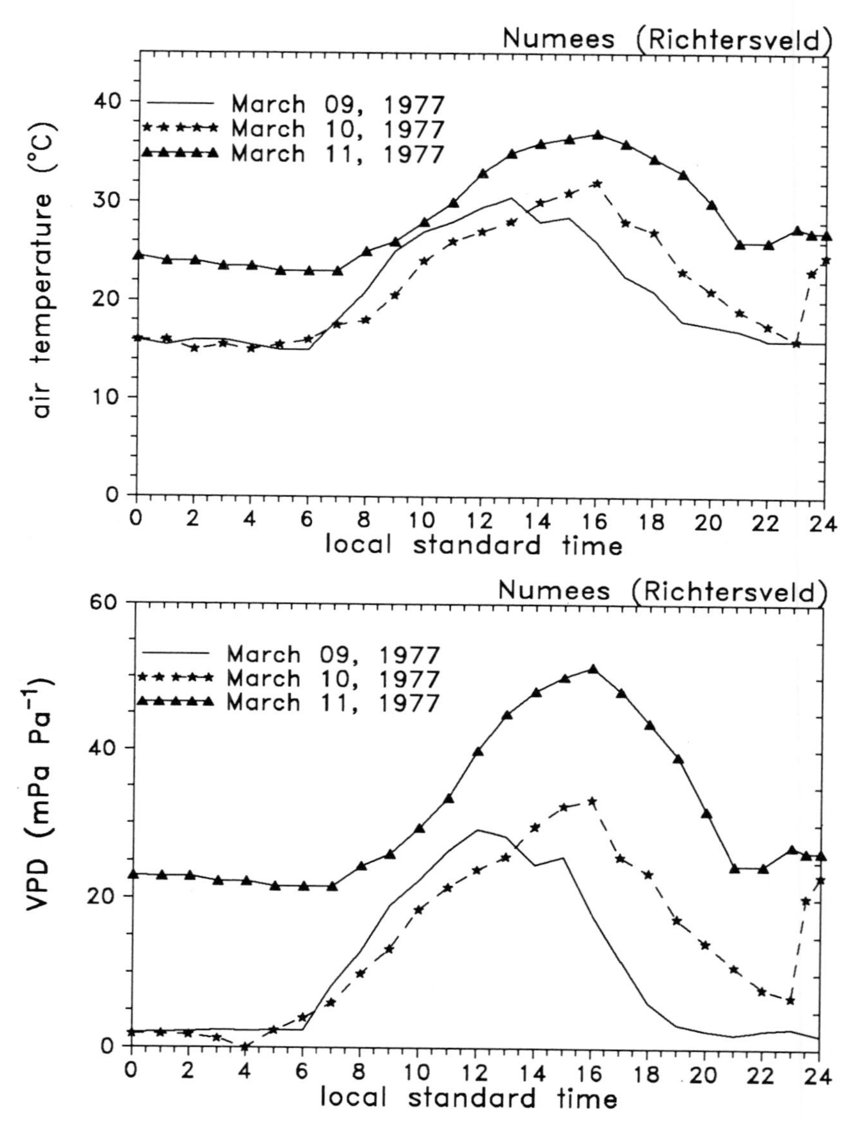

Fig. 4.81 Air temperature and VPD on three consecutive days at Numees (Richtersveld). Air temperature and relative air humidity were recorded with a thermohygrograph. VPD was calculated from these data according to Section 2.1.4. As thermohygrographs cannot measure air humidity lower than 20% accurately, VPD during bergwind is higher than given here.

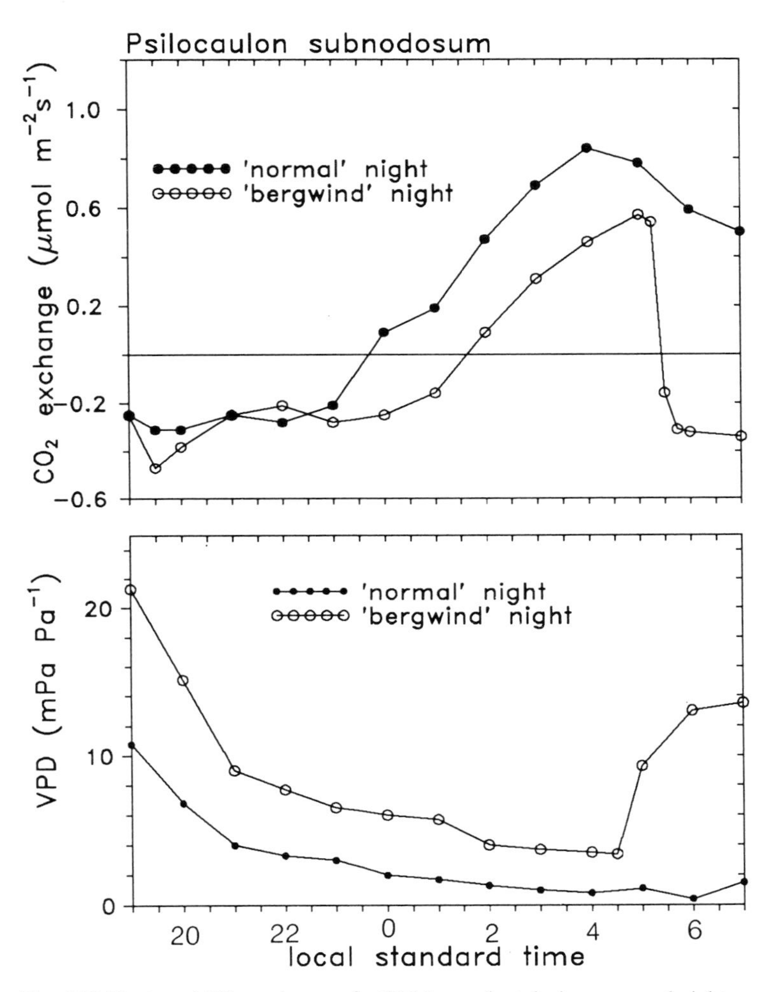

Fig. 4.82 Nocturnal CO_2 exchange of a CAM succulent during a normal night and a dryer night with an additional bergwind arising at 4:30. the actual VPD is given in the lower graph.

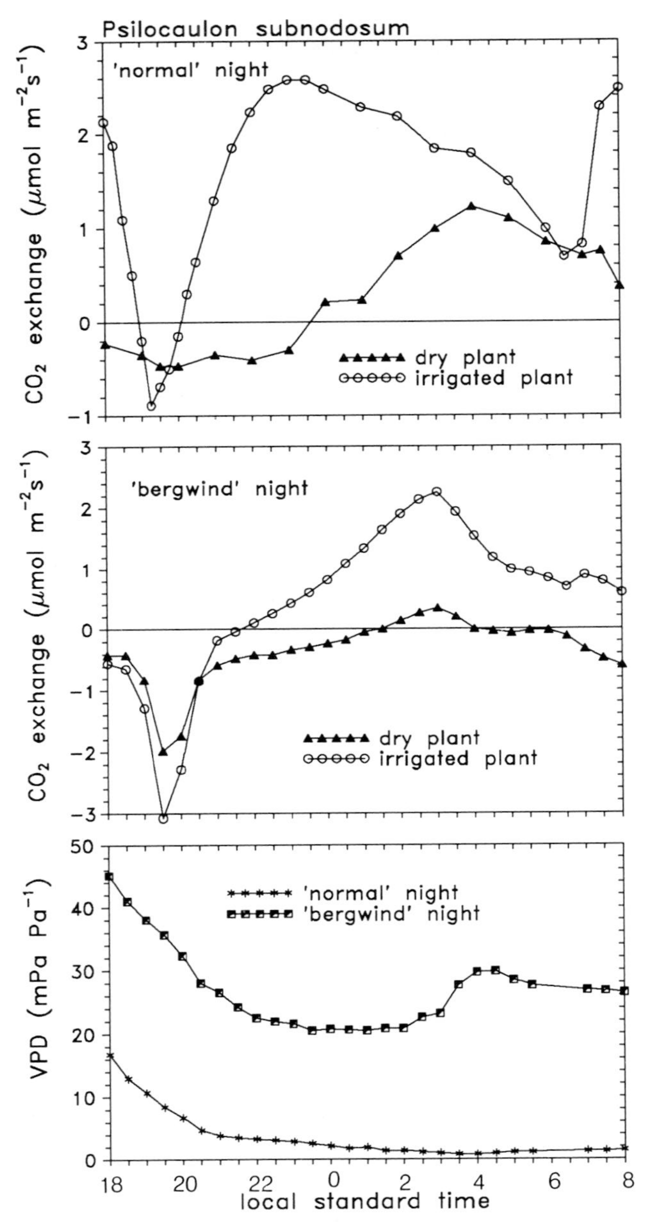

Fig. 4.83 Nocturnal CO_2 exchange of a CAM succulent during a normal night and a night under bergwind conditions with an additional increase of the bergwind at 3:00. The graphs compare the CO_2 exchange of a dry plant and a plant which was irrigated several times with an equivalent of 20 mm of precipitation. The lowest graph gives the VPD for both nights.

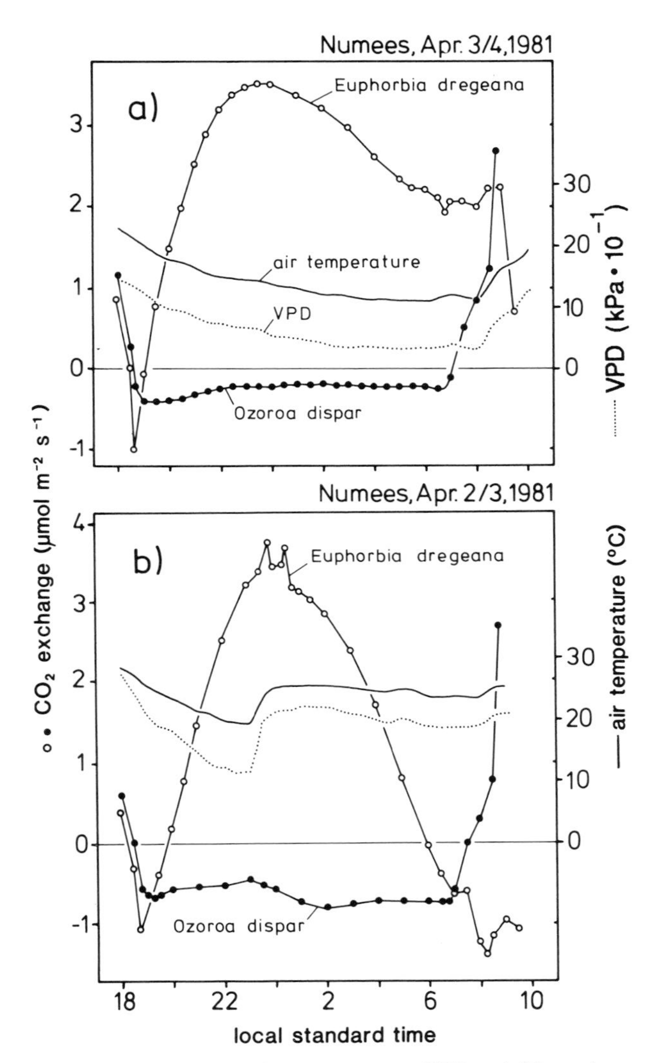

Fig. 4.84 Course of air temperature, VPD and CO_2 exchange of *Euphorbia dregeana* (CAM succulent) and *Ozoroa dispar* (sclerophyllous C_3 plant) two weeks after an abundant rainfall event in a cool and humid night (upper graph) and a night with a bergwind arising at 23:30. Sunset was at 19:00 and sunrise at 6:30 local standard time.

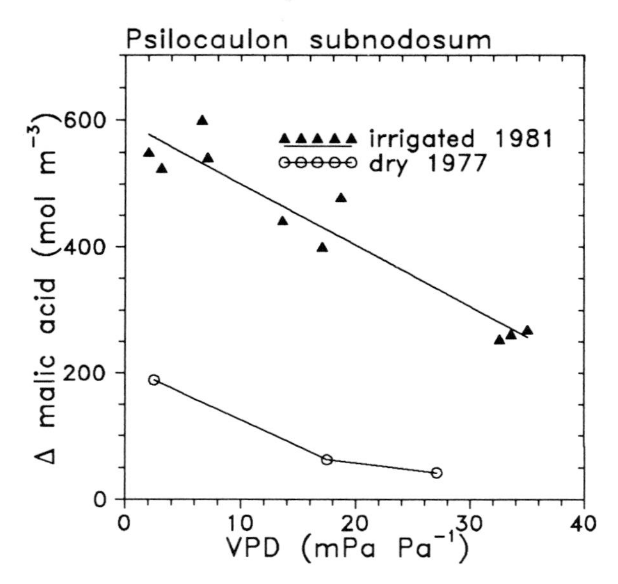

Fig. 4.85 Relation between the overnight accumulation of malic acid and the mean VPD of the same night. The relation is presented for a non-watered CAM succulent at the end of the dry season in March 1977 and for an irrigated plant in February 1981.

reduction of CO_2 uptake and malic acid accumulation occurred (von Willert *et al.* 1983, 1985). This means that the quality of the response to atmospheric drought is not controlled by the availability of water in the soil. Even when the fraction of utilizable water is at its maximum and the plant has the additional possibility of refilling its water storage organs via uptake by the roots, the plant does not use its utilizable water to bridge such short-term droughts. This shows impressively that water saving is the overruling necessity in arid habitats. As soon as water loss exceeds water gain the stomata are closed independent of the existence and quantity of utilizable water.

We can summarize these results in a graph (Fig. 4.85) showing the reduction of nocturnal acid accumulation of the same CAM succulent *Psilocaulon subnodosum* with increasing mean VPD between sunset and sunrise. The water status of the plant determines the total amount of acid accumulation but not the sensitivity against atmospheric drought.

So far we have dealt only with the effect of a short-term drought on dark fixation of CO_2 and nocturnal acid accumulation of CAM succulents. We shall now briefly address the question of how daytime CO_2 uptake is

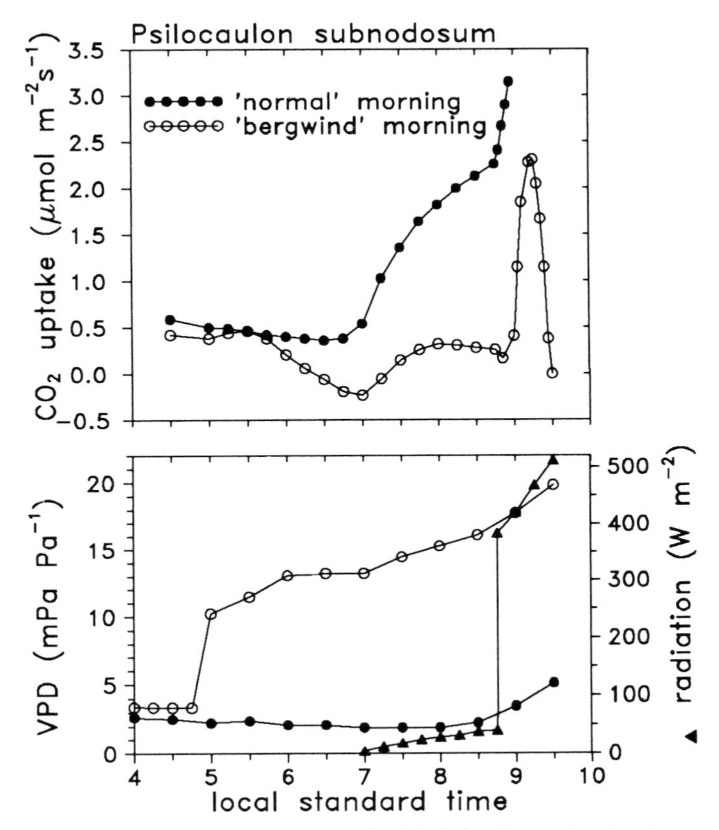

Fig. 4.86 CO_2 exchange of a CAM succulent during the dawn period on a normal day and on a morning with bergwind conditions. The lower graph shows the VPD and the solar radiation on both mornings. Plants came into bright sunlight at 8:45 local standard time.

influenced. We have data to compare a CAM succulent with a number of non-succulent species. A well-irrigated *Psilocaulon subnodosum* plant starts off at dawn with a steadily increasing CO_2 uptake. Bright sunlight then causes a further gulp of CO_2 uptake (Fig. 4.86). On a bergwind morning (bergwind arose at 5:00), CO_2 uptake in the phase of diffuse light is markedly reduced, by 85%, and bright sunlight leads to an only very temporary CO_2 gulp. Non-succulent C_3 plants behave quite differently as is shown for one prominent and representative species *Forsskaolea candida* in Fig. 4.87. With dawn a bergwind arose but CO_2 uptake in the morning was enhanced instead of reduced when compared with a 'normal' morning on which, additionally, moist air from the coast was blown in keeping the VPD

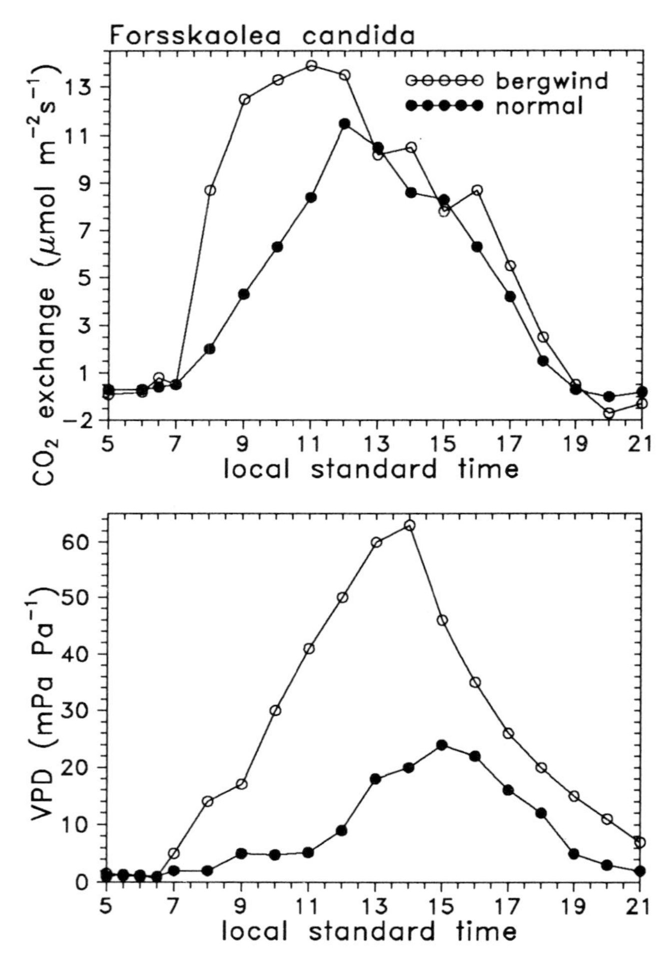

Fig. 4.87 CO_2 exchange of the non-succulent *Forsskaolea candida* for a normal day and a day on which a bergwind arose just at dawn and ceased at 14:00 in the afternoon. The lower graph gives the VPD of the two days.

low until noon. Table 4.11 shows that other C_3 plants also behave similarly. CO_2 uptake between 7:00 and 11:00 is compared for two days as is the total CO_2 uptake from dawn to dusk. All four species show a much higher CO_2 uptake during bergwind conditions than on a cool and humid morning. There is no doubt that water vapour loss is much higher at high VPDs. When comparing the total CO_2 uptake from dawn to dusk bergwind conditions either cause a decrease (*Codon royenii*) or an increase (*Forsskaolea candida*) or leave the CO_2 budget more or less unchanged.

Table 4.11. *CO_2 budget, water budget (H_2O) and water use efficiency (WUE) of four non-succulent C_3 plants and a C_4 grass (*Fingerhuthia africana*) of the Numees area (Richtersveld) over a period of four hours in the morning and over the whole period from dawn to dusk on two different days (one day with a bergwind starting just before dawn and one 'normal' day). Measurements were done on the same leaf with a porometer in October 1983*

	'Bergwind' day						'Normal' day					
	7:00–11:00			7:00–19:00			7:00–11:00			7:00–19:00		
Plant species	CO_2	H_2O	WUE	CO_2	H_2O	WUE	CO_2	H_2O	WUE	CO_2	H_2O	WUE
Codon royenii	122	47	2.6	315	141	2.2	71	12.2	5.8	472	104	4.5
Forsskaolea candida	151	47	3.2	371	183	2.0	57	10.1	5.6	236	63	3.7
Solanum incanum	77	22	3.5	174	62	2.8	29	7.6	3.8	178	41	4.3
Monechma mollissimum	50	9.6	5.2	138	27	5.1	23	63	3.7	114	32	3.6
Fingerhuthia africana	83	21	4.0	275	82	3.4	—	—	—	—	—	—

Note:
CO_2 budget in mmol m^{-2}; H_2O budget in mol m^{-2}; WUE in mmol mol^{-1}.

Solanum incanum and *Monechma mollissimum* also differ in their transpirational water loss from the other two species. Transpiration is rather similar on both days. This might be due to the fact that the leaves of these two species are densely covered with hairs.

Unfortunately, no data for C_3 succulents of that area are available. Nevertheless, we have to conclude that the utilizable water of succulents does not mean any advantage for a short-term atmospheric drought stress. In fact the case seems to be just the opposite; succulents seem to be better water savers than non-succulent species.

4.3.4.2 The long-term drought

In contrast to a short-term drought stress a long-term drought stress develops slowly with desiccation of the soil. This leaves enough time for a plant to adjust gradually to the environment that continuously becomes harsher. A number of reactions are known to be initiated by drought stress. They range from biochemical reactions to changes in the morphology and the habitus of a plant.

In areas where drought coincides with a distinct season adaptations generally aim at reducing the plant's surface during the dry season giving rise to seasonally deciduous plants. Where seasonal droughts do not occur, drought deciduous plants may have evolved. These are plants that carry leaves only for a short period after rainfall. Since growth requires water, the growing season will always be the rainy season. And as the amount of rainfall in arid areas varies greatly from year to year and remains unpredictable, growth should either be rapid or water should be stored as long as it is available from the soil. We have already seen that the rather high investment necessary to build up a suitable water storage tissue prohibits succulents from growing fast. Therefore the abundance of succulents in many dry areas implies that their advantage may lie in the careful use of stored water during a long-term drought stress.

There was an early attempt to show that the CO_2 budget of a succulent is less sensitive to desiccation than the CO_2 budget of a non-succulent species (Kluge 1976). The species compared were *Kalanchoe daigremontiana* and *Coleus* sp. Within two days after stopping irrigation the CO_2 budget of the non-succulent *Coleus* became negative while that of the succulent *Kalanchoe daigremontiana* remained positive over the whole period of five days of measurements. In nature these species never grow together. We therefore repeated such desiccation experiments with numerous plants growing in the Richtersveld but never got qualitatively different results (von Willert and Brinckmann 1986).

Fig. 4.88 Young plants of (a) *Pachypodium namaquanum* and (b) *Tylecodon paniculatus*. Both species are growing in the same area at Numees (Richtersveld). Both have succulent stems with a chlorenchyma underneath the periderm and fairly succulent leaves which are seasonally deciduous. The leaves of *Pachypodium namaquanum* perform a C_3 photosynthesis while the leaves of *Tylecodon paniculatus* exhibit a CAM.

Very often the attribute succulence is brought into connection with CAM, rather as if all succulents perform a CAM. This is not the case. However, the above-mentioned difference in the CO_2 budget during desiccation between succulent and non-succulent species can easily be misinterpreted as a difference between C_3 and CAM. In doing so the relevance of succulence, which also occurs with C_3 plants, is neglected. Of course there is still the possibility of an additional benefit from CAM augmenting the effect of succulence *per se* and giving CAM succulents a further advantage over C_3 succulents. To answer this question requires a proper comparison of two succulents differing only in their mode of CO_2 fixation. A pair of succulents fulfilling this requirement are *Tylecodon paniculatus* (CAM) and *Pachypodium namaquanum* (C_3). Both species have a succulent stem and succulent leaves which are strictly seasonally deciduous (Fig. 4.88). Leaf shedding and flush of leaves are photoperiodically controlled and not triggered by drought and rainfall. The modes of photosynthesis exhibited by the leaves of both species differ significantly from each other. The *Tylecodon paniculatus* leaf performs a CAM and shows a well-expressed overnight acid accumulation while the *Pachypodium namaquanum* leaf performs a C_3 photosynthesis (Fig. 4.89). The stems of both species have a subcortical chlorenchyma enabling the stems to refix respiratory CO_2 but not to allow for a net CO_2 uptake in light. Though quantitatively different both stems show qualitatively the same gas exchange pattern (Fig. 4.89).

These species together with another pair of succulents allow a judgement about the contribution of CAM to the CO_2 budget during desiccation. The other two species are *Plectranthus marrubioides* (CAM) and *Delosperma cooperi* (C_3). Both species have succulent leaves and a more or less woody shoot.

The changes in the CO_2 budget over 24 hours during proceeding desiccation of the soil was followed for these four succulents and was additionally compared with a non-succulent species from the Richtersveld, *Cucumis meeusei.* The results are presented in Fig. 4.90. As expected for a non-succulent species the CO_2 budget of *Cucumis meeusei* dropped quickly. Already on the third day after irrigation was stopped, *Cucumis* lived from its reserves. Even the leaf succulent CAM plant *Plectranthus marrubioides* showed an immediate decrease in the CO_2 budget but this decline slowed down on the following days and stabilized at about 30% of the initial CO_2 gain. The C_3 succulent *Delosperma cooperi* also stabilized at about 30% of the initial CO_2 budget but the drop was delayed by three days. The stem and leaf succulent species showed, regardless of their modes of photosynthesis, at first an increase in their CO_2 gain. After one week of desiccation the CO_2 budget was as good as in wet soil and even after 12 days without watering the CO_2 budget was still at about 50% of the initial value. Consequently it is not CAM which makes a succulent temporarily independent from external water supply but succulence. The higher the amount of utilizable water the higher is the degree of independency.

So far we have dealt with the costs to envelop a unit of water and with the benefit of succulence for the CO_2 budget. We have now to turn to the water costs that are inevitably involved in a carbon gain.

Increasing soil drought causes, irrespective of plants being succulent or not, stomatal closure during the driest time of the day. This is referred to as the midday depression of photosynthesis. During very severe droughts stomatal opening and hence photosynthesis can be restricted to the phase just after dawn. For the rest of the day plants then operate either at or even markedly below the CO_2 compensation point. An often observed behaviour of stomata from succulent C_3 species is a rapid opening shortly after sunrise. At that time of the day photosynthesis is limited by the photon flux

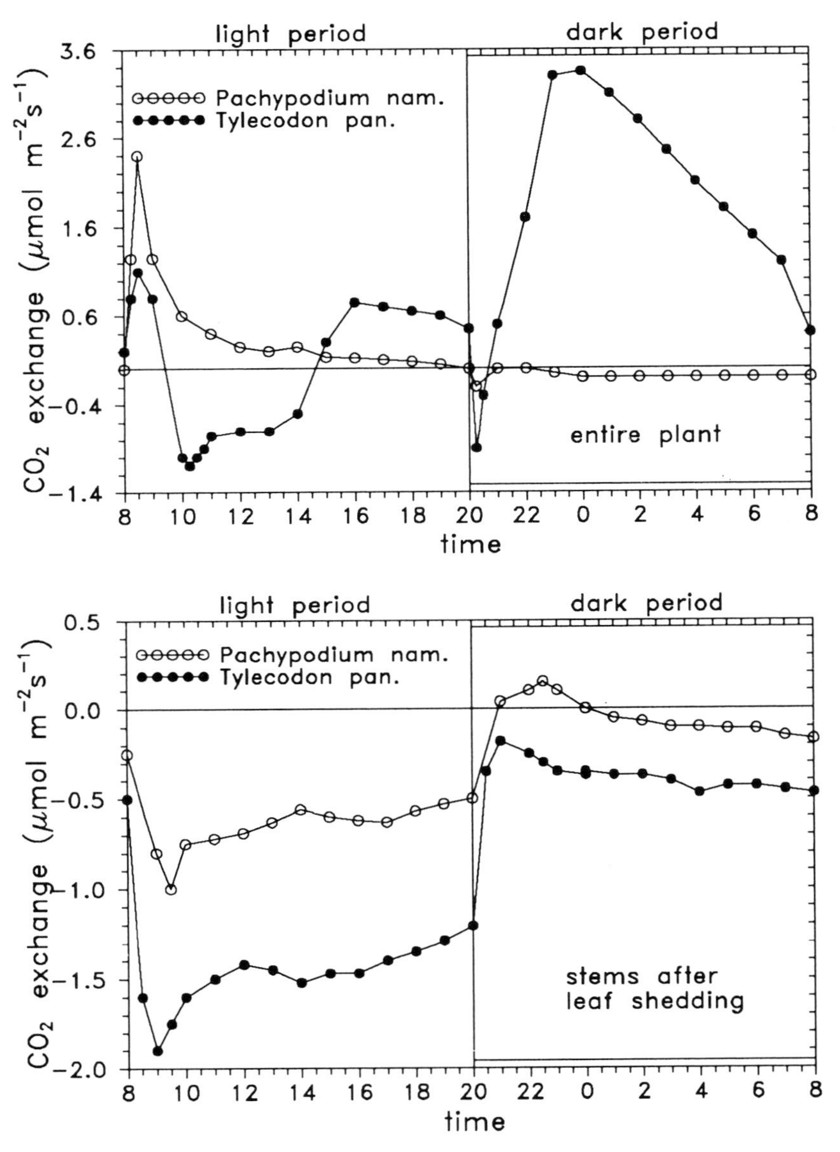

Fig. 4.89 Comparison of the CO_2 exchange of the two succulents *Pachypodium namaquanum* (C_3) and *Tylecodon paniculatus* (CAM) in a laboratory experiment. The upper graph shows the gas exchange of the entire plant. The lower graph shows the gas exchange of the same plants after naturally occurring leaf shedding. Conditions during the gas exchange measurements were: photoperiod 12 to 12 hours, day temperature 27 °C, night temperature 15 °C, dewpoint temperature 13 °C throughout day and night, photon flux density 470 μmol m^{-2} s^{-1}.

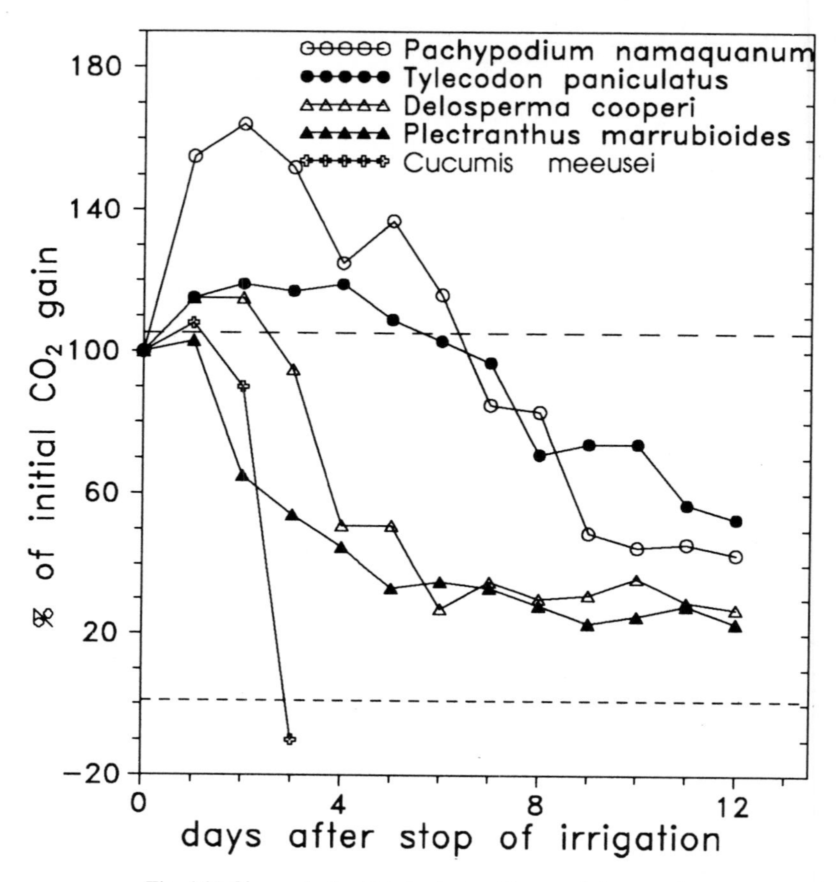

Fig. 4.90 Change in the CO_2 budgets of four succulents and one non-succulent species during desiccation of the soil. Succulent species that are compared were strictly selected for identical growth form and differed only in their modes of photosynthesis, either C_3 or CAM. One pair is formed by the stem and leaf succulents *Pachypodium namaquanum* and *Tylecodon paniculatus*, the other pair are the leaf succulents *Delosperma cooperi* and *Plectranthus marrubioides*. The non-succulent C_3 species is *Cucumis meeusei*. CO_2 gas exchange of these species was measured with a minicuvette system in a temperature and humidity controlled chamber with a photoperiod of 12 to 12 hours. Other conditions were: day temperature 27 °C, night temperature 15 °C, dewpoint temperature 13 °C throughout day and night, photon flux density 470 μmol m^{-2} s^{-1}.

density and hence stomata are wider open than would be necessary for the observed rate of CO_2 uptake. Maximum photosynthesis is reached when stomata are already in their closing movement (Fig. 4.91) which leads to the strange fact that photosynthesis increases with decreasing leaf conductance. Fig. 4.91 compares net CO_2 exchange, leaf conductance and internal CO_2 concentration of two rather short-lived succulents, *Augea capensis* and

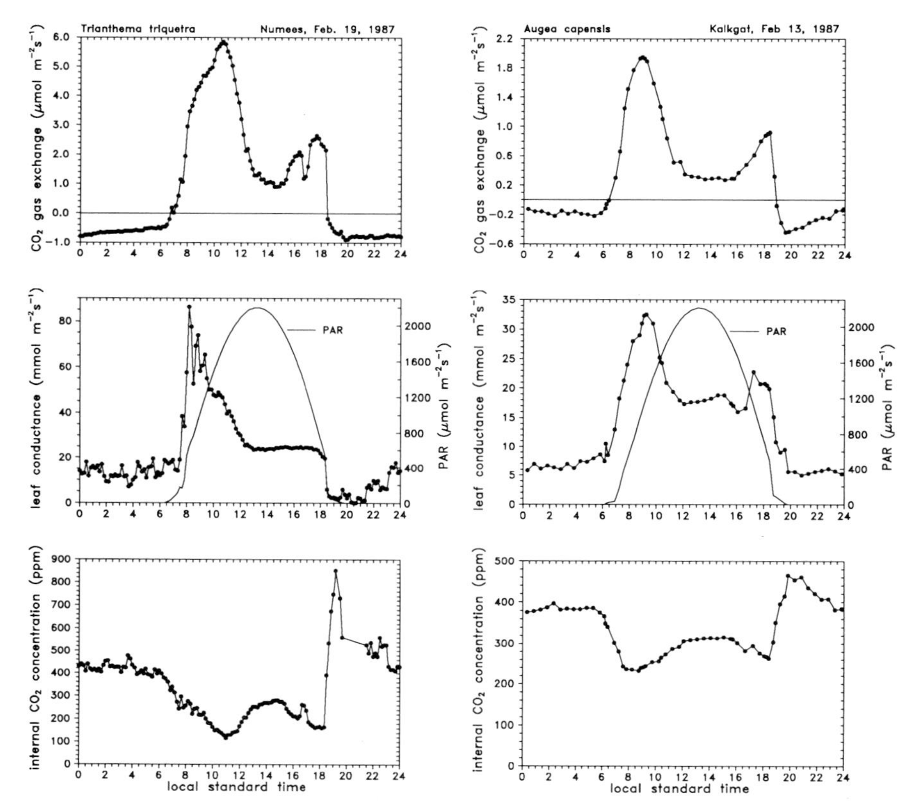

Fig. 4.91 Diurnal course of net CO_2 exchange, leaf conductance, photosynthetically active radiation (PAR) and internal CO_2 concentration of two C_3 succulents growing in the southern Namib, *Trianthema triquetra and Augea capensis*. Transpiration and further environmental factors have been given in Fig. 4.45.

Trianthema triquetra, showing two types of stomatal response. The *Augea* stomata open slowly with increasing photon flux density in the morning and hence photosynthesis follows exactly stomatal opening. *Trianthema triquetra*, on the other hand, opens the stomata rapidly during dawn so that increase in photosynthesis coincides with stomatal closure. This different behaviour of the two succulents becomes clearer in Fig. 4.92 which shows the relation between net CO_2 exchange on the one side and leaf conductance, internal CO_2 concentration and transpiration on the other.

Since very little information concerning CO_2 exchange of C_3 succulents is available we want to present data for another species, *Galenia dregeana*, which reacts in a similar way to *Trianthema triquetra* with regard to early morning opening of stomata (Fig. 4.93).

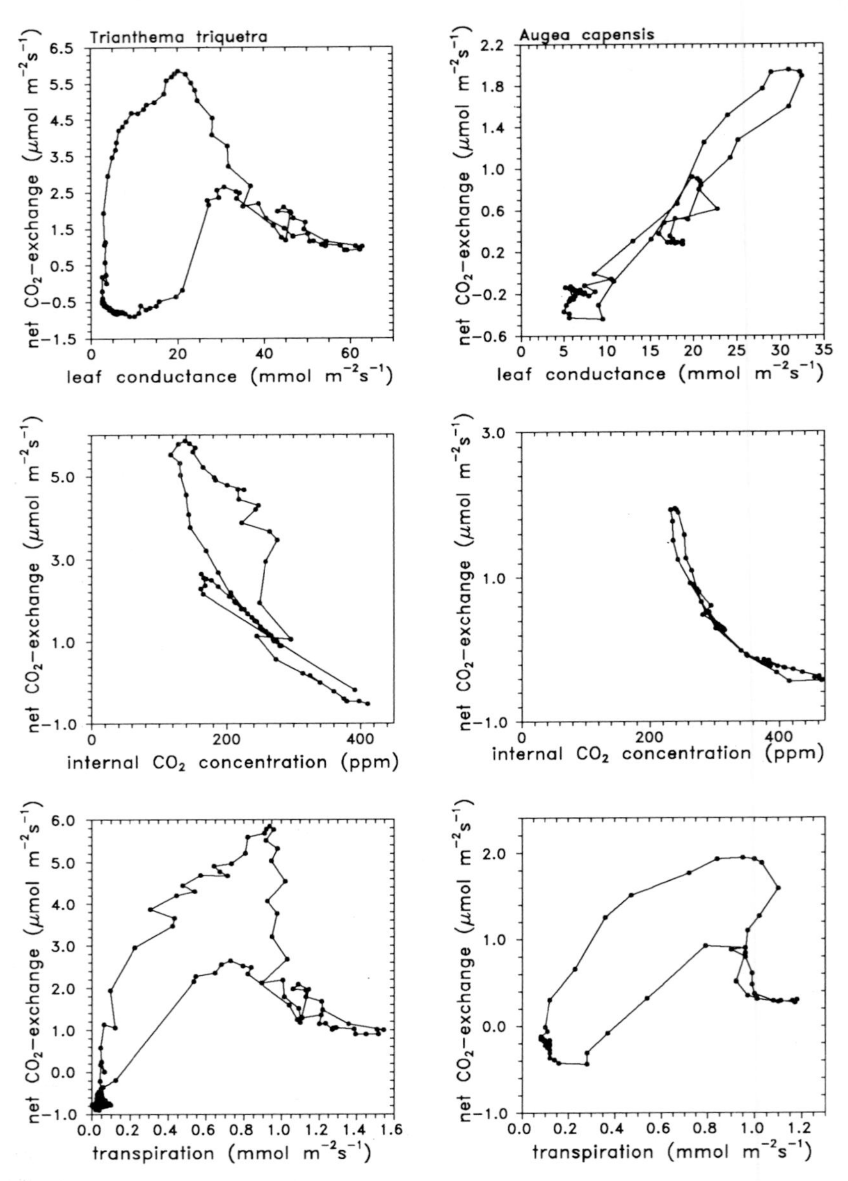

Fig. 4.92 Plots of net CO_2 exchange of two C_3 succulents, *Trianthema triquetra* and *Augea capensis*, over leaf conductance, internal CO_2 concentration and transpiration. While for *Augea capensis* CO_2 exchange is directly correlated with stomatal aperture this relationship is much more complex for *Trianthema triquetra*. Here photosynthesis increases while stomata close by about two-thirds. This behaviour, i.e. that stomata are much wider open in the morning than necessary for the corresponding photosynthesis, was repeatedly observed in C_3 succulents.

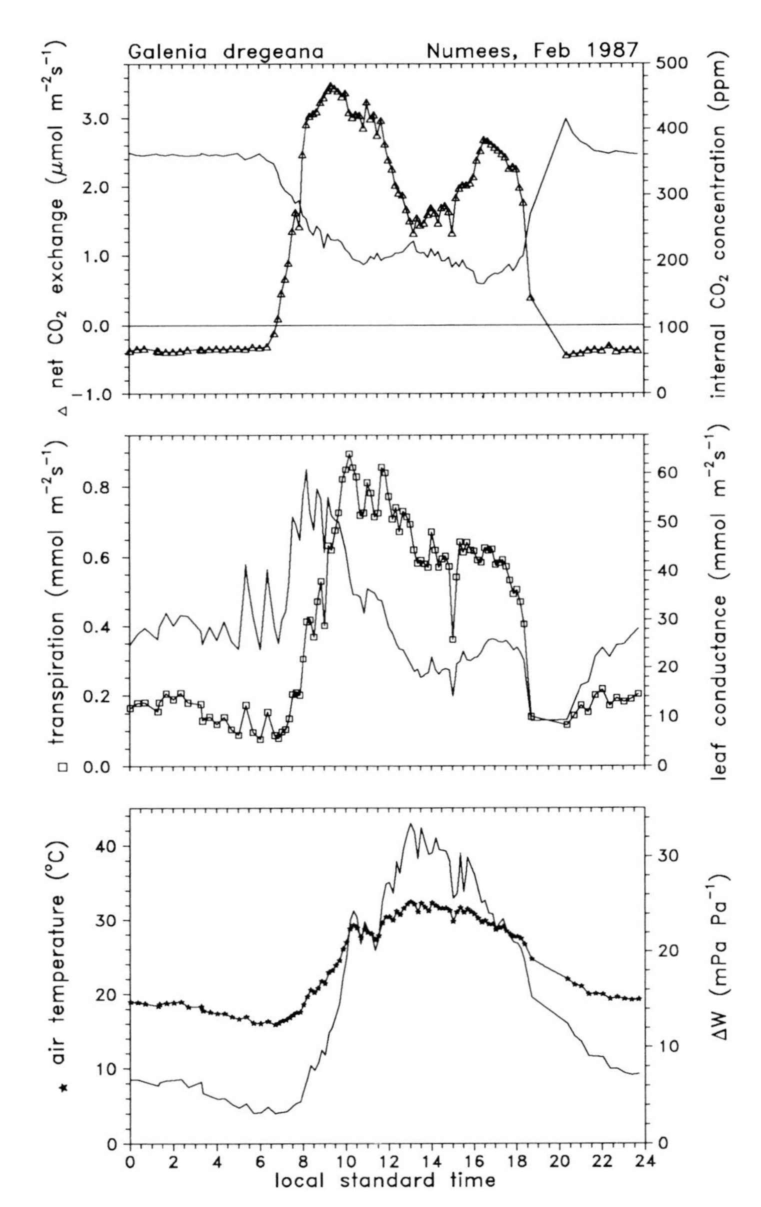

Fig. 4.93 Diurnal course of net CO_2 exchange, internal CO_2 concentration, transpiration leaf conductance, air temperature and leaf to air water vapour gradient ($\triangle W$) for the C_3 succulent *Galenia dregeana*. Note again the widely open stomata in the early morning and the partial closure with increasing photosynthesis.

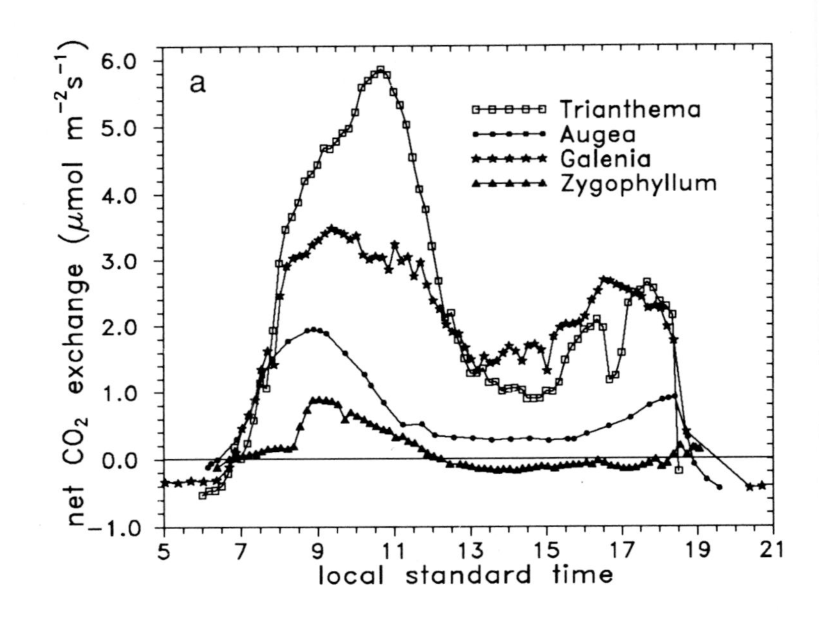
a
Trianthema
Augea
Galenia
Zygophyllum
net CO_2 exchange (μmol $m^{-2}s^{-1}$)
6.0
5.0
4.0
3.0
2.0
1.0
0.0
−1.0
5
7
9
11
13
15
17
19
21
local standard time

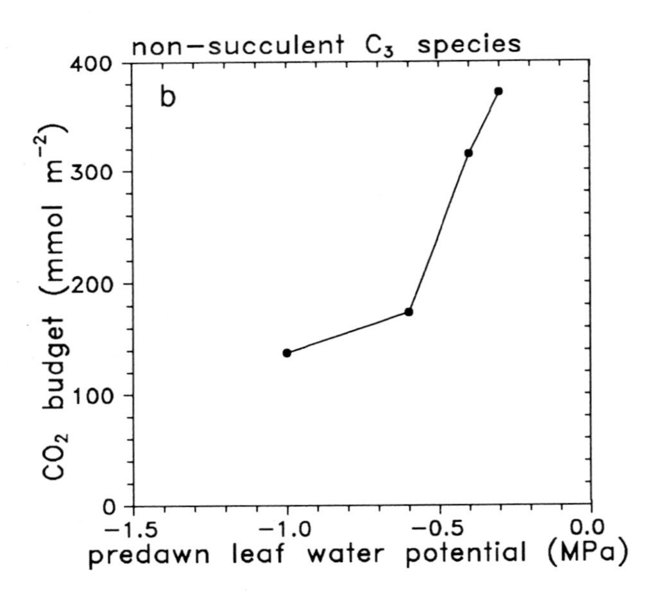
non−succulent C_3 species
b
CO_2 budget (mmol m^{-2})
400
300
200
100
0
−1.5
−1.0
−0.5
0.0
predawn leaf water potential (MPa)

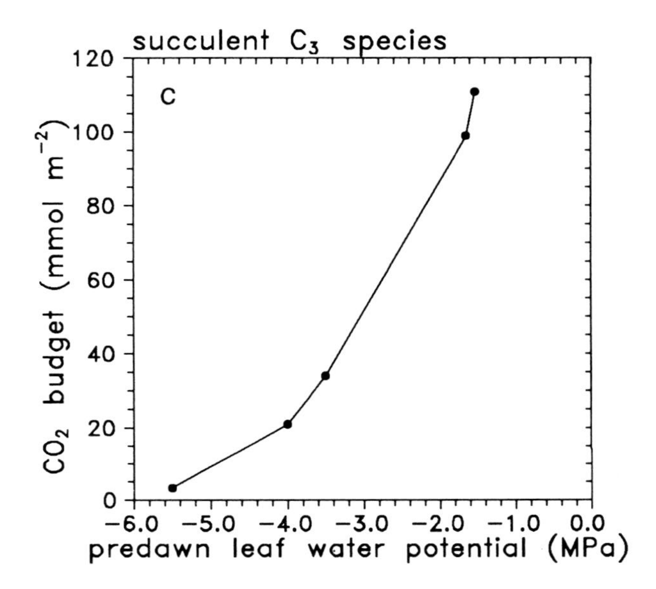

Fig. 4.94 (a) Comparison of the net CO_2 exchange of four different C_3 succulents at the end of the dry season. Data are given for days with fairly comparable climatic factors. The species are *Trianthema triquetra*, *Augea capensis*, *Galenia dregeana* and *Zygophyllum prismatocarpum*. (b) and (c) Correlation between the CO_2 budget from dawn to dusk of (b) four non-succulent C_3 species and (c) four succulent C_3 species in the southern Namib. The budget of the non-succulent species was determined at the beginning of the dry season, that of the succulents at the end of the dry season. The non-succulent species are: *Codon royenii*, *Forsskaolea candida*, *Solanum incanum* and *Monechma mollissimum*. The succulents are *Augea capensis*, *Galenia dregeana*, *Trianthema triquetra* and a dry and irrigated *Zygophyllum prismatocarpum*.

From the limited databank available it is very difficult to judge if and how any factor expressing succulence determines net CO_2 exchange. Necessary for such a comparison is that the climatic prehistory and the prevailing environmental conditions during the measurement are reasonably identical. We are able to present such data but only for four different C_3 succulents all measured within ten days in the same area. Unfortunately, only for two of them is a measure of our newly defined 'succulence quotient' available. Net CO_2 exchange during the day for these four succulents is given in Fig. 4.94a. *Zygophyllum prismatocarpum* is by far the least succulent species and shows indeed the lowest CO_2 gain from dawn to dusk while *Trianthema triquetra* ranks much higher and has the best CO_2 gain. Whether the other two species fit into this sequence is unknown. Although these are very few data they still allow a plot of the CO_2 budget of the C_3

succulents over their predawn water potential. As Fig. 4.94c shows there might be a relationship between these two parameters. This is not at all surprising but needs further clarification. Interestingly, non-succulent species of the same area that were investigated a few weeks after an abundant rainfall event exhibit a similar relationship but with much higher CO_2 gains and predawn water potentials (Fig. 4.94b).

The water use efficiency (WUE) is defined by the relation

$$\mathrm{WUE} = \frac{CO_2 \text{ budget}}{H_2O \text{ budget}} \qquad (4.62)$$

WUE can be given as an instantaneous value by simply dividing the two gas flux densities or can be given for a distinct period of time, a part of a day, a day, a month, a season or a year. The instantaneous WUE of C_3 succulents shows the same daily pattern as known for non-succulent species. Examples are given in Fig. 4.95. Values for the period from dawn to dusk are presented in Table 4.12 which also allows a comparison with the WUE of non-succulent species given in Table 4.11. Towards the end of the dry season the WUE of C_3 succulents is markedly lower than for non-succulent species at the beginning of the summer drought. Again much more information throughout the season and for more species is required.

The relief from a long-term drought is a rewetting of the soil. We will end this section with a comparison of the reactions of a C_3 and a CAM succulent to such a relief. At the end of the summer drought the C_3 succulent *Zygophyllum prismatocarpum* had a negative CO_2 budget over 24 hours. One abundant irrigation event restored daytime uptake of CO_2 and led to a positive CO_2 budget within days (Fig. 4.96). This reaction was accompanied by a marked improvement of the WUE (Fig. 4.97 and Table 4.12). At the end of a long drought period we measured the CAM succulent *Ruschia* sp. which looked as shown in Fig. 4.38. Such a plant did not show a net CO_2 uptake in the dark, but a rainfall event re-established it within days (Fig. 4.98). Parallel to the development of nocturnal CO_2 uptake restoration of photosynthetic CO_2 uptake also proceeded and even more rapidly than nocturnal CO_2 uptake (von Willert *et al.* 1983).

Considering the whole section concerning the interaction between carbon and water we have to conclude that although the generation of succulent organs is costly, it is ecologically relevant to make this investment. A high amount of utilizable water is a valuable possession for a plant and it is normally handled with utmost frugality. It is not spoilt even if storage organs can be refilled by water uptake from the soil. For many botanists it might be surprising that CAM succulents and C_3 succulents perform rather similarly with regard to the CO_2 budget and that obviously

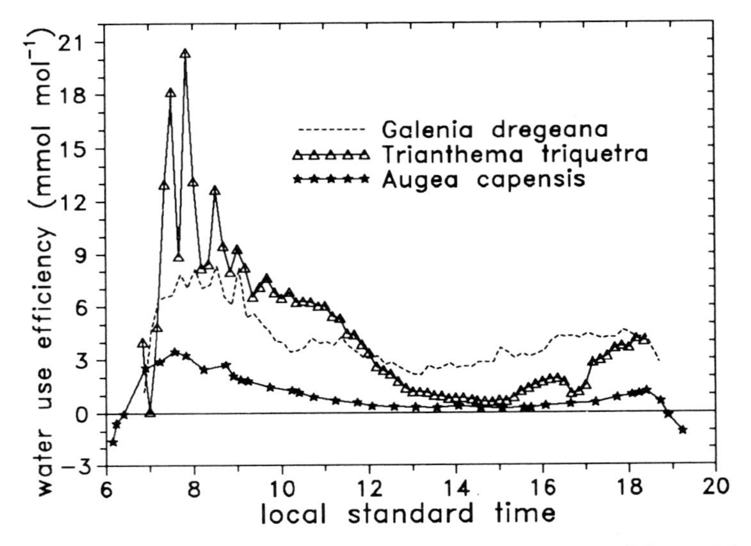

Fig. 4.95 Diurnal course of the instantaneous water use efficiency of three C_3 succulents in the Richtersveld.

succulence is the overruling quality, not the mode of photosynthesis. One must stress, however, that much more work is required in this field to evaluate appropriately the various ecophysiological differences between plants of the same natural environment.

4.3.4.3 *The role of CAM*

We have seen that the borderlines separating the plants into C_3, C_4 and CAM plants become more and more blurred. Plants can shift from one mode of photosynthesis to another and backwards or show features which characterize them as intermediates. We have also mentioned that the first euphoric idea of CAM being a biochemical adaptation to drought was repeatedly subject to critical comments. The extension of CAM from succulents of arid areas to underwater-living plants forces reconsideration of the whole complex of CAM. It is not the scope of this book to do this. But, since CAM is or seems to be the dominating metabolism in succulents especially in the winter rainfall areas of southern Africa, we have to deal with CAM in more detail.

There is full agreement that CAM has a polyphyletic origin. This might imply that it developed for different purposes in different groups of plants. Hence it is not strictly necessary to come up with a unifying theory of CAM. It may be more reasonable to define carefully and differentiate the various

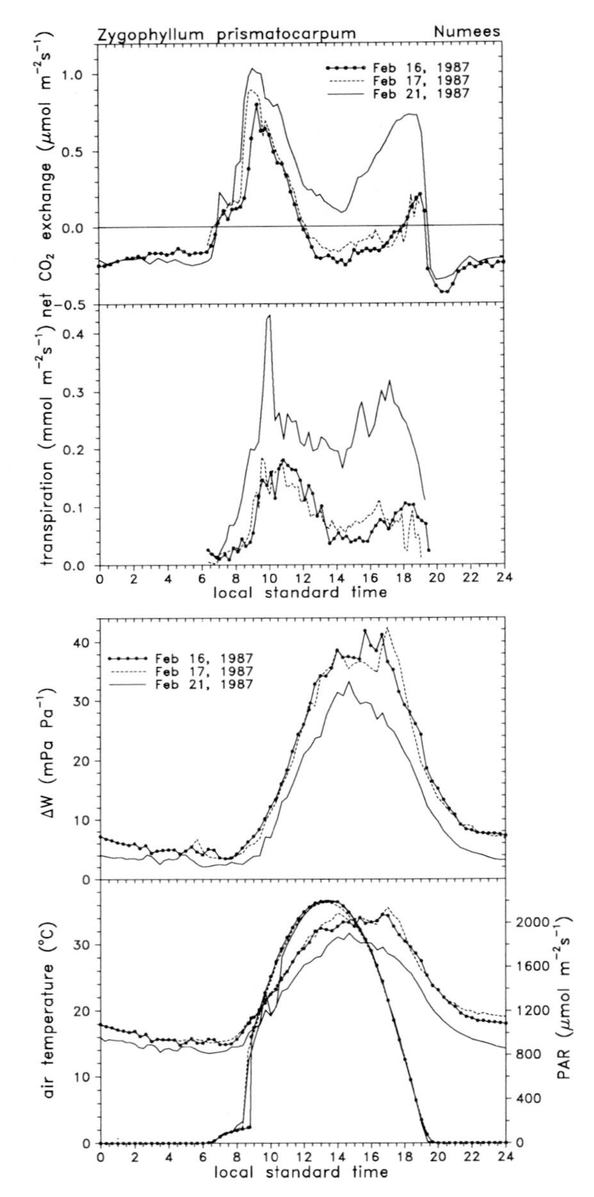

Fig. 4.96 Diurnal course of net CO_2 exchange and transpiration of the C_3 succulent *Zygophyllum prismatocarpum* at the end of the dry summer season (16 Feb.). At the end of that day the plant was abundantly watered and the gas exchange was followed over the next days. The dotted line represents the gas exchange on the day after irrigation (17 Feb.), the solid line that of the fifth day after irrigation. The lower graphs give the air temperature and photon flux density of the three days and the leaf to air water vapour gradient ($\triangle W$).

Table 4.12. *CO_2 budget, water budget and water use efficiency of four C_3 succulents of the Numees area (Richtersveld) over a period of four hours in the morning and over the whole period from dawn to dusk. Measurements were done with a minicuvette system in February 1987*

	7:00–11:00			7:00–19:00		
Plant species	CO_2	H_2O	WUE	CO_2	H_2O	WUE
Augea capensis	21	11	1.9	34	41	0.8
Galenia dregeana	40	8	4.0	99	25	4.0
Trianthema triquetra	56	8	7.0	111	38	2.9
Zygophyllum prismatocarpum (dry)	5.3	1.2	4.4	3.3	3.8	0.9
Zygophyllum prismatocarpum (irrigated)	9.4	2.9	3.2	21	9.6	2.2

Note:
CO_2 budget in mmol m^{-2}; H_2O budget in mol m^{-2}; WUE in mmol mol^{-1}.

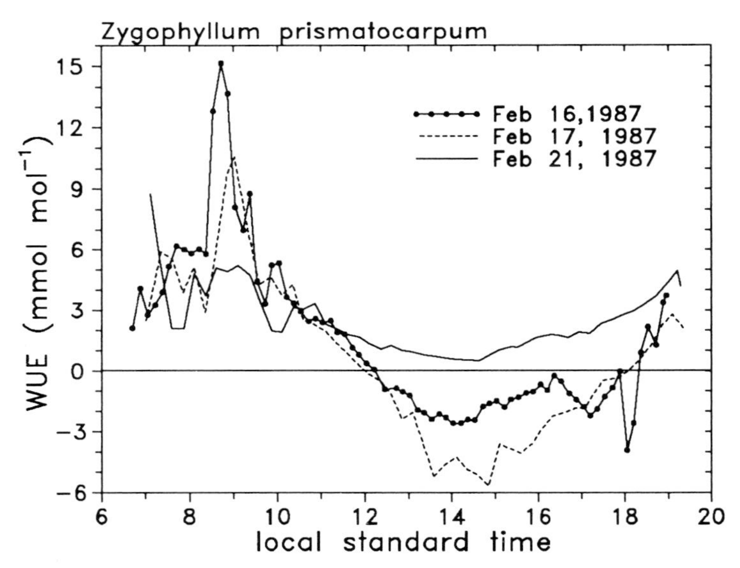

Fig. 4.97 Diurnal course of the instantaneous water use efficiency of *Zygophyllum prismatocarpum* showing the effect of irrigation.

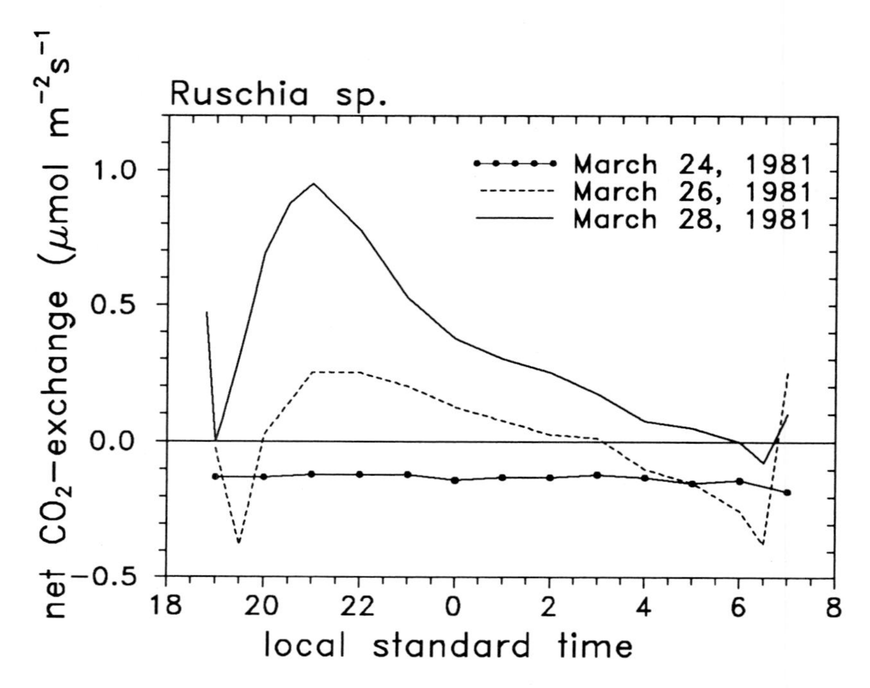

Fig. 4.98 Nocturnal net CO_2 exchange of the CAM succulent *Ruschia* sp. at the end of a prolonged drought prior to a rainfall event (24 March), one day after the rainfall (26 March) and three days after the rainfall (28 March).

expressions of CAM in terms of its possible or actual ecological function and relevance.

CAM might have been an 'invention' for an unknown and very different purpose but then proved also to have advantages for other purposes which it now serves. Or it may be that it has no advantage any longer but as it has no disadvantages CAM is still performed by many plants without obeying relevances botanists are always looking for.

CAM has many facets and CAM performance in cacti is not necessarily identical with that in Mesembryanthemaceae. CAM within an area or a plant family also is not necessarily the same. For evergreen *Crassula* species it may be an essential part of their life strategy; for a seasonally deciduous *Tylecodon* it may have another meaning.

A series of facts suggest that many cells possess the ability to perform CAM-like reactions. PEPC is present in all green and non-green living cells. Only the way in which a cell makes use of this enzyme is different and seems to be a matter of a metabolic regulation that so far is not understood. Non-green tissues, especially roots, can build up high amounts of malic acid which function as counterions for cations. Many guard cells of the epidermal tissues have high levels of malic acid which again balance the positive charge of potassium. Synthesis and degradation of malic acid accompany stomatal movements. In contrast to the mesophyll PEPC the guard cell PEPC is stimulated by potassium even in CAM succulents (Springstubbe 1990). Leaf slices of spinach leaves in aqueous solutions accumulate large quantities of malic acid (Böcher and Kluge 1977), a reaction which is not found in the intact leaf. Consequently, succulence, though generally associated with malic acid accumulation, is not a must for it. Cutting a leaf into strips does not change its succulence but triggers acid accumulation.

Essential for a valid definition of CAM is an understanding of its relevance. What is it for? Is it a mechanism to keep a CO_2 budget positive when the environment gets harsh? If so, why then are large quantities of CO_2 taken up during the night released on the subsequent day, as is generally shown by many members of the Crassulaceae? On the other hand, if conditions do not allow a net CO_2 uptake during the day, will the refixation of respiratory CO_2 during the night not save utilizable biomass and hence positively contribute to the CO_2 balance of the plant? Or is CAM, like C_4, a mechanism to outmanoeuvre photorespiration by increasing the CO_2 concentration at the site of the rubisco? CAM reduction under very bad conditions might then be seen from the aspect that an operating photorespiration is essential to keep the photosynthetic apparatus alive. This then should be found predominantly in evergreen CAM succulents. But what about CAM in annual succulents, are they all 'shifters' like *Mesembryanthemum crystallinum*, which generates CAM with age? Does CAM contribute at all to growth? Or is CAM a mechanism which must be closely linked to water relations? Certainly not for submerged plants. CAM via the production of solutes in fact facilitates water uptake into a cell, into a tissue or drives water movements between tissues and cells by the generation of osmotic gradients. For this task only the property of being able to accumulate and subsequently degrade solutes is essential and the release of CO_2 during the day and a negative CO_2 budget can be tolerated. There can be no doubt that transferring parts of the daily CO_2 uptake into the more humid night with its much lower evaporating demand is an important tool in saving water. CAM may also play a role in water uptake from deposits on the outer surface.

So far attempts to define CAM try to characterize it by its performance. This ends up in classifications like weak, full or super CAM, facultative or obligate CAM. We would like to suggest classification of CAM by its function rather than by its performance. This is, of course, much more difficult and we are far away from presenting a suitable system. Coming from the functional direction the five criteria given in Section 4.3.2.3 might be judged very differently.

Net CO_2 uptake in the night

With respect to the CO_2 gas exchange CAM-people like to differentiate between CAM, CAM idling and CAM cycling. The difference can be explained by the CO_2 exchange reactions. At least during parts of the night, 'real' CAM plants exhibit a net CO_2 uptake. If such 'real' CAM plants suffer from a severe drought they reduce and often cease their CO_2 exchange with the environment but they still show day–night oscillations of acids which indicate a refixation of respiratory CO_2 during the night. This performance is termed 'CAM idling' and is characterized by acid fluctuations in the absence of CO_2 exchange reactions. 'CAM-cycling', finally, means a more or less normal C_3-like photosynthetic CO_2 uptake during the day but a performance at the CO_2 compensation point during the night. This again indicates refixation of respiratory CO_2 and means acid fluctuations, but a net CO_2 uptake at night is never achieved. 'CAM idling' and 'CAM cycling' are different in their performance but functionally and from their result they are similar. Among southern Namib succulents the number of species performing CAM cycling might be rather large though it seems to be restricted to succulents with drought or seasonally deciduous leaves. An example is given in Fig. 4.99 for *Monilaria globosa*.

We do not support the terminology of weak, full and super CAM since it has been shown recently (Herppich 1989) that one and the same species can exhibit all types of CO_2 gas exchange. The performance depends on the prehistory of conditions and markedly on the conditions during the measurement. Within a few days the succulent CAM plant *Plectranthus marrubioides* can shift from a plant that shows actually very little net CO_2 uptake in the night to a plant exhibiting CO_2 uptake only in the night. This process is fully reversible. Nevertheless, during all treatments the accumulation of malic acid continues, showing that *Plectranthus* is a CAM plant. Since gas exchange reactions can easily be manipulated and are highly controlled by environmental factors, as was outlined earlier, we suggest that net CO_2 uptake in the night should be left out of the list of essential criteria. Nevertheless a proper knowledge of the gas exchange of a CAM succulent may help evaluate the function of CAM in the plant. In particular, the question of whether net CO_2 uptake in the night contributes

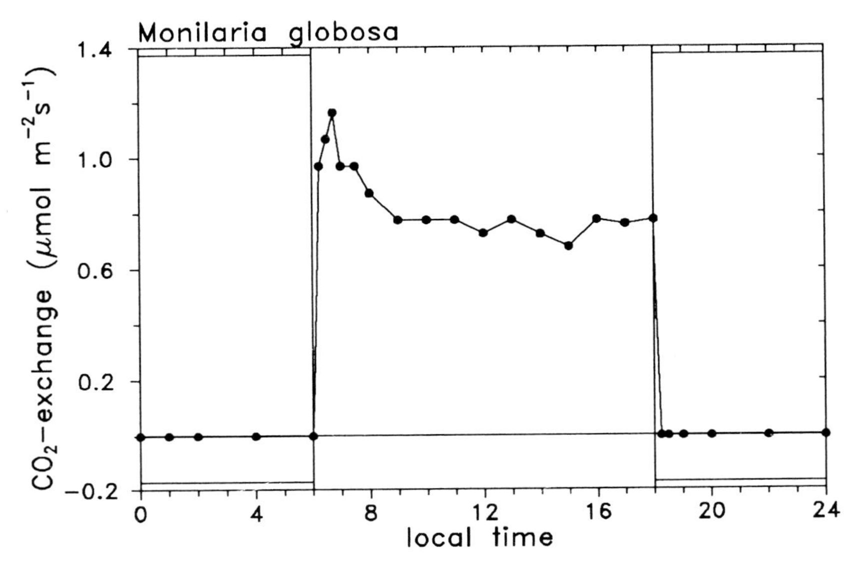

Fig. 4.99 CO_2 gas exchange of the seasonally deciduous succulent *Monilaria globosa*. The light period was from 6:00 to 18:00, the irradiance was 340 W m^{-2}, the temperature was 25 °C during the light period and 15 °C during the night, and the dewpoint temperature was 12 °C throughout the day and night. The succulent was sufficiently watered. Nocturnal accumulation of malic acid was 1356 μmol and that of citric acid 51 μmol g^{-1} of dry matter.

to growth can only be answered from gas exchange measurements accompanying growth analysis.

The presentation of responses of CAM succulents to environmental factors cannot be finalized without dealing with the influences of temperature and humidity on net CO_2 uptake in the dark. With respect to these two parameters great confusion exists in older literature. In many cases where the effect of temperature had been investigated either no care was taken of the VPD, or the relative humidity of the air was kept constant or the humidity conditions were not mentioned. That is why the results obtained either mix up temperature with VPD and hence do not allow an evaluation or they cannot be used at all. For a proper measurement the VPD at night must be kept constant while the temperature is varied. While doing this one must be aware of the fact that the amplitude of the temperature change from day to night may have an effect on nocturnal CO_2 exchange *per se*. For the curve given in Fig. 4.100 day time temperature was 30 or 35 °C and night temperature varied from 5 to 35 °C. VPD at night was kept in the range 5 ± 3 mPa Pa^{-1}. Nocturnal CO_2 uptake of the CAM succulent *Plectranthus*

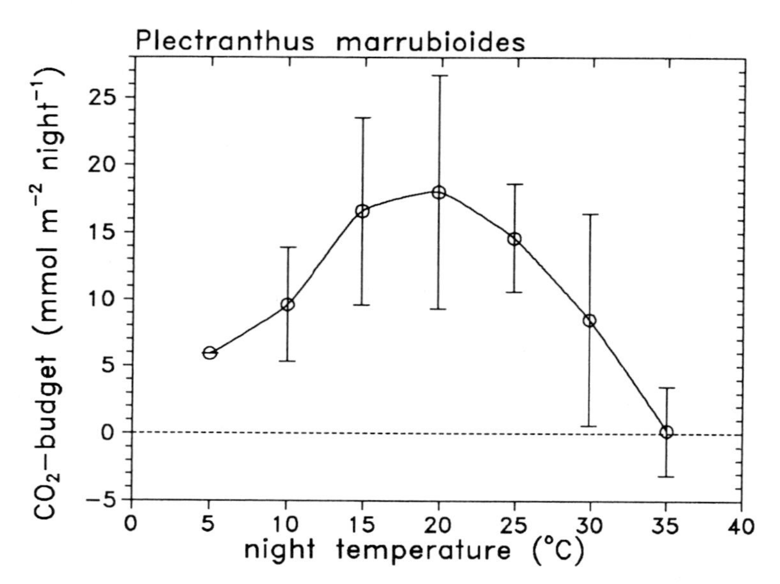

Fig. 4.100 Temperature optimum of the nocturnal CO_2 budget of *Plectranthus marrubioides*. The temperature during the light phase was 30 to 35 °C. Each point represents the average of at least nine measurements. Data were selected for comparable VPDs (Herppich 1989).

marrubioides is highest between 15 and 25 °C with a broad optimum around 20 °C (Herppich 1989). The high variance of values is due to the different soil water potentials the plants faced during the measurements. The decline towards lower temperatures may reflect slow-running biochemical reactions whereas the decline towards higher temperatures may result from a faster accelerating release of CO_2 from respiration. The CO_2-fixing enzyme PEPC itself has a temperature optimum far above 20 °C when measured in an *in vitro* test.

The effect of the VPD in CO_2 exchange has been presented in many figures and will not be repeated here. One must, however, state that in the given examples temperature also changed with increasing VPD but, judged from Fig. 4.100, the temperature effect should not be significant and should not be responsible for the observed effect.

The better measure for a humidity effect on nocturnal reactions is leaf conductance. It has been shown for *Opuntia inermis* and *Kalanchoe pinnata* that leaf conductance during the night increases with decreasing VPD (Osmond *et al.* 1979, Medina 1982). The same relationship was found for *Plectranthus marrubioides* for night temperatures ranging from 15 to 30 °C

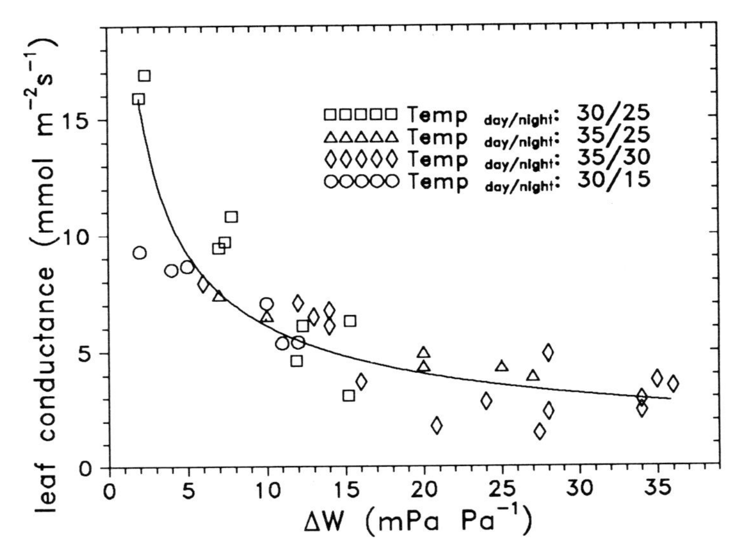

Fig. 4.101 Relationship between leaf conductance and leaf to air water vapour gradient (ΔW) for various night–day temperatures as given in the graph. ΔW represents the average value over the whole night. All *Plectranthus marrubioides* plants were sufficiently watered. The curve represents the best-fit polynomial. (Data taken from Herppich 1989.)

(Fig. 4.101). The fast decrease of the leaf conductance at rather low VPD values and the approximation to a threshold with further increasing water deficits are both characteristic (Herppich 1989).

Nocturnal acid accumulation

Nocturnal acid accumulation is common to all CAM performances. In our opinion one should not differentiate for the origin of CO_2 that is used for the acid formation, be it either external atmospheric CO_2 or endogenously released metabolic CO_2. The result of the fixing of CO_2 is of interest and is always an organic acid. Generally malic acid is the one which is predominantly formed overnight and stored in the vacuole for re-use on the following day. Other acids that can also be synthesized and accumulated are citric acid and isocitric acid.

Fig. 4.102 presents the day–night oscillation of malic acid in the leaf succulent *Sphalmanthus trichotomus* at the end of the rainy winter period when there is still sufficient water in the soil (upper graph). Malic acid is the dominating acid but citric acid is also present in rather high amounts. Clear diurnal oscillations in citric acid are not exhibited. This performance alters

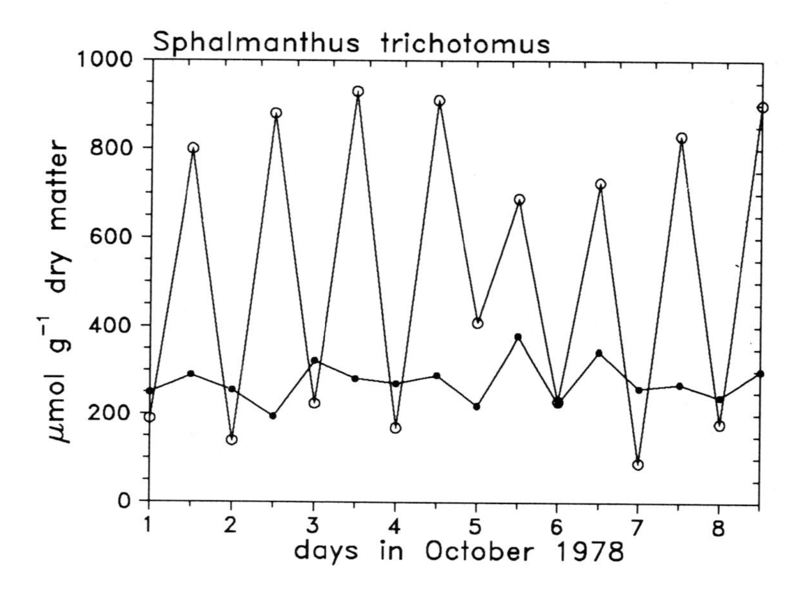

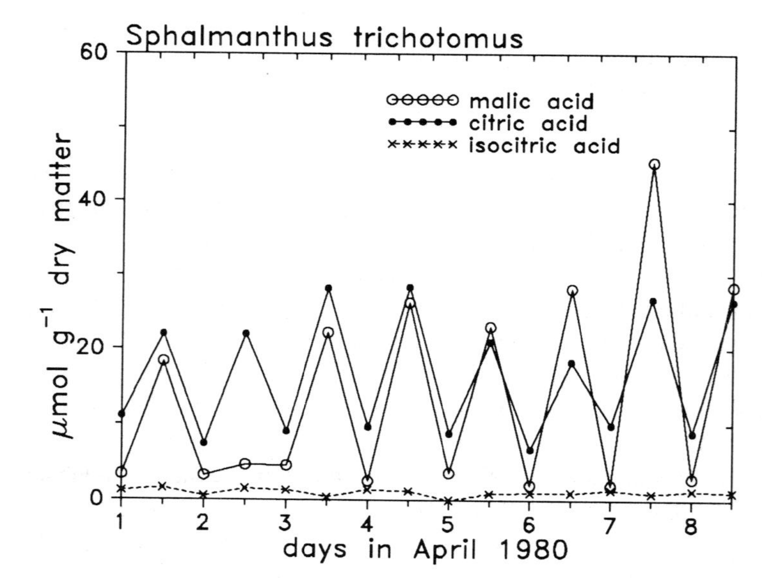

Fig. 4.102 Malic acid, citric acid and isocitric acid content in leaves of *Sphalmanthus trichotomus* at the beginning of the dry season (upper graph) when there was still sufficient moisture in the soil and at the climax of a prolonged drought at the end of the dry season (lower graph). Measurements started on the evenings of 1 October and 1 April, hence morning values are between the figures. Acids were determined enzymatically in the field (Numees, Richtersveld).

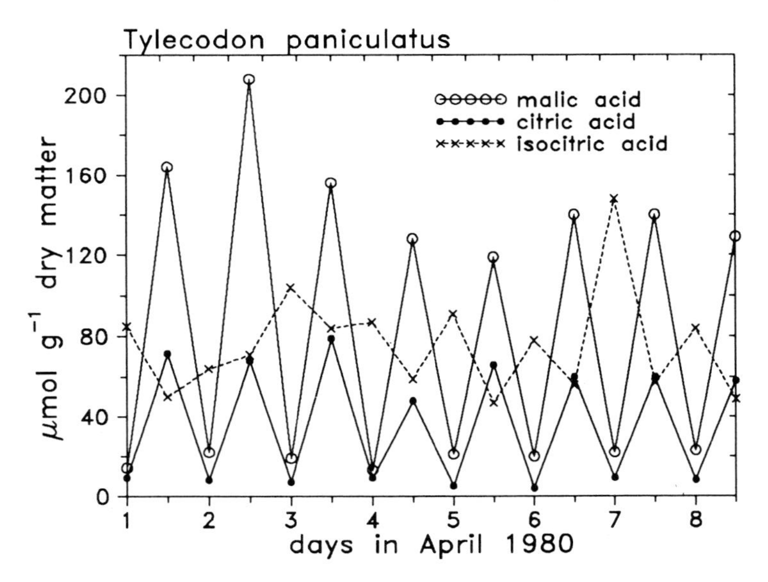

Fig. 4.103 Malic acid, citric acid and isocitric acid content in leaves of the seasonally deciduous *Tylecodon paniculatus*. The plant started to grow leaves during the severe drought in early March. Evening contents stand above the figures on the x-axis, morning contents between the figures. Acids were determined enzymatically in the field (Numees, Richtersveld).

significantly at the climax of a prolonged drought (lower graph). Now malic and citric acid both oscillate with about the same amplitude, but the total amount has dropped dramatically to about one-tenth for citric and one-twentieth for malic acid. CO_2 gas exchange of *Sphalmanthus* was negligible at that time indicating that citric acid synthesis might result from refixation of respiratory CO_2 via the path outlined in Section 4.3.1.

The leaves of *Tylecodon paniculatus* start growing at the beginning of March independent of rainfall. Here, too, citric acid oscillates together with malic acid but with an amplitude that is about 50% lower (Fig. 4.103). *Tylecodon paniculatus* also has significant amounts of isocitric acid which sometimes oscillate counter to citric acid. This phenomenon has been found for many *Tylecodon* species but so far not in any other succulent. Its meaning is unknown. After rainfall malic acid oscillation is augmented substantially and reaches values of 900 μmol g^{-1} of dry matter and more (von Willert *et al.* 1985).

When water is available in the soil malic acid is generally the main acid which is accumulated overnight but many members of the Crassulaceae and Portulacaceae exhibit day–night oscillations of citric acid too. With

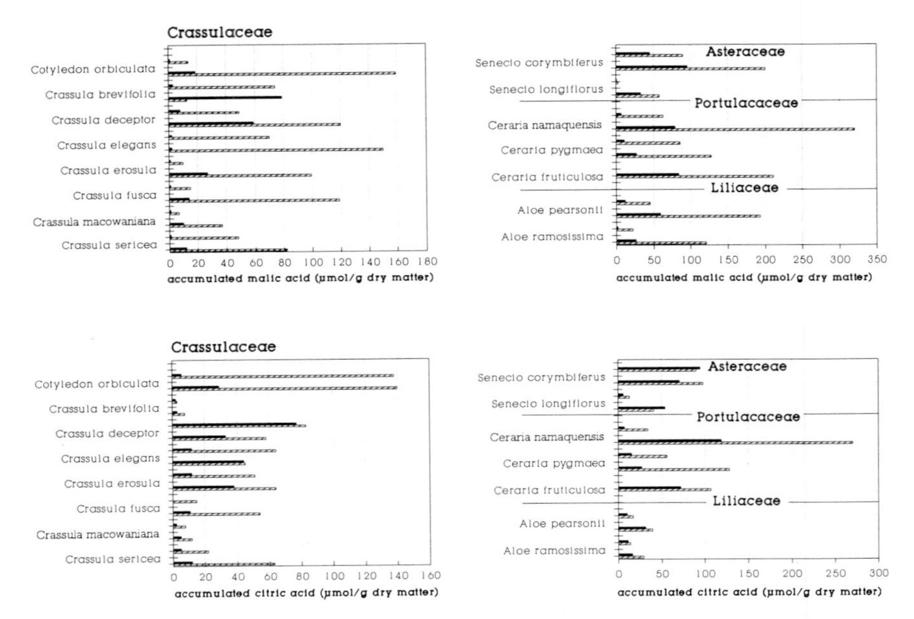

Fig. 4.104 Malic acid and citric acid content in the evening (black bar) and morning (dashed bar) prior to an abundant rainfall event at the end of a severe drought period and two weeks after the rainfall event. Bars above the species names represent the contents before the rainfall, bars below the species name the respective contents after the rainfall. Data are summarized for plant families and the most prominent members within these families are shown here.

decreasing soil water malic acid oscillation in these CAM succulents declines significantly while oscillations of citric acid either persist at the same magnitude or decrease only slightly. Examples are given in Fig. 4.104 for a number of different CAM succulents of the Numees area (Richtersveld).

Members of the Mesembryanthemaceae follow about the same lines but features are more diverse and clear generalizations cannot be made. Fig. 4.105 compares Mesembryanthemaceae of the all-cell succulent type with those of the partially succulent type.

While there is consensus about the role of malic acid in CAM, little attention has been paid to the role of citric acid. First hints that citric acid can overtake the role of malic acid in severely drought-stressed CAM succulents reach back to observations of von Willert (1979). Measurements in CAM succulents in the southern Namib revealed that high but not oscillating citric acid contents start to oscillate during a severe bergwind causing night conditions that do not allow stomatal opening. High and

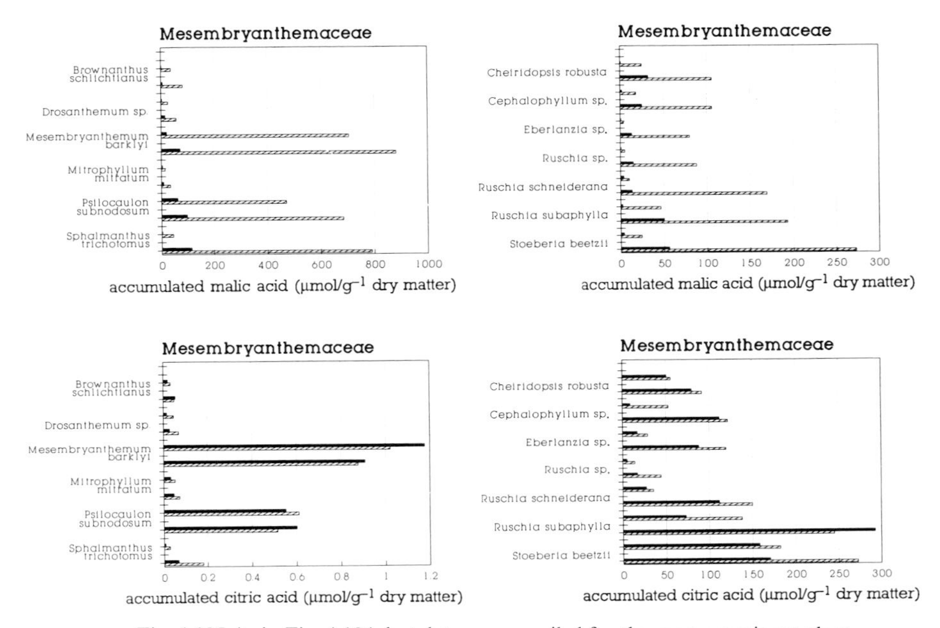

Fig. 4.105 As in Fig. 4.104, but data are compiled for the most prominent plant family of the Richtersveld, the Mesembryanthemaceae. The graphs on the left-hand side show the contents of succulents belonging to the all-cell succulent type, the graphs on the right-hand side show the contents of plants of the partially succulent type.

sometimes oscillating contents of citric acid were reported for many Mesembryanthemaceae. Since extensive work has been done in this field in the following years it seems worthwhile to present some more details here.

In the Mesembryanthemaceae citric acid seems to undergo seasonal oscillations which are triggered by photoperiod. For the winter rainfall area of the southern Namib increase of citric acid content in CAM succulents happens in April over a period of a few weeks. We have no idea whether this is an expression of an activation of reserves to be well prepared for a rainfall event which is now more likely to occur or whether this reflects a protection against low temperatures which are also expected to come. Fig. 4.106 shows this increase for four different Mesembryanthemaceae in the Richtersveld area. Citric acid content stays high throughout the winter season. The decline of citric acid in October is even more rapid in *Psilocaulon subnodosum* (Fig. 4.107) but happens in many other Mesembryanthemaceae too. Fig. 4.106 also shows that at the climax of the drought citric acid is by far the dominating acid, that oscillations in malic acid are obscure and nearly

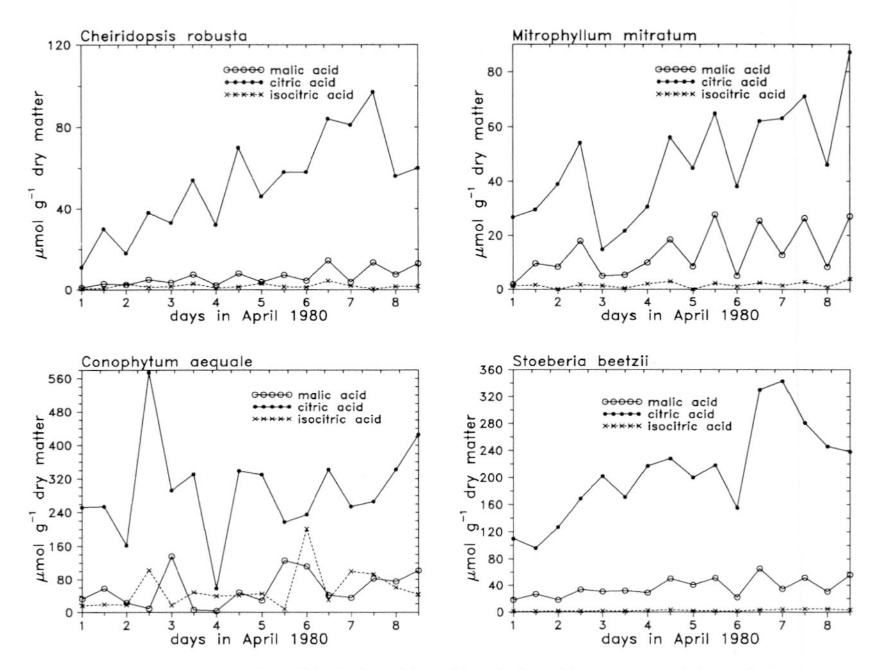

Fig. 4.106 Malic acid, citric acid and isocitric acid content and day–night oscillation of the same acids in leaves of four different members of the Mesembryanthemaceae during a severe drought period. Evening contents are above the figures on the x-axis, morning values are between the figures. Acids were determined in the field (Numees, Richtersveld).

negligible and that isocitric acid plays no role at all. With the exception of *Cheiridopsis robusta* no clear and sustained oscillation in the content of citric acid occurs. Where oscillations occur they are as expected with high values in the morning and lower values in the evening. In such an extreme stress situation each branch and leaf of a succulent must be regarded as an independent individuum which means that even leaves on the same branch can differ significantly from each other with respect to ionic composition. Though sampling took into account statistical demands, clear oscillations might still be masked by highly diverse leaf material. We will give a proof for that shortly.

An interesting feature was found to accompany the shift from C_3 to CAM in salt-treated *Mesembryanthemum crystallinum*. We have already mentioned that increase of PEPC precedes nocturnal accumulation of malic acid by days (see Fig. 4.76). A detailed analysis of the complete set of organic acids, however, revealed that citric acid starts to oscillate when

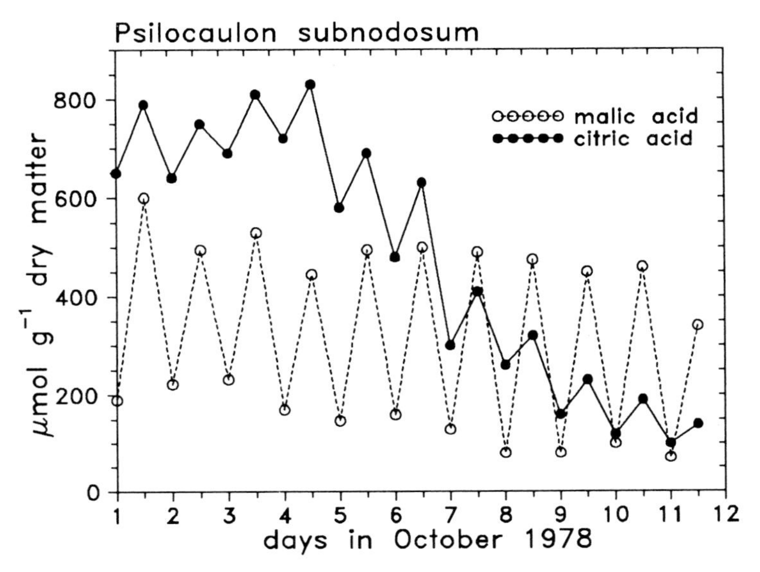

Fig. 4.107 Malic acid and citric acid content and day–night oscillation of the same acids in leaves of *Psilocaulon subnodosum* at the beginning of the dry summer season. Note the rapid decline in citric acid content (Numees, Richtersveld).

PEPC increases. This oscillation is markedly reduced and finally disappears when malic acid starts day–night oscillations. Concomitant gas exchange measurements suggest that citric acid oscillations reflect refixation of respiratory CO_2 at night (M. Herppich, unpublished).

The results presented here for citric acid substantiate the opinion of Lüttge (1988) who on a much smaller database speculated about the role of citric acid in CAM. From our experience one must be very careful in mixing investigations in the field with those under artificial conditions. The results are seldom identical and great efforts are necessary to verify similarities and discrepancies. At a first glance the results we obtained for isocitric acid are similar to those that we have just presented for citric acid. Succulents belonging to the Liliaceae and Crassulaceae contain rather large amounts of isocitric acid. This acid seems also to oscillate in a day–night rhythm as might be deduced from the figures for *Aloe* (Fig. 4.108) or various *Crassula* species (Fig. 4.109). A careful analysis and comparison, however, showed that isocitric acid and its fluctuations have nothing to do with CAM performance and CAM function. In both genera isocitric acid is used rather to chelate calcium. It became obvious that the same oscillations were also

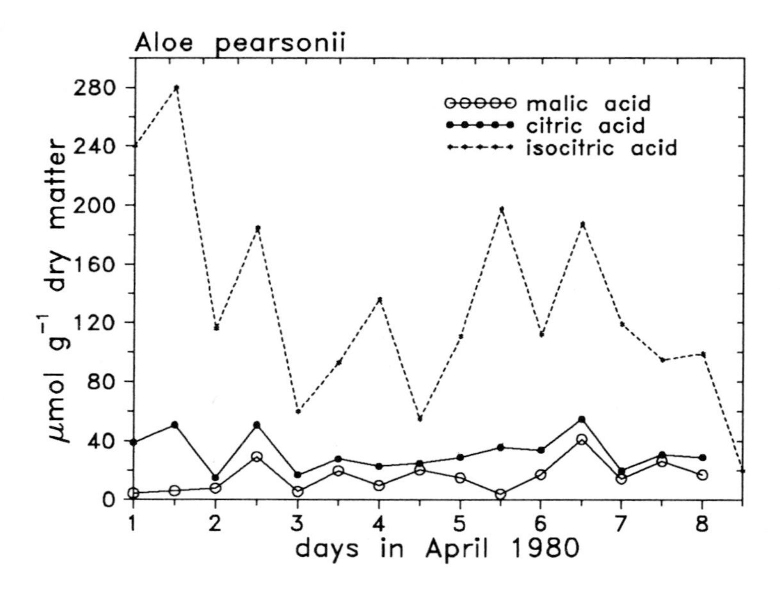

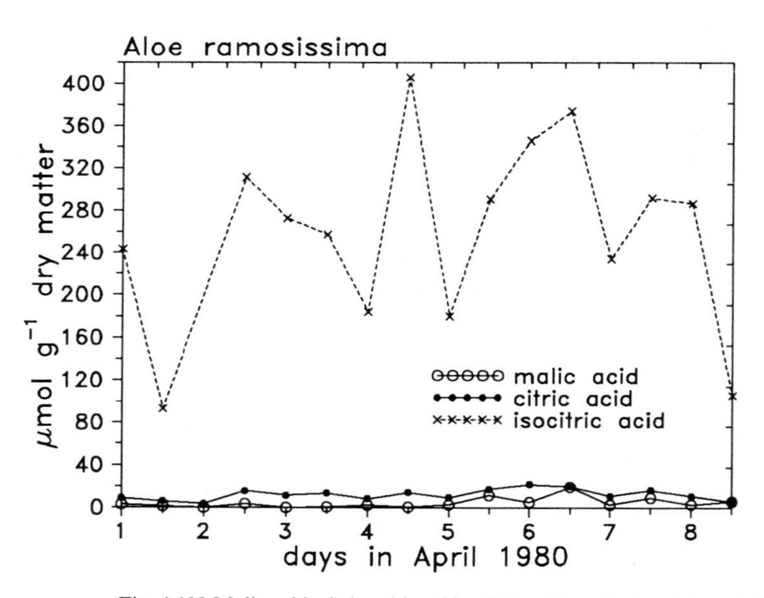

Fig. 4.108 Malic acid, citric acid and isocitric acid content and day–night oscillation of the same acids in leaves of two Aloes during a severe drought period. Other conditions as in Fig. 4.106.

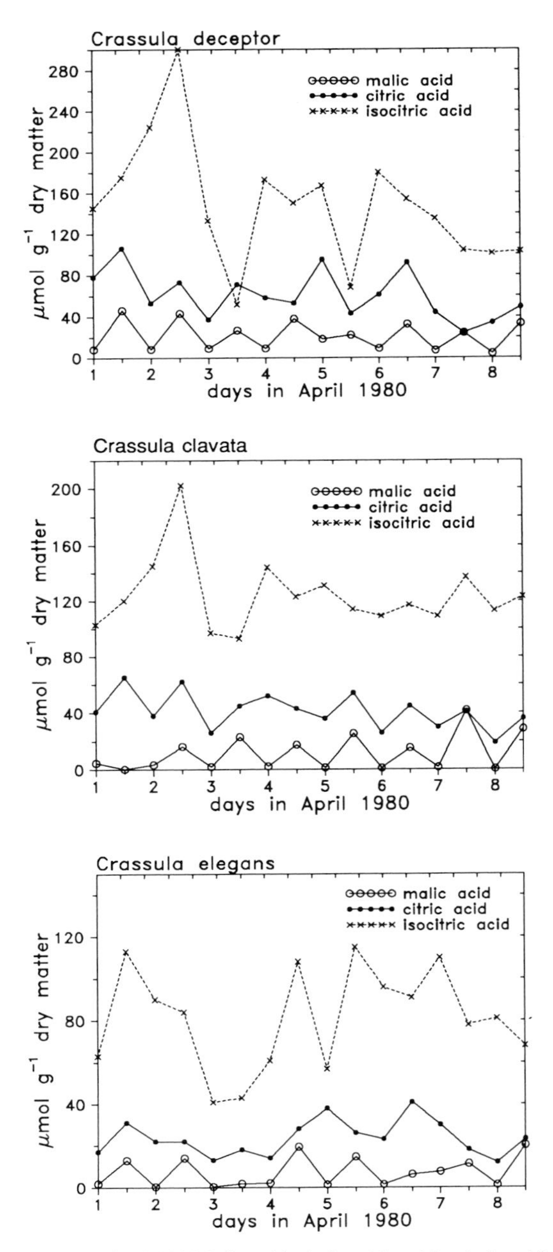

Fig. 4.109 Malic acid, citric acid and isocitric acid content and day–night oscillation of the same acids in leaves of three crassulas during a severe drought period. Other conditions as in Fig. 4.106.

exhibited by calcium. Consequently, these oscillations reflect inhomogeneity of the calcium and isocitric acid contents between leaves of the same plant and between plants. A plot of the calcium content over the content of isocitric acid makes this clear (Fig. 4.110). There is a significant difference between aloes and crassulas. Aloes have more isocitric acid than calcium whereas crassulas have more calcium than isocitric acid. The meaning of this is unknown. The correlation given here between calcium and isocitric acid impressively supports earlier findings by Rössner and Popp (1986).

During a severe drought CAM succulents hardly behave like CAM plants as has been well documented above. From all the criteria mentioned only acid fluctuations persist at a very low and sometimes even negligible level. At that stage CAM succulents do not differ from C_3 succulents; many C_3 succulents exhibit features very similar to CAM succulents when they are subjected to the same severe drought. Fig. 4.111 gives two examples for the C_3 succulents *Othonna opima* and *Pachypodium namaquanum* and additionally for a sclerophyllous C_3 plant *Cucumis meeusei*. The C_3 succulents can show significant day–night oscillations of malic and citric acid which do not differ from CAM succulents in amplitude. It must be stressed here that both species do not show any acid oscillation when there is no water shortage. They also do not show 'CAM cycling' but start with acid oscillations only under severest drought stress. We assume that this reversible acid accumulation which is triggered or induced by drought stress serves water translocation from the central water tissue to the chlorenchyma (*Othonna*) and from the stem to the leaf (*Pachypodium*). Consequently, from a functional point of view both species might also be put in the group of CAM succulents. We have to conclude that even the borderline between CAM and C_3 succulents becomes blurred.

Overnight malic acid accumulation declines with decreasing water content of the plant. This feature seems to be common to all terrestrial CAM succulents regardless of their life form and regardless of whether measurements were done in the natural habitat or in a growth chamber. This is impressively demonstrated by a comparison of malic acid accumulation in *Opuntia basilaris* (Fig. 4.112) in the Sonora desert, *Plectranthus marrubioides* in growth chamber experiments (Fig. 4.113) and in a number of succulents investigated over a period of five years at Numees (Fig. 4.114). In all cases acid accumulation decreases with decreasing leaf water.

Succulence

The last important criterion which is thought to be necessary for CAM is succulence. We have dealt with succulence throughout this book, hence it does not need special reference here. But a few comments seem

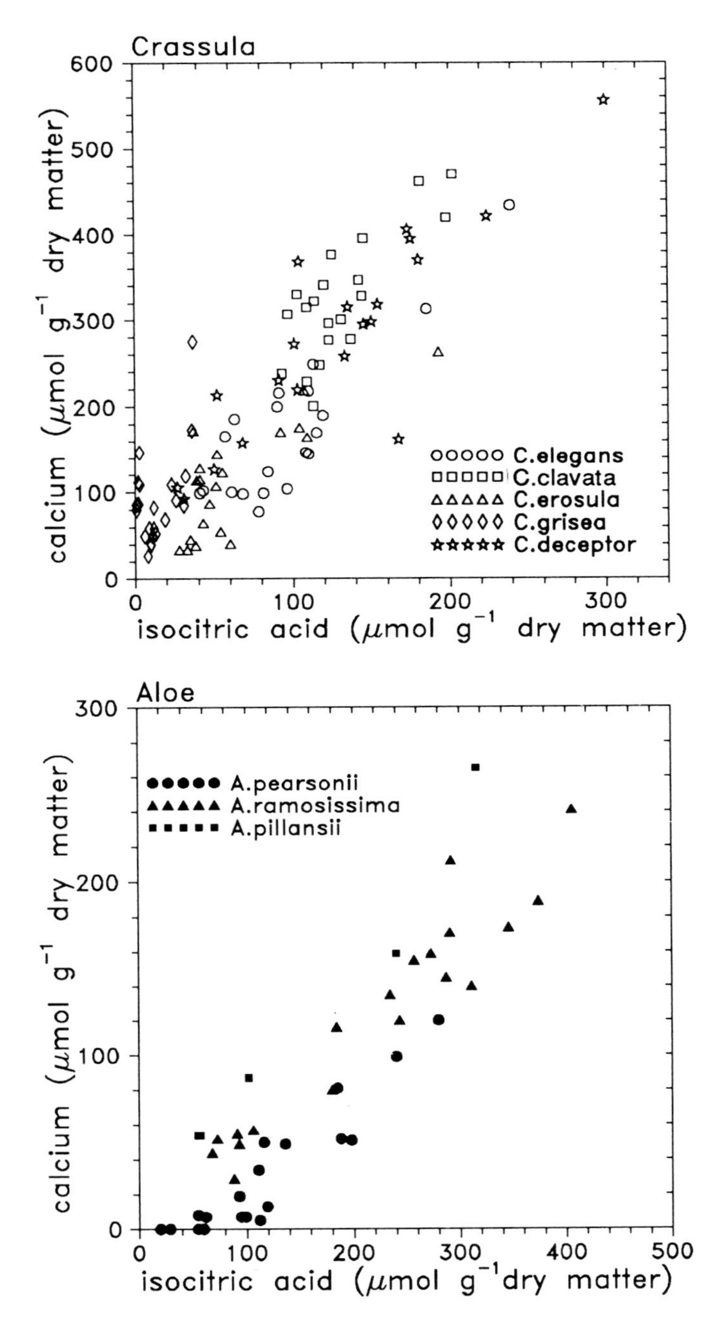

Fig. 4.110 Correlation between the calcium content and the content of isocitric acid in the genera *Crassula* and *Aloe*. There is a strong correlation between the contents of these two constituents.

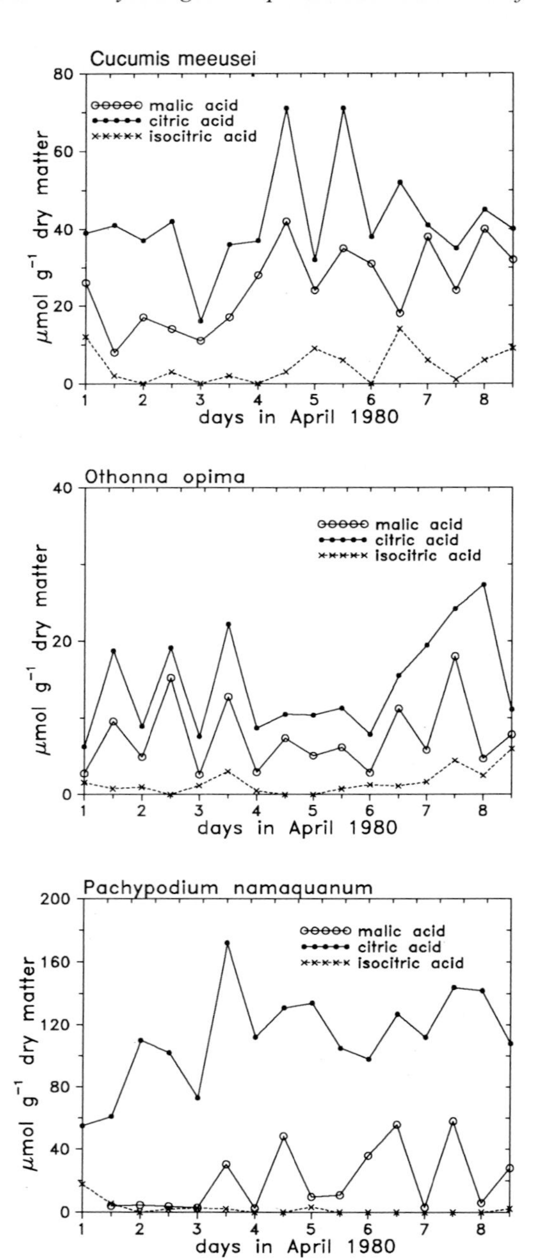

Fig. 4.111 Malic acid, citric acid and isocitric acid content and day–night oscillation of the same acids in leaves of two C_3 succulents, *Othonna opima* and *Pachypodium namaquanum*, and a sclerophyllous C_3 plant, *Cucumis meeusei*, during a severe drought period. Other conditions as in Fig. 4.106.

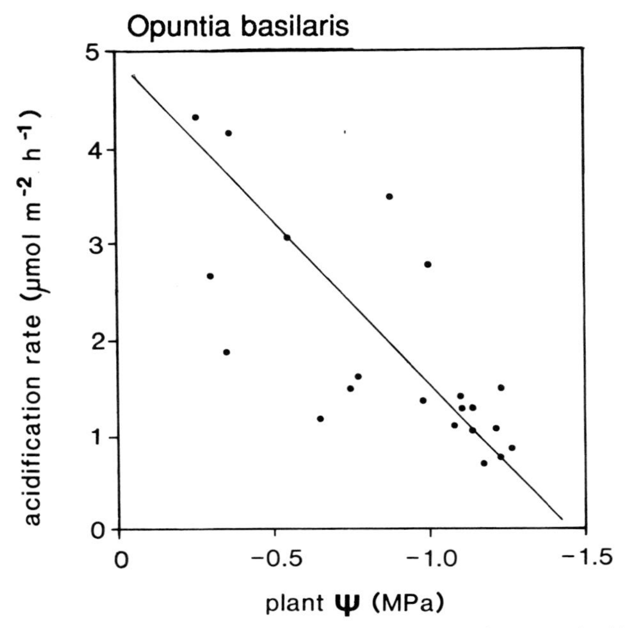

Fig. 4.112 Correlation between the rate of nocturnal acidification and plant water potential in the cactus *Opuntia basilaris*. (Data taken from Szarek and Ting 1974.)

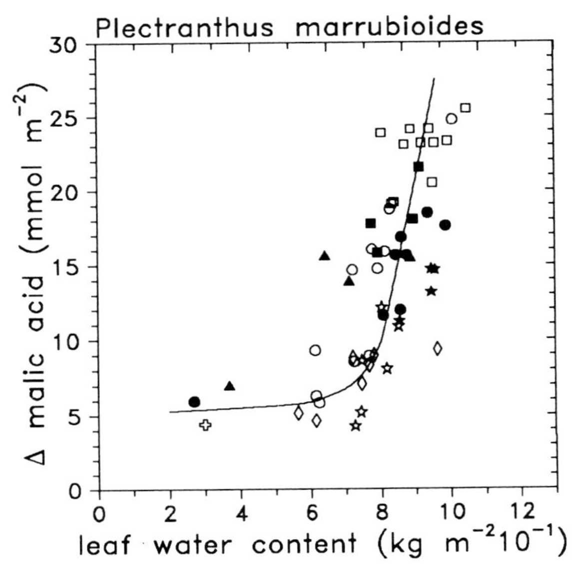

Fig. 4.113 Relationship between the overnight accumulation of malic acid and the corresponding apparent leaf water content. The different symbols represent means from measurements under very different temperature and humidity conditions which are of no interest here but best reflect natural conditions. (Data are taken from Herppich 1989.)

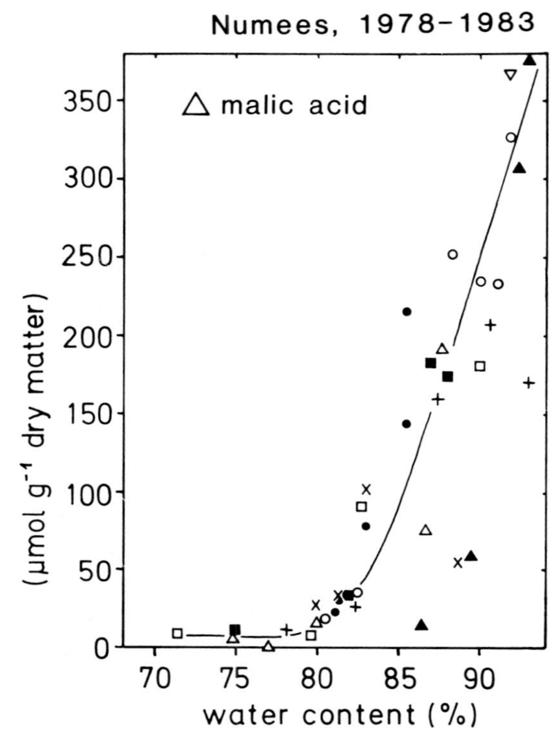

Fig. 4.114 Correlation between the amount of malic acid accumulated overnight and the corresponding content of leaf water in naturally growing (●) *Brownanthus schlichtianus*, (▲) *Sphalmanthus trichotomus*, (○) *Stoeberia beetzii*, (+) *Ruschia* sp., (△) *Cheiridopsis robusta*, (X) *Crassula deceptor*, (□) *Senecio longiflorus* and (■) *Aloe pearsonii*. Determinations were done in September 1978, March 1980, February 1981, April 1981, September 1982 and October 1983 at Numees (Richtersveld).

worthwhile. There is full agreement that the degree of succulence is a bad description of succulence. Evidence has been given earlier. A better description of succulence which is not only based on morphological features but also takes into account physiological criteria is the so called 'mesophyll succulence' (S_m). This feature is defined as (Kluge and Ting 1978)

$$S_m = \frac{\text{water content (g)}}{\text{chlorophyll content (mg)}} \tag{4.63}$$

Starnecker (1984) investigated 31 succulents of the genus *Peperomia*. When he ranked these species according to their degree of succulence he got a mix up of CAM and C_3 succulents whereas ranking them according to their mesophyll succulence separated the CAM succulents clearly from the C_3 succulents. Earlier in the book we presented a new quotient, the succulence quotient, which does not aim at separating C_3 from CAM succulents but which tries to include growth patterns into succulence providing a more valuable measure of ecological relevance of succulence.

CAM and life form

The definition of CAM as a biochemical adaptation to drought would imply that CAM predominates in evergreen leaves or stems. Findings that stems of succulents which carry deciduous leaves perform a CAM while the leaves exhibit a C_3 photosynthesis support this idea (Lange and Zuber 1977). However, *Mesembryanthemum crystallinum* and other annual CAM succulents of the same or other genera show that CAM does not only occur in long-living plants or plant organs. Can the functional concept that CAM is a 'matter of maintenance', also be applied to annuals? With the information available this question cannot be answered but it seems more likely that succulence rather than CAM is the important feature. This also proved to be the case for evergreen leaf succulent C_3 and CAM plants in the Richtersveld (von Willert and Brinckmann 1986).

The occurrence of CAM in numerous seasonal and drought deciduous succulent leaves of perennial succulents does not support the idea of CAM functioning as an adaptation to drought. We shall come back to this matter in the following chapter.

To our knowledge there is only one life form that is strictly linked to CAM. (According to the definitions that will be given in Chapter 5 we should better speak of a life strategy than of a life form.) Succulents belonging to that life strategy are characterized by completely recycling leaves. They grow only one pair of leaves per growing season even in the absence of sufficient rainfall. High amounts of utilizable biomass and utilizable water are required for this life strategy. Growth is extremely slow and CAM might play a key role in water translocation from the old leaves that are dying back to the growing ones. Unfortunately very little is known about the exciting biology of these succulents.

The more facts we know about succulents with and without CAM the more difficult it becomes to evaluate the ecological relevance of CAM. As outlined earlier we do not aim at a unifying theory. However, the link between CAM and succulence and hence utilizable water raised the question: can utilizable water be used beneficially in keeping a CO_2 balance less negative? In this context we can say that CAM in all its expressions, i.e. including all cycling and idling phenomena, may be a helpful tool to achieve this.

5

Life strategies of succulents

The remarkable rhythm of growth and the astonishing variety in form of desert plants have always attracted the attention of man. They soon aroused the suspicion that form and growth rhythm were essentially linked to the survival of the plants in that harsh environment. Early accounts of such views can be found in the work of Theophrastos of Eresos, writing in the fourth century BC and referring to plants in the Persian deserts. In time patterns in plant form and plant functioning were detected and were found to correspond with certain features of the environment in which the plants live. Ultimately this led to the development of classification systems in which plants can be assigned to types defined in functional and/or morphological terms and which are considered useful in summarizing the overwhelming variety around us. These are the systems of growth forms, life forms or plant strategy types classifying plants according to characters in their architectural designs, their functioning or their ways of exploitation of and survival in their habitats.

An early and interesting classificatory system of desert plants according to their functioning is that of Abu Hanifa, an Arabian scientist of the ninth century. He stated that desert plants belong to one of three classes: (a) plants that survive winter with their roots and their stems, (b) plants where the stem is killed by the winter but the roots survive and the plants sprout again from the roots in the favourable season and (c) plants where both stems and roots die in winter, so that the plants have to re-establish from seeds. It is amazing that this system is so similar to the well-known Raunkiaer (1907) system of life forms and went completely unnoticed though it antedates Raunkiaer's by over a thousand years (Silberberg 1910, 1911, Werger 1983c).

Classification of plants by growth forms or life forms has received much attention since the second half of the nineteenth century. Various approaches have been adequately reviewed by Du Rietz (1931) and

Barkman (1988). They show that most authors used a mixture of morphological and functional criteria to define their categories. Today the life form system of Raunkiaer, or modifications thereof, is the most commonly used system for rapid general evaluation of the plant cover of an area on a mixture of functional and morphological criteria. Recent contributions by Barkman (1988) and Rauh (1988) considerably clarified the confusion about the mixture of criteria. Barkman designed a growth form system purely on architectural features and free from hypotheses about the adaptational value of those features. Rauh recognized the principle that growth form refers to the genetically fixed morphological potential of a plant, while he considered life form the actual plant form modified in response to environmental (climatic) constraints within the limits set by the inherent morphological plan of the plant.

More recently the dynamical aspects of plant growth and plant functioning in response to the environment have been included in the evaluation of plant form and plant performance. This led to the concept of life strategy. It refers to the suite of integrated or perhaps co-adapted plant characteristics that together maximize the chance that the plant successfully exploits its habitat. This set of features can be considered at the level of the life cycle of the individual plant, referred to in the concept of plant strategy, or it can be considered at the level of population dynamics, referred to in the concept of life history strategy. These concepts thus integrate the constraints of the plant's inherent morphological plan and its physiology with the static and dynamic constraints put to the plant or population by its environment along the time scale (Fig. 5.1). Thus the dynamic response of the plant, or the population, to its environment is addressed.

Grime (1977, 1979) used this approach in a simple model explaining the evolution of plant strategies as a result of environmental pressures and natural selection. He stated that stress and disturbance (which are loosely defined by Grime) always have been the two main environmental pressures plants have had to cope with, and that this has led to the evolution of three basic plant strategies. Stress-tolerators evolved under severe and prolonged stress and limited or no disturbance, while ruderals evolved under the opposite constellation of environmental pressures; in habitats of limited or no stress nor disturbance competitors evolved, but no plants could evolve that were adapted to both severe stress and severe disturbance, according to Grime. This seems illogical (cf. Werger 1983c), and also not correct, as can be seen in a number of species of particular desert habitats.

It is the merit of Grime's model, however, that the dynamics in the pattern of investment of photosynthates in the plant is implicitly empha-

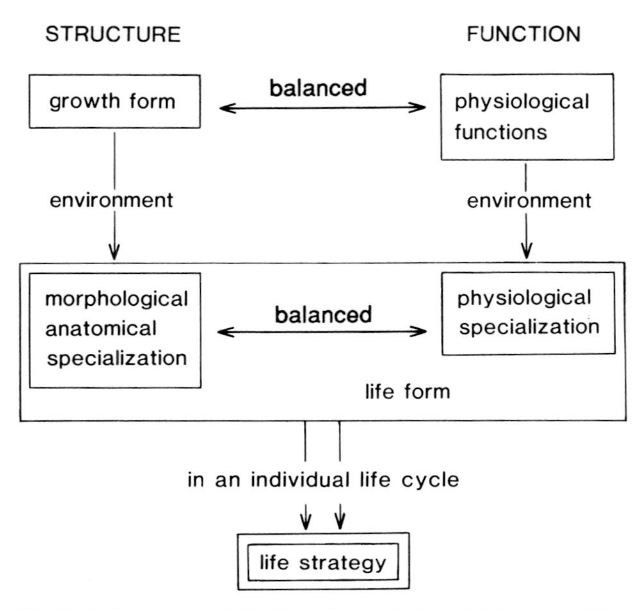

Fig. 5.1 Scheme of the definition of growth form, life form and life strategy.

sized as a crucial feature determining the plant's success within specific environmental constraints. However, in the desert environment the carbon balance of a plant and the carbon allocation pattern cannot be considered independent of the plant's water balance. Energy inflow in desert plants is very tightly coupled to and controlled by the water flow through the plant (Noy-Meir 1973). It is therefore important that we first address the relevant aspects of the water and carbon balance within the life cycle of desert plants before we return to an outline of the life strategies of desert succulents.

5.1 The water and carbon balance within the life cycle

A plant can persist only when over a period of time after germination its total of gains outweighs its total of losses. This is quite independent of a plant's life form. It leads to an increase in biomass of the plant over time. More exactly, it is not the change in total biomass that matters in this respect. Only the change in a certain fraction of the total biomass, i.e. the utilizable biomass, is relevant. It is this fraction that determines how long the plant will survive when suddenly the plant's photosynthetic carbon gain would halt, and whether or not the plant can ripen off its seeds with its available resources.

The relevance of the amount of utilizable biomass becomes clear even during germination and establishment. Seeds with a small fraction of utilizable biomass have to achieve a positive CO_2 balance soon after germination. Only then is growth assured. The situation is analogous for water. Here too we should distinguish between the utilizable and the non-utilizable fraction of the total water content (see Section 4.2). A special case in this respect are those seeds that hold too little utilizable water to realize germination. Here seed swell results in a rapidly improving positive water balance by increasing the level of utilizable water and this triggers germination.

The distinction between utilizable and non-utilizable biomass and water is relevant for all plants, and it is of particular importance for succulents, especially ephemeral succulents such as *Mesembryanthemum crystallinum* and annual succulents such as *Dorotheanthus* species. We would like to present the basic principles and interpretations of these points before we embark on particular explanations for various examples.

The total biomass of a plant consists of utilizable and non-utilizable biomass:

$$M_c = M_{cu} + M_{cs} \tag{5.1}$$

where M_c is the total, M_{cu} the utilizable and M_{cs} the structural, non-utilizable biomass.

Losses by leaf fall, herbivory or exudation concern both kinds of biomass:

$$L_c = L_{cu} + L_{cs} \tag{5.2}$$

where L_c is the total loss, L_{cu} the loss of M_{cu}, and L_{cs} the loss of M_{cs}. Thus, over a specific period of time (Δt) from t_1 to t_2, total loss is expressed by:

$$\Sigma L_c\,(t_1 - t_2) = \Sigma L_c = \Sigma L_{cu} + \Sigma L_{cs} \tag{5.3}$$

where t_1 is either the time of germination or any other time within the life cycle of the plant and t_2 marks the end of the period of time in question.

The production and dispersal of fruits and seeds also constitutes a loss of biomass, but this loss is qualitatively different from the losses discussed so far. We can, however, express these losses (F_c) over a specified period of time (Δt) in a similar way:

$$\Sigma F_c\,(t_1 - t_2) = \Sigma F_c = \Sigma F_{cu} + \Sigma F_{cs} \tag{5.4}$$

where F_{cu} and F_{cs} are the utilizable and structural biomasses respectively in inflorescences, fruits and seeds.

During the time interval from t_1 to t_2 the plant has realized a certain rate of CO_2 exchange, J_c(net). The carbon turnover, M_c(net), over this interval can be expressed as:

$$M_c(\text{net}) = J_c(\text{net})\ \mathrm{d}t = M_{cu}(\text{net}) + M_{cs}(\text{net}) \tag{5.5}$$

Consequently, the carbon-allocation pattern determines whether the carbon turnover concerns primarily utilizable (M_{cu}) or non-utilizable (M_{cs}) biomass.

When $M_{cs} > M_{cu}$ and M_c(net) is positive, the newly gained carbon is predominantly invested in structural elements, e.g. thick cell walls. A xeromorphic succulent shows a high carbon investment per unit area and unit volume. Non-xeromorphic plants invest much less carbon for the same amount of volume. Accordingly, with equal M_c(net) they will grow faster than xeromorphic plants. In Section 4.3.3 we dealt in detail with the different aspects of carbon investment per unit leaf area or unit water volume in succulents of the Richtersveld.

Let us now consider the change in biomass (ΔM_c) over an interval Δt and its partitioning between utilizable and non-utilizable biomass. Equation (5.1) can be written as

$$\Delta M_c = \Delta M_{cu} + \Delta M_{cs} \tag{5.6}$$

To some extent during interval Δt there can be a shift from utilizable biomass M_{cu} to non-utilizable biomass M_{cs} independent of the size of ΔM_c. This part, M_{cd}, represents growth, the formation of flowers, etc. and the shift can occur without ΔM_c being necessarily positive during this time interval. This is a crucial point, together with the analogous changes in utilizable water, in the understanding of the meaning of succulence. It concerns the basic strategy of the succulents. We have shown this in detail in Section 4.2.5.

From equations (5.3), (5.4) and (5.5) follows

$$\Delta M_{cu} = M_{cu}(\text{net}) - M_{cd} - \Sigma L_{cu} - \Sigma F_{cu} \tag{5.7}$$

i.e. the change in total utilizable biomass over a given period of time is the result of the net change of utilizable biomass over this period diminished by the amount of utilizable biomass that has been withdrawn for structural purposes and diminished by losses and fruit formation.

In analogy we can write:

$$\Delta M_{cs} = M_{cs}(\text{net}) + M_{cd} - \Sigma L_{cs} - \Sigma F_{cs} \tag{5.8}$$

Accordingly, the utilizable and non-utilizable biomasses, respectively, at time t_2 are adequately given by the following two equations, taking changes in M_{cd} into account:

$$M_{cu}(t_2) = M_{cu}(t_1) + M_{cu}(\text{net}) - M_{cd} - \Sigma L_{cu} - \Sigma F_{cu} \quad (5.9a)$$

$$M_{cs}(t_2) = M_{cs}(t_1) + M_{cs}(\text{net}) + M_{cd} - \Sigma L_{cs} - \Sigma F_{cs} \quad (5.9b)$$

The utilizable biomass at t_2 equals the sum of all gains minus all losses since t_1. In the first year of the life cycle $M_{cu}(t_1)$ is the amount of utilizable biomass available in the seed at the time of germination.

In perennial plants M_{cu} will initially rapidly increase as the plant grows and then more or less asymptotically reach a saturation value. However, during the life cycle of perennials periods of negative ΔM_{cu} values can be survived and the length of these periods depends on the total amount of M_{cu}. It is thus clear that the amount of M_{cu} at the beginning of a period of stress crucially determines the survival of the plant.

M_{cu}(net) represents the amount of carbon gained as utilizable biomass over Δt. This value can be positive or negative. For example, at the beginning of its dormant period a bulb has a larger amount of utilizable biomass than at the time of sprouting. The losses originate from the negative carbon turnover as a result of respiration, and thus M_{cu}(net) is negative.

M_{cd} represents the amount of utilizable biomass that is transferred into non-utilizable biomass and thus has been utilized. This concerns germination, sprouting, growing of flowers and seeds and all other growth processes at a zero or negative carbon balance.

ΣL_{cu} comprises all losses of a plant which occur because of herbivory, deciduousness, exudation (including nectar), removal of flowering parts, etc. The equation shows that this factor can attain crucial magnitude.

ΣF_{cu} comprises that part of the utilizable biomass that is used directly for reproduction. In ephemerals and annuals it is the largest sink.

Entirely analogous, and we do not derive it here in full detail, we arrive at the following equations concerning the plant's water:

$$M_{wu}(t_2) = M_{wu}(t_1) + M_{wu}(\text{net}) - M_{wd} - \Sigma L_{wu} - \Sigma F_{wu} \quad (5.10a)$$

$$M_{ws}(t_2) = M_{ws}(t_1) + M_{ws}(\text{net}) + M_{wd} - \Sigma L_{ws} - \Sigma F_{ws} \quad (5.10b)$$

where M_{wu} is the amount of utilizable water, M_{ws} is the amount of non-utilizable water (structural and solute water), M_{wu}(net) and M_{ws}(net) are analogous to the parameters in equation (5.5), and M_{wd} is the displaced water, shifted from M_{wu} to M_{ws}.

Thus, M_{wd} is an important parameter of succulents. It shows that when M_{wu} is sufficiently large, part of it can be used for growth without the necessity to stop the other processes that use part of the reserves in M_{wu}. *Prenia sladeniana* demonstrates this point very impressively (Fig. 4.66). Here water is consistently transported from source to sink independent of whether, relative to the plant's main stem, the source or sink are at the upper or lower end of the plant.

It is necessary to explicitly point out that an increase of M_{cs} refers to the primacy of growth. Growth implies water use. The plant will use this water from internal resources that are replenished from the environment. But as soon as the plant cannot replenish its water resources from the environment growth will draw water at the costs of the pool of utilizable water and then an increase of M_{cs} implies a decrease in M_{wu}.

This should suffice as to general considerations and we will now address the special cases.

5.1.1 Ephemerals and annuals

In an annual that successfully completes its life cycle from germination (t_1) till seed ripening (t_2), $M_{cu}(t_2)=0$. As $M_{cu}(t_1)$ is very small compared to M_{cu}(net) it can be disregarded. Usually M_{cd} is also not important, because utilizable biomass is used virtually entirely in the inflorescence and thus is covered by the term ΣF_c. Consequently, equation (5.9a) simplifies to

$$\Sigma F_{cu} = M_{cu}(\text{net}) - \Sigma L_{cu} \tag{5.11}$$

Similarly, equation (5.10a) leads to

$$\Sigma F_{wu} = M_{wu}(\text{net}) - \Sigma L_{wu} \tag{5.12}$$

Thus, the investment available for seed solely depends on growth conditions and loss through herbivory or, in other words, on environmental conditions.

We made a preliminary estimate that the utilizable biomass in a succulent ephemeral such as *Mesembryanthemum crystallinum* amounts to over 80% of the total biomass. (We cannot as yet estimate the amount of utilizable water in this species.) Based on this assumption, the development of the utilizable biomass and utilizable water during the life cycle of an ephemeral or an annual should approximate the pattern shown in Figs. 5.2a and 5.2b.

In a dry seed M_{wu} is nearly zero, and M_{cu} is by far the largest component of the biomass. As the seed swells M_{wu} rapidly rises and continues to increase as long as water can be drawn from the soil and the plant grows. M_{wu} of the whole plant decreases as soon as M_{wu}(net) is negative for more than 24 hours and no additional precipitation falls. At that point daily water use and water needs for the growth of fruits and seeds determine whether or not and how many seeds can be grown with the available amount of utilizable water. Measurement of these variables will allow an

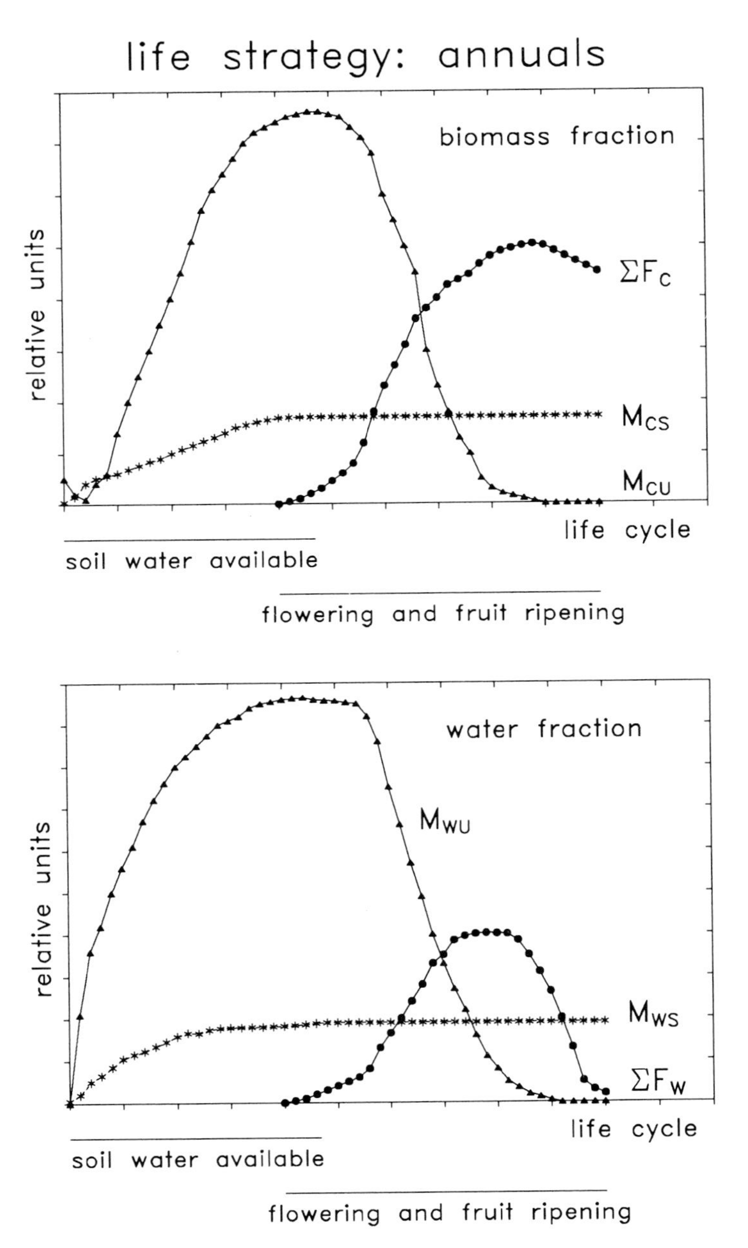

Fig. 5.2 Development and relation of the various fractions of biomass and water within the life cycle of succulent annuals. Data are given in relative units and are based on estimations for *Mesembryanthemum crystallinum*. M_{cs} = non-utilizable, structural biomass, M_{cu} = utilizable biomass, $\sum F_c$ = total biomass of fruits and seeds, M_{ws} = structural water, M_{wu} = utilizable water and $\sum F_w$ = total water of fruits and seeds.

estimate of M_{wu}. Even at this stage plants with CAM can realize a positive CO_2 balance over a 24-hour period. This may slow down or even briefly reverse the decrease in utilizable biomass. The pattern in M_{cu} will be very similar to that in M_{wu} except for the seed swell and germination phases. Should the result of either equation (5.11) or (5.12) become zero before the seeds have ripened that would imply that the plant has died before it could reproduce. And should this happen in a number of consecutive years then the survival of the population of ephemerals or annuals in that environment depends on the longevity of the seeds, the seedbank and the regulation of the germination of those seeds. Such a succession of events is typical for summer annuals growing in a winter rainfall area, such as *Zygophyllum simplex*. This species is rarely seen in the Richtersveld where it meets the southwestern limit of its distribution area.

If biologically effective amounts of precipitation fall intermittently even annuals can resume growth once lack of moisture provisionally has forced them to stop. Under such circumstances M_{cu} and M_{wu} show stepped rises during the life cycle.

The amount of M_{cu} and M_{wu} built up by annuals as a result of a rainfall event determines whether or not the plant can persist till the next event. In this respect succulent annuals appear to have an advantage over non-succulent annuals. They have a better rain gap endurance. This explains why succulent annuals often are so conspicuously abundant in arid areas such as the southern Namib, where rainfall events are few and temporally far between. Instructive examples are provided by *Mesembryanthemum pellitum*, *Opophytum aquosum* and *Dorotheanthus* spp.

5.1.2 Hapaxanthous plants

In principle there is little difference between the life cycles of hapaxanthous plants and annuals. In hapaxanthous plants the critical phase of establishment is longer. Survival is solely determined by the amounts of M_{cu} and M_{wu} reached at the beginning of the dry period and by the duration of the dry period.

Until the inflorescence is formed ΣF_{cu} stays zero. If losses are presumed to be negligible compared to $M_{cu}(t_1)$ and if there is only growth when M_c(net) is positive, the equation for a 'normal' year reads

$$M_{cu}(t_2) = M_{cu}(t_1) + M_{cu}(\text{net}) - M_{cd} \qquad (5.13)$$

This implies that growth ($M_{cs}(t_2) > M_{cs}(t_1)$) depends on gain of utilizable biomass (M_{cu}(net) from the plant's CO_2 budget (equation (5.5)) and that is controlled by the water balance. Thus, the environmental conditions during the year are crucial for growth. In the year of flowering (t_1) the situation is similar to annuals:

$$F_{cu}(t_n) = M_{cu}(t_{n-1}) + M_{cu}(\text{net}) - L_{cu} \qquad (5.14)$$

When the seeds turn ripe $M_{cu}(t_n) = 0$. We do not embark here on a separate consideration of M_{wu} since it is analogous in all relevant aspects to what has been said about M_{cu}.

Obviously a hapaxanthous plant can have more losses than gains in the course of the year. Then $M_{cu}(t_2) < M_{cu}(t_1)$. The length of the unfavourable period that can be survived will be proportional to the size of the plant, or better, the amount of $M_{cu}(t_1)$ just before the start of the unfavourable period. Thus, the time courses of available amounts of utilizable biomass and utilizable water in hapaxanthous plants will show stepped increases and decreases as will be the case for all perennial plants. Apparently a critical threshold value of M_{cu} and M_{wu} has to be reached in hapaxanthous plants before reproduction is initiated. How this value is measured and reproduction is triggered in, for example *Agave deserti*, remains unknown. We will return to *Agave deserti* in the following pages.

5.1.3 Geophytes

A geophyte establishing from seed must build up large reserves of water and biomass in the first year. These reserves should suffice not only to bear the unavoidable carbon and water losses during the subsequent period of dormancy but also to guarantee the renewed flush of the plant in the next growth season until a positive CO_2 and water balance (over a 24-hour period) is achieved. Thus, a very large proportion of M_c(net) must flow into the M_{cu}(net) pool, so that M_{cu} is considerably larger than the M_{cd} needed to flush. Since geophytes have a short growth season, it seems clear that they have to refrain from flowering until their M_{cu} and M_{wu} are sufficiently large. It is possible that for reasons argued in Chapter 4 leaf succulent geophytes here have an advantage over non-succulent geophytes. It is also possible that geophytes that flower during the time they have leaves or immediately after leafy period refrain from flowering when unfavourable weather conditions force them to abandon their leaves prematurely. But relevant observations are lacking. However, if they did not do this, they might exhaust their reserves when $F_{cu} > M_{cu}(t_1) + M_{cu}(\text{net})$.

Apart from the work of Pate and Dixon (1981), rather little is known about the physiology of desert geophytes, presumably because of their short and rather uncertain growth season. It seems that as well as C_3 photosynthesis, CAM or at least CAM cycling occurs. We conclude this from our own measurements in the southern Namib (e.g. *Trachyandra falcata*, *Crassula* sp., *Pelargonium* sp., *Anacampceros comptonii*, etc.).

5.1.4 Deciduous perennials

In this category we consider those plants that possess perennial above-ground organs with deciduous leaves. For these deciduous perennials the situation is basically the same as for geophytes. The only difference is that their perennial above-ground organs during the leafless period burden the carbon and water balances of the plants still more strongly.

Deciduous above-ground perennating succulents are characteristic for the southern Namib. Therefore we want to show the annual time courses of M_{cu} and M_{wu} for a typical representative of this group, *Tylecodon paniculatus*.

Figs. 5.3a and 5.3b show the same individual, leafless in February 1981 and in full leaf in August 1981. The following analysis of this individual was carried out in August and October 1982. About 90% of the stem's fresh weight consisted of water and contained about 77 litres. In full leaf this individual contained 612 leaves with a total transpirational surface area (stomata on both sides) of 1.83 m^2. The leaves contained 2.2 l of water. This dropped to 1.5 l in mid-October due to persistent bergwind conditions. The number of leaves did not change in the period from August until October but the total leaf area dropped from 1.83 to 1.32 m^2 due to shrinkage. Leaf drop started at the end of October. Apparently water loss from the leaves was not immediately replaced by water from the stem when the soil did not any longer provide sufficient water. We measured transpiration during a day with bergwind and during a foggy day. Under the extreme conditions of bergwind (the VPD of the air averaged 55 mPa Pa^{-1} over the 24 hours of measurement, cf. Fig. 3.11) 0.448 l of water was lost in 24 hours which equals 20% of the total leaf water quantity.

When fully turgid the total amount of water in the leaves is only 3% of the amount in the stem. Nevertheless there seems to be an intensive reallocation of water from the leaves to the stem before the leaves are dropped. Leaf flush is strictly photoperiodically controlled. It occurs during the hot season of the year, months before the winter rains are to resume. In the beginning only a few rosettes with few, small leaves are grown. The rosettes make up to 50% of the total maximal leaf area developed in the rainy season. Thus, for the development of these early leaf rosettes only about 2% of the stem's water is needed. Until the first rains the transpiration losses from the leaves also have to be drawn from the stem's water reserves. As a result of the rapid changes from bergwind to 'normal'

Fig. 5.3 The seasonal deciduous *Tylecodon paniculatus* (a) in the leafless stage during the summer drought and (b) in full leaf during winter.

conditions, as from March (cf. Fig. 3.10) the average daily (24 hours) water use of the winter foliage amounts to 0.1–0.2 l. When the winter rains start late a hundred or more days can pass before it is possible for the plant to replenish its water reserves (compare the years 1979 and 1980, Fig. 3.8). Depending on the amount of leaf area carried by the plant and the weather conditions this may imply a considerable loss of water from the stem.

Leaf drop in October and renewed leaf flush in February before the winter rains allow the plant to save a considerable amount of water as compared to the alternative of staying evergreen and fully leaved. Should the plant keep its leaves during the whole summer, it would use 40 l of water which amounts to 60% of the stem's water. It certainly is doubtful whether so high a percentage of the stem's water content is utilizable water. Leaf flush at the beginning of autumn (February) and maintenance of these leaves till the rains start take about 10 to 20 l of water, or maximally 30% of the stem's water reserves. One can easily notice the water loss from the stem during the dry summer season: while the stem is rigid and swollen during the rainy season, it is soft and not fully turgid just before the rains start.

Tylecodon paniculatus flowers in summer after the leaves have been dropped. That means that all investments in the inflorescence are taken

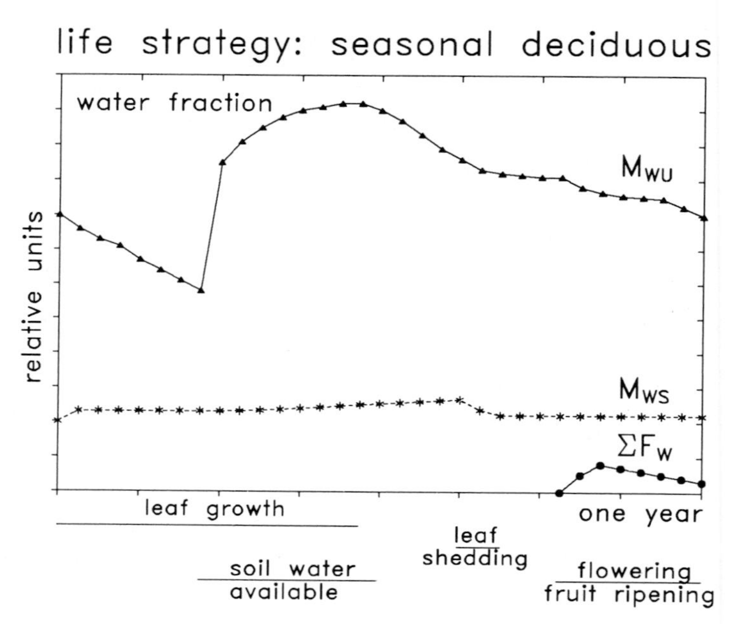

Fig. 5.4 The possible course of the development of the various fractions of water in *Tylecodon paniculatus* during an 'ideal' year. Curves were calculated from field data given in the text and are presented in relative units. M_{wu} = utilizable water, M_{ws} = non-utilizable, structural water and $\sum F_w$ = total water of the inflorescence and seeds.

from the pools of utilizable biomass and water in the stem. Unfortunately, we do not know of observations that show that *Tylecodon* refrains from flowering when winter rains in the season before were absent or very poor. We do know, however, that *Tylecodon paniculatus* grew new leaves in February 1980 and 1981 despite rather poor winter rains in the years before.

Fig. 5.4 shows the possible course of M_{wu} in *Tylecodon paniculatus* during an 'ideal' year. Under such conditions $M_{wu}(t_n) < M_{wu}(t_{n+1})$. The lengths of the various phases of leaf growth, leaf shedding, flowering and soil water availability are based on our observations in the Richtersveld during long periods of time. The proportions of M_{wu} and M_{ws} in the total amount of water were set at 80:20. Water reserves in the root system were ignored. Based on these estimates and the values for transpiration, leaf water content, leaf flush, etc., we calculated the various water fractions for a growth of 5% (Fig. 5.4). Biomass values are not separately shown, because they resemble the curve for water values at all important points.

We tried to determine the sizes of M_{cu} for several succulents of the Richtersveld and compare these values with that for the leaves of a sclerophyllous tree. The results are shown in Table 5.1.

Table 5.1. *Estimates of utilizable biomass* (M_{cu}) *in percentage of total biomass of five perennial species of the Richtersveld (four evergreen succulents and one sclerophyllous deciduous tree,* Ozoroa dispar*) and a mangrove (*Pelliciera rhizophoreae*) from the coastal swamps of Colombia*

Prenia sladeniana	48%	CAM
Othonna opima	38%	C_3
Cotyledon orbiculata	36%	CAM
Senecio corymbiferus	25%	CAM
Ozoroa dispar	4–5%	C_3
Pelliciera rhizophoreae	11–16%	C_3

In this respect the results obtained with *Agave deserti* are interesting (Nobel 1977b). *Agave deserti* is hapaxanthous. The dry mass of its inflorescence averages 1.25 kg. The annual photosynthetic production of a full-grown leaf is about 19 g, which means that *Agave deserti* needs at least 60 to 80 'leaf years' before it can grow an inflorescence from its photosynthetic production. *Agave deserti* needs about 10 to 25 years to grow the number of leaves needed. The inflorescence constitutes an enormous sink for carbon and water: from the initiation of the inflorescence until seed ripening the plant loses 77% of its fresh weight and 35% of its dry weight. In this example a plant of *Agave deserti* with 68 leaves loses 24.9 kg of its fresh weight and it dies with seed ripening. Thus $M_{wu}(t_1)$ must have been about this amount, and $M_{wu}(t_2)=0$. Transpiration losses of the plant are 7.1 kg ($M_{wu}(\text{net})=7.1$). The development of the inflorescence uses 3.1 kg ($M_{wd}=3.1$), and 14.7 kg are lost by the inflorescence until seed ripening ($\Sigma F_{wu}=14.7$).

Though the analogy is perhaps not complete we can use these values to calculate the various amounts of utilizable water:

$$M_{wu}(t_2)=M_{wu}(t_1)-M_{wu}(\text{net})-M_{wd}-\Sigma F_{wu}$$
$$0=24.9-7.1-3.1-14.7 \quad (5.15)$$

We could use the same reasoning to calculate the amount of utilizable biomass but assessment of the various compartments is not as easy as in the case of the amount of utilizable water.

The *Agave* plant had 68 leaves which produced about $68\times 0.019=1.29$ kg of photosynthates. An unknown part of this was needed for respiration, i.e. $M_{cu}(\text{net})=(1.29-x)$. The investments in the inflorescence including the seeds (?) amount to 1.25 kg, thus $M_{cd}=(1.25-y)$ where y symbolizes the amount of utilizable biomass in the seeds. The inflorescence respires a total of 0.59 kg until the time when the seeds are ripe, thus $\Sigma F_{cu} = 0.59 + y$. The value of y is unknown but that does not matter since y is lost through addition. The amount of the initial utilizable biomass $M_{cu}(t_1)$ plus the turnover of utilizable biomass $M_{cu}(\text{net})$ have to be 1.84 kg, or

$$[M_{cu}(t_1) + M_{cu}(\text{net})] = M_{cd} + \Sigma F_{cu} = 1.25 + 0.59 = 1.84 \text{ (kg)} \tag{5.16}$$

$M_{cu}(\text{net})$ must be a part of the total of 1.29 kg of produced photosynthates and $M_{cu}(t_1)$ is thus 1.84 minus that part.

Ferocactus acanthodes, a barrel cactus that flowers every year, increases its total biomass by 10% per year. That is only a quarter of the annual rate of biomass increase of *Agave deserti*. But *Ferocactus* uses only 6% of its pool of utilizable water for its annual flowering and seed ripening. Unfortunately we do not have any data on its carbon investments.

We sincerely wish that similar analyses will be carried out for other areas with succulents and for all kinds of life forms. They will provide a basic understanding of the life strategies of these plants.

5.2 Life strategies

In the early chapters we discussed the multitude of challenges that characterize arid habitats. Much of the environmental variation in arid regions is stochastic; favourable and unfavourable conditions are impossible to predict in space and time. This forces plants to develop diverse adaptive strategies, each of them fit to face a certain set of conditions but probably never suited to overcome all constraints. Such adaptive strategies could develop as a result of natural selection preserving a rich array of modified and specialized features in the plants' basic sets of structural and functional characteristics in response to the environment. These modifications are to some extent specific responses to specific environmental constraints. In this way a variety of life forms evolved that impressively document diversity as well as convergence. Careful analysis of the different life forms and their characteristic functional and structural specializations shown during various phases of their life cycles reveals that certain modifications, e.g. succulence, can occur in various life forms but differ considerably in relevance. We defined the life strategy of a plant as the plant's set of integrated and perhaps co-adapted features that are suitable, in the context of the whole life cycle of the plant, to maximize the chances to successfully meet the specific ecological problems set by its habitat, to exploit the resources in its habitat and thus to maximize its chances for survival. Consequently, a life strategy comprises a set of elements that each have a certain rank value, a weighted importance in the total of the integrated set of features that we call life strategy. There is a hierarchical ranking in the importance of the different elements of the life strategy. The elements can be causally linked but they can also be independent. Strictly photoperiodically induced leaf flush of a perennial plant in an arid area with seasonal rainfall requires the availability of a considerable amount of utilizable water and thus is necessarily linked with the life strategy element of succulence. We discussed the example of *Tylecodon paniculatus* in detail in the previous section. Acquisition or loss of an adaptive feature can increase or decrease the value of another element of the life strategy. As a result of this we find the same element in different strategies at different ranks of importance in the hierarchical system of life strategy elements. For example, CAM is of primary importance in winter annuals whereas it is of little significance in perennials which are seasonally deciduous (see Table 5.3).

Considering life form in the context of the life cycle we arrive at four basic life strategies among the succulents, solely based on the lengths of their cycles (Table 5.2). We purposely distinguished ephemerals apart from annuals because they constitute very different strategies. An ephemeral germinates after every rain storm independent of the time of the year. An annual reacts to precipitation only when it falls in the proper season. We are not sure whether or not ephemeral succulents really occur, but our own observations persuade us to consider *Mesembryanthemum crystallinum* as perhaps belonging to this group. Its occurrence in arid areas with only erratic precipitation also suggests this.

Among the annuals the mode of photosynthesis is an important feature. Whereas C_3 photosynthesis is equally common among summer and winter annuals, C_4 photosynthesis is mainly found in summer annuals and CAM in winter annuals. The number of succulents is larger among the annuals in winter rainfall areas than in summer rainfall areas. We do not need to emphasize the importance of the strategy element annuality. The economic use of the available pools of utilizable water and biomass in the seed is very effective. Succulence, particularly in combination with CAM, favours survival in rainless periods as we have shown repeatedly.

The pauciennial strategy is very interesting. There are both C_3 succulents (e.g. *Augea capensis*) and CAM succulents (e.g. *Mesembryanthemum barklyi*) among the pauciennials. Most of the species we know are able to realize a neotenic growth pattern: they flower in their juvenile stage

Table 5.2. *Life strategies of succulents*

1. Ephemerals

2. Annuals
- 2.1. summer annuals
- 2.2. winter annuals

3. Pauciennials
- 3.1. seasonal
- 3.2. aseasonal

4. Perennials
- 4.1. plant below-ground persistent, above-ground deciduous (geophytes)
  - 4.1.1. flowers and leaves in the same season
  - 4.1.2. flowers and leaves in different seasons
- 4.2. entire plant below-ground persistent, only flowers above ground
  - 4.2.1. plant without leaves
  - 4.2.2. plant persistently leaved
    - 4.2.2.1. leaves annually replaced
    - 4.2.2.2. leaves perennial
- 4.3. plant above-ground persistent
  - 4.3.1. plant without green leaves (except cotyledons)
    - 4.3.2. plant with evergreen leaves
    - 4.3.3. plant with deciduous (annually replaced) leaves

(provisional flowering) but can start regrowth, often from their inflorescences, when new rains come in time. This often leads to the situation that a plant carries ripe seeds and flowers simultaneously. Such a growth pattern is also typical for *Salicornia europaea*. Succulent pauciennials necessarily are evergreen. Retranslocation of water and biomass is well developed and the utilizable fractions are very high. CAM seems to positively influence the maximal age of the pauciennial plants. It also seems to improve their competitive vigour as against C_3 pauciennials along a gradient of decreasing precipitation. Based on the germination response we can distinguish between seasonal pauciennials, which germinate only when rains fall in the appropriate season, and aseasonal pauciennials, which germinate upon precipitation in any season (e.g. *Augea capensis*). Though we have no data available, we presume that pauciennials use all precipitation, independent of the season, to grow and to increase their M_{cu} and M_{wu}.

Outside the rainy season perennials are the most commonly encountered plants in arid regions though even some of them are hard to find. We distinguish three substrategies in the category of perennials. The first substrategy comprises the geophytes in the sense of Raunkiaer. These plants perennate with their subterraneous organs only; above ground they resemble annuals. Very little is known about this interesting group of plants, the so-called 'soil succulents'. We want to refer to the comprehensive description of Pate and Dixon (1981) of the geophytes of West Australia.

The second substrategy comprises plants that are usually small and either grow entirely subterraneously or just reach the soil surface and hardly stick out. Only when flowering are they easily spotted. Examples are species of soil cacti, *Lithops*, *Frithia*, *Oophytum*, etc. Many of the species in this substrategy are window plants, discussed in Section 4.1. A majority of those plants have a short stem and only two succulent leaves (e.g. *Lithops*) that are replaced, seasonally or rain-induced, by two new ones. The growth of the new leaves is photoperiodically controlled and occurs at the cost of the old leaves. Consequently, the new leaves are always smaller than the old ones and only following sufficient rain do they fully expand. Sometimes they swell till they burst. Most of these species can be provoked to a more rapid replacement of their leaves but this soon leads to the death of the plant. This suggests that a period of dormancy is required. The strategy to replace one pair of leaves annually is very common among the perennials in arid southern Africa, though it seems to be restricted to the Mesembryanthemaceae. Examples are found in the genera *Argyroderma*, *Conophytum*, *Cheiridopsis*, *Lithops*, etc.

It is impressive to see the use of the pools of utilizable biomass and utilizable water in these plants, but only qualitative observations as presented for *Prenia sladeniana* in Fig. 4.66 and for *Brownanthus schlichtianus* in fig. 4.72 are available; no quantitative data are available.

The third substrategy comprises those succulents that have aboveground perennating organs. In this category a group can be distinguished that has no green leaves except for the cotyledons. All the leaf's functions have been taken up by stem and twigs. Clear examples are found in the Cactaceae and in several Euphorbiaceae and Asclepiadaceae.

Apart from this group there are two other groups with leaves, one evergreen and the other deciduous. The deciduous species somewhat resemble, at least qualitatively, the annuals. Above-ground persistent perennial with deciduous leaves is a frequent strategy in the flora of southern Africa. That is why we have made a detailed hierarchical classification of strategy elements as known to us for plants in this category (Table 5.3 and Fig. 5.5). This classification emphasizes that the element of annuality in the perennial plant is again of great importance in its life

Table 5.3. *Detailed presentation of the life strategies of perennial succulents which persist above ground. The listing shows the hierarchical structure of the elements of the strategies. Compare also with Fig. 5.5*

4. Perennials
- 4.3. above-ground persistent
  - 4.3.3. plant with deciduous (annually replaced) leaves
    - 4.3.3.1. leaves drought deciduous
      - 4.3.3.1.1. plant with woody non-succulent shoots
        - 4.3.3.1.1.1. leaves with C_3 photosynthesis
        - 4.3.3.1.1.2. leaves with CAM
      - 4.3.3.1.2. plant with succulent shoots
        - 4.3.3.1.2.1. shoots with CAM features
          - 4.3.3.1.2.1.1. leaves with C_3 photosynthesis
          - 4.3.3.1.2.1.2. leaves with CAM
        - 4.3.3.1.2.2. shoots without CAM features
          - 4.3.3.1.2.2.1. leaves with C_3 photosynthesis
          - 4.3.3.1.2.2.2. leaves with CAM
    - 4.3.3.2. leaves seasonally deciduous
      - 4.3.3.2.1. plant with succulent shoots
        - 4.3.3.2.1.1. shoots without CAM features
          - 4.3.3.2.1.1.1. leaves with C_3 photosynthesis
          - 4.3.3.2.1.1.2. leaves with CAM
        - 4.3.3.2.1.2. shoots with CAM features
          - 4.3.3.2.1.2.1. leaves with C_3 photosynthesis
          - 4.3.3.2.1.2.2. leaves with CAM
      - 4.3.3.2.2. plant with woody non-succulent shoots (?)

strategy. While in the life strategy 'annuals' the mode of photosynthesis is the hierarchically next most important strategic element, in the perennials with deciduous (annual) leaves other elements are more important.

What triggers leaf drop and leaf flush is very important. This can be either drought and precipitation or photoperiod. It sometimes is difficult to verify this with certainty in the field, because drought in the winter rainfall areas of the southern hemisphere is always coupled with an increasing photoperiod. Seasonality can be clearly assessed, however, when leaves flush before the start of the rainy season, as for example in *Tylecodon paniculatus*. Otherwise greenhouse experiments under controlled light and moisture conditions are necessary.

Drought deciduousness and seasonal deciduousness are very different strategy elements that require very different preconditions. Leaf flush without rain necessarily requires succulence in stem or roots, whereas

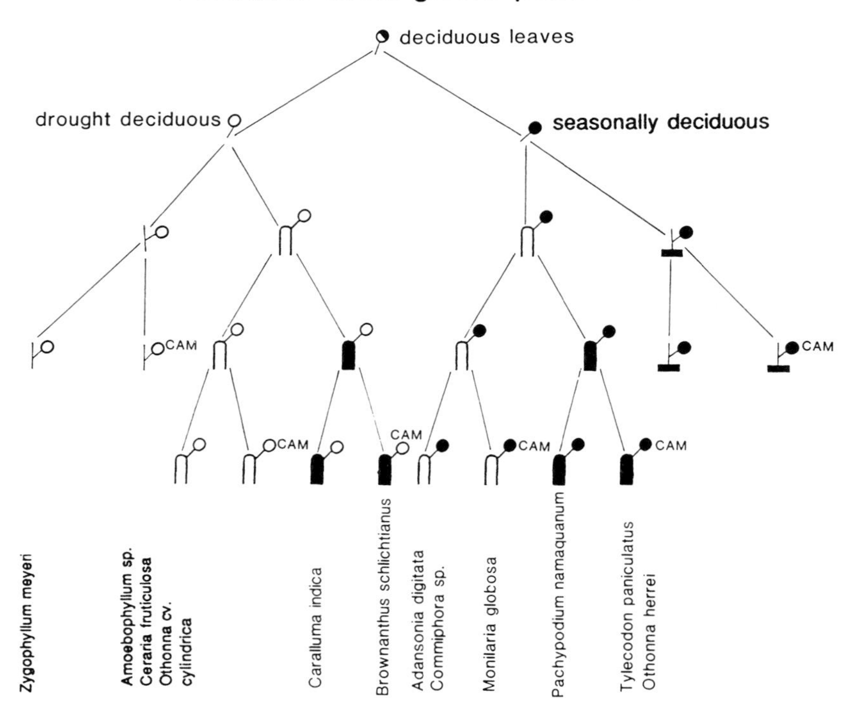

Fig. 5.5 Scheme showing the hierarchical structure of the elements of the life strategy 'above-ground persistent perennial succulents with deciduous leaves'. Examples for the various combinations of elements of the different life strategies are also given. Open circles represent drought deciduous leaves, filled circles seasonally deciduous leaves. A vertical line stands for a woody non-succulent stem, the vertical bar for a succulent shoot either without CAM activity in the dry season (open bar) or with CAM activity in the dry season (filled bar). A filled horizontal bar stands for a succulent root. So far no examples of seasonally deciduous succulents with a woody stem and succulent roots are known. The scheme illustrates Table 5.3.

drought deciduousness and leaf flush following rain is possible without perennating succulent organs. In this latter group (4.3.3.1.1) it is, according to our observations, rather irrelevant whether or not the succulent leaves possess the CAM mode of photosynthesis, as there is no measurable advantage to the plant. And if the stem is succulent (group 4.3.3.1.2) the mode of photosynthesis of the leaves is even less relevant.

A comparison between *Tylecodon paniculatus* and *Pachypodium namaquanum* is illustrative. Both are perennial (4), above-ground persistent (4.3), with annually deciduous leaves (4.3.3), and the leaves are dropped seasonally (4.3.3.2). Their shoots are succulent (4.3.3.2.1) and possess CAM features (4.3.3.2.1.1). The only difference is that their leaves possess the CAM (*Tylecodon*) or C_3 (*Pachypodium*) mode of photosynthesis. Both species grow well in the same habitat and survived the severe drought of 1979–1980 without damage. Somehow we got the impression that *Pachypodium* could survive even more extreme sites than *Tylecodon*.

A scheme of the groups of above-ground persistent perennial succulents with deciduous leaves is presented in Fig. 5.5 and examples as far as are known are given for the various life strategies.

Comparative studies into the analysis of the adaptive significance of strategy elements strictly require that the plants differ in only the element under study. If this condition is not met incorrect interpretations are easily obtained.

It is tempting to continue this path into more philosophical reflections, for instance to consider the question whether or not the development of annuality late in evolution has made all earlier adaptations to aridity insignificant or futile. However, we do not want to go down this road. We confine ourselves to this presentation of life strategies. Many more observations and measurements are needed and will lead to many corrections of our concept and presentation of life strategies. We believe, however, that our concept is suitable to stimulate discussion and that our proposal offers the basis for an understanding of the various ways of life of succulents in the desert.

References

NB References with the same first author and two or more other authors (cited as *et al.* in the text) are listed in date order.

Ackerman, T. L. and S. A. Bamberg 1974. Phenological studies in the Mojave desert at Rock Valley (Nevada Test Site). In: H. Lieth (ed.): *Phenology and Seasonality Modelling*, Springer, Berlin, New York, pp. 215–226.

Acocks, J. P. H. 1953. Veld types of South Africa. *Memoirs of the Botanical Survey of South Africa*, **28**:1–192.

Adamson, R. S. 1938. *The vegetation of South Africa*, British Empire Vegetation Commission, London.

André, M., D. A. Thomas, D. J. von Willert and A. Gerbaud 1979. Oxygen and carbon dioxide exchange in Crassulacean-acid-metabolism-plants. *Planta*, **147**:141–144.

Bange, G. G. J. 1953. On the quantitative explanation of stomatal transpiration. *Acta Botanica Neerlandica*, **2**:255–297.

Barbosa, L. A. G. 1970. *Carta fitogeografica de Angola*. Instituto de Investigação Cientifica de Angola, Luanda.

Barbour, M. G. 1981. Plant-plant interactions. In: D. W. Goodall and R. A. Perry (eds.), **2**:33–49.

Barkman, J. J. 1988. New systems of plant growth forms and phenological plant types. In: M. J. A. Werger *et al.* (eds.): *Plant Form and Vegetation Structure*, SPB Academic Publishers, The Hague, pp. 9–44.

Beadle, N. C. W. 1981. *The Vegetation of Australia*. Cambridge University Press, Cambridge, 690 pp.

Beatley, J. C. 1967. Survival of annuals in the northern Mojave Desert. *Ecology*, **48**:745–750.

Benzing, D. H. 1984. Epiphytic vegetation: a profile and suggestions for future inquiries. In: E. Medina, H. A. Mooney and C. Vazquez-Yanes (eds.): *Physiological Ecology of Plants of the Wet Tropics*, Junk, The Hague, Boston, pp. 155–171.

Böcher, M. and M. Kluge 1977. Der C_4-Weg der C-Fixierung bei *Spinacea oleracea*. I. ^{14}C-Markierungsmuster suspendierter Blattstreifen unter dem Einfluß des Suspensionsmediums. *Zeitschrift für Pflanzenphysiologie*, **83**:347–361.

Brandt, A. B. and S. V. Tageeva 1967. *Optical Parameters of Plant Organisms*, Nauka Publishers, Moscow.

Brinckmann, E., M. Wartinger and D. J. von Willert 1985. Turgoränderungen in Blasenzellen von Mesembryanthemaceen. *Berichte der Deutschen Botanischen Gesellschaft*, **98**:447–454.

Brulfert, J., D. Guerrier and O. Queiroz 1973. Photoperiodism and enzyme activity: balance between inhibition and induction of the CAM. *Plant Physiology*, **51**:220–222.

Brunnhöfer, H., H. Schaub and K. Egle 1968a. Der Verlauf des CO_2-und O_2- Gaswechsels bei *Bryophyllum daigremontianum* in Abhängigkeit von der Temperatur. *Zeitschrift für Pflanzenphysiologie*, **59**:285–292.

Brunnhöfer, H., H. Schaub and K. Egle, 1968b. Die Beziehungen zwischen den Veränderungen der Malat- und Stärkekonzentration und dem CO_2- und O_2-Gaswechsel bei *Bryophyullum daigremontianum. Zeitschrift für Pflanzenphysiologie,* **60**:12–18.

Caldwell, M. M. and J. H. Richards 1989. Hydraulic lift: water efflux from upper roots improves effectiveness of water uptake by deep roots. *Oecologia*, **79**:1–5.

Cameron, R. J. 1970. Light intensity and the growth of *Eucalyptus* seedlings. II. The effect of cuticular waxes on light absorption in leaves of *Eucalyptus* species. *Australian Journal of Botany*, **18**:275–284.

Cannon, W. A. 1911. *The Root Habitats of Desert Plants. Publications of the Carnegie Institute*, No. 131.

Chen, S.-S. and C. C. Black. 1983. Diurnal changes in volume and specific tissue weight of Crassulacean acid metabolism plants. *Plant Physiology*, **71**:373–378.

Cloudsley-Thompson, J. L. 1977. *Man and the Biology of Arid Zones*. Edward Arnold, London.

Coetzee, B. J. and M. J. A. Werger 1975. A west–east vegetation transect through Africa south of the Tropic of Capricorn. *Bothalia*, **11**:539–560.

Coetzee, C. G. 1969. The distribution of mammals in the Namib desert and adjoining inland escarpment. *Scientific Papers of the Namib Research Station*, **40**:23–36.

Cole, D. T. 1988. *Lithops: Flowering Stones*. Acorn Books, Randburg, Republic of South Africa.

Compton, R. H. 1929. The vegetation of the Karoo. Journal of the Botanical Society of South Africa, **15**:13–21.

Cornet, A. F., J. P. Delhome and C. Montana 1988. Dynamics of striped vegetation patterns and water balance in the Chihuahuan desert. In: H. J. During *et al.* (eds.): *Diversity and Pattern in Plant Communities*, SPB Academic Publishers, The Hague, pp. 221–231.

Cowling, R. M. and B. M. Campbell 1983. A comparison of fynbos and non-fynbos coenoclines in the lower Gamtoos River valley, SE Cape, South Africa. *Vegetatio*, **53**:161–178.

Cowling, R. M. and P. W. Roux (eds.) 1987. The Karoo biome: a preliminary synthesis, Part 2. *South African National Science Programme*, Report No. 142, 134 pp.

Cowling, R. M., G. E. Gibbs Russel, M. T. Hoffman and C. Hilton-Taylor 1989. Patterns of plant species diversity in southern Africa. In: B. J. Huntley (ed.): *Biotic Diversity in Southern Africa*, Oxford University Press, Cape Town, pp. 19–50.

Cutler, J. M., Rains, D. W. and R. S. Loomis 1977. The importance of cell size in the water relations of plants. *Physiologia Plantarum*, **40**:255–260.

Dainty, J. 1976. Water relations of plant cells. In: U. Lüttge and M. G. Pitman (eds.): *Transport in Plants. II. Encyclopedia of Plant Physiology*, NS, vol. II A, Springer-Verlag, Berlin etc. pp. 12–35.

Delf, E. M. 1912. Transpiration in succulent plants. *Annals of Botany*, **26**:409–442.

Denius, H. R. and P. H. Homann 1972. The relation between photosynthesis, respiration, and CAM in leaf slices of *Aloe arborescens* Mill. *Plant Physiology*, **49**:873–880.

Dixon, M. and J. Grace 1983. Natural convection from leaves at realistic Grashof numbers. *Plant, Cell and Environment*, **6**:665–670.

Du Rietz, G. E. 1931. Life-forms of terrestrial flowering plants. *Acta Phytogeographica Suecica*, **3**:1–95.

Dusinberre, G. M. 1961. *Heat Transfer Calculations by Finite Differences*, International Textbook Company, Scranton/USA.

Edwards, G. and D. A. Walker 1983. *C_3, C_4 Mechanisms and Cellular and Environmental Regulation of Photosynthesis*, Blackwell Scientific Publications, Oxford.
Ehleringer, J. 1981. Leaf absorptance of Mohave and Sonoran desert plants. *Oecologia*, **49**:366–370.
Ehleringer, J. and O. Björkman 1978. Pubescence and leaf spectral characteristics in a desert shrub, *Encelia farinosa*. *Oecologia*, **36**:137–148.
Ehleringer, J. R. and H. A. Mooney 1978. Leaf hairs: effects on physiological activity and adaptive value to a desert shrub. *Oecologia*, **37**:183–200.
Ehleringer, J. R. and K. S. Werk 1986. Modification of solar-radiation absorption and implications for carbon gain at the leaf level. In: T. J. Givnish (ed.): *On the Economy of Plant Form and Function*, Cambridge University Press, Cambridge, pp. 57–82.
Ehleringer, J., O. Björkman and H. A. Mooney 1976. Leaf pubescence: effects on absorptance and photosynthesis in a desert shrub. *Science*, **192**:376–377.
Ellenberg, H. 1981. Ursachen des Vorkommens und Fehlens von Sukkulenten in den Trockengebieten der Erde. *Flora*, **171**:114–169.
Eller, B. M. 1979. Die strahlungsökologische Bedeutung von Epidermis-Auflagen. *Flora*, **168**:146–192.
1982. Strahlungsabsorption von *Agyroderma pearsonii* (N.E.Br.) Schw. in der Vegetations- und Ruheperiode. *Berichte der Deutschen Botanischen Gesellschaft*, **95**:333–340.
Eller, B. M. and N. Grobbelaar 1982. Geophylly: consequences for *Ledebouria ovatifolia* in natural habitat. *Journal of Experimental Botany*, **33**:366–375.
1986. Diurnal temperature variation in and around a *Lithops lesliei* plant growing in its habitat on a clear day. *South African Journal of Botany*, **52**:403–407.
Eller, B. M. and A. Nipkow 1983. Diurnal course of the temperature in a *Lithops* sp. (Mesembryanthemaceae Fenzl) and its surrounding soil. *Plant, Cell and Environment*, **6**:559–565.
Eller, B. M. and B. Ruess 1982. Water relations of *Lithops* plants embedded into the soil and exposed to free air. *Physiologia Plantarum*, **55**:329–334.
Eller, B. M. and P. Willi 1977. Die Bedeutung der Wachsausblühungen auf Blättern von *Kalanchoe pumila* Baker für die Absorption der Globalstrahlung. *Flora*, **166**:461–474.
Eller, B. M., E. Brinckmann and D. J. von Willert 1983a. Optical properties and succulence of plants in the arid Richtersveld (Cp., Rep. South Africa). *Botanica Helvetica*, **93**:47–55.
Eller, B. M.,N. van Rooyen, G. K. Theron and N. Grobbelaar 1983b. Spectral properties of some plant species of the Sourish Mixed Bushveld. *South African Journal of Botany*, **3**:43–49.
Eller, B. M., D. J. von Willert, E. Brinckmann and R. Baasch 1983c. Ecophysiological studies on *Welwitschia mirabilis* in the Namib desert. *South African Journal of Botany*, **2**:209–223.
Evenari, M. 1985a. The desert environment. In: M. Evenari, I. Noy-Meir and D. W. Goodall (eds.), vol. 12A, pp.1–22.
1985b. Adaptations of plants and animals to the desert environment. In: M. Evenari, I. Noy-Meir and D. W. Goodall (eds.), vol. 12A, pp. 79–92.
Evenari, M. and Y. Gutterman 1976. Observations on the secondary succession of three plant communities in the Negev desert, Israel. I. Artemisietum herbae-albae. In: R. Jacques (ed.): Etudes de biologie végétale, CNRS, Paris, pp. 57–86.
Evenari, M., I. Noy-Meir and D. W. Goodall (eds.) 1985. *Hot Deserts and Arid Shrublands*, Ecosystems of the world, vols. 12A + B, Elsevier, Amsterdam, Oxford, New York, Tokyo.
Farrar, J. F. and O. P. Mapunda 1977. Optical properties of the leaves of some African crop plants. *Applied Optics*, **16**:248–251.

Flach, B. M.-T. 1986. Strahlungsangebot, Strahlungsgenuß und Photosynthese nicht ebener Blattflächen in Abhängigkeit vom Tagesgang der Sonne. Thesis, University of Zürich. (Abstr.)

Gates, D. M. 1965. *Energy Exchange in the Biosphere*, Harper and Row, New York.

1980. Biophysical ecology, Springer-Verlag, New York.

Gates, D. M. and Ch. M. Benedict 1962. Convection phenomena from plants in still air. *American Journal of Botany*, **50**:563–573.

Gates, D. M. and L. E. Papian 1971. *Atlas of Energy Budgets of Plant Leaves*, Academic Press, London.

Gates, D. M. and W. Tantraporn 1952. The reflectivity of deciduous trees and herbaceous plants in the infrared to 25 microns. *Science*, **115**:613–616.

Gates, D. M., H. J. Keegan, J. C. Schleter and V. R. Weidner 1965. Spectral properties of plants. *Applied Optics*, **4**:11–20.

Gibson, A. C. and P. S. Nobel 1986. *The Cactus Primer*, Harvard University Press, Cambridge.

Giess, W. 1968. A short report on the vegetation of the Namib coastal area from Swakopmund to Cape Frio. *Dinteria*, **1**:13–29.

1971. A preliminary vegetation map of South West Africa. *Dinteria*, **4**:1–14.

1974. Zwei Fahrten zur *Jensenobotrya lossowiana* Herre. *Dinteria*, **10**:3–12.

1981. Die in der Zentralen Namib von Suedwestafrika/Namibia festgestellten Pflanzenarten und ihre Biotope. *Dinteria*, **15**:32–61.

Givnish, T. J. 1986. Biomechanical constraints on crown geometry in forest herbs. In: T. J. Givnish (ed.): *On the Economy of Plant Form and Function*, Cambridge University Press, Cambridge, pp. 525–583.

Givnish, T. J. and G. J. Vermeij 1976. Sizes and shapes of liane leaves. *The American Naturalist*, **110**:743–778.

Goodall, D. W. and R. A. Perry 1979. *Arid-land Ecosystems: Structure, Functioning and Management*, vol. 1, IBP 16, Cambridge University Press, Cambridge.

1981. *Arid-land Ecosystems: Structure, Functioning and Management*, vol. 2, IBP 17, Cambridge University Press, Cambridge.

Goudie, A. 1972. Climate, weathering, crust formation, dunes and fluvial features of the Central Namib Desert, near Gobabeb, South West Africa. *Madoqua Series*, **2**:15–31.

Grace, J. 1975. Wind damage to vegetation. *Current Advances in Plant Science*, **6**:883–894.

Grime, J. P. 1977. Evidence for the existence of three primary strategies in plants and its relevance to ecological and evolutionary theory. *The American Naturalist*, **111**:1169–1194.

1979. *Plant Strategies and Vegetation Processes*, Wiley, New York.

Grisebach, A. 1872. *Die Vegetation der Erde nach ihrer klimatischen Anordnung*, 2 vols, Engelmann, Leipzig.

Guralnick, L. J., P. A. Rorabaugh and Z. Hanscom 1984a. Influence of photoperiod and leaf age on Crassulacean acid metabolism in *Portulacaria afra* (L.) Jacq. *Plant Physiology*, **75:**454–457.

1984b. Seasonal shifts of photosynthesis in *Portulacaria afra* (L.) Jacq. *Plant Physiology*, **76**:643–646.

Gutterman, Y. 1980/81. Influences on seed germinability: phenotypic maternal effects during seed maturation. *Israel Journal of Botany*, **29**:105–117.

Haberlandt, G. 1909. *Physiologische Pflanzenanatomie*, 4th edition, Verlag W. Engelmann, Leipzig.

Harmse, H. J. M. v. 1978. Schematic soil map of southern Africa south latitude 16° 30′. In: M. J. A. Werger (ed.), pp. 71–75.

Harper, J. L. 1977. *The Population Biology of Plants*, Academic Press, London.

Hartmann, H. 1982. Monographien der Subtribus Leipoldtiinae. III. Monographie der

Gattung *Fenestraria* (Mesembryanthemaceae). *Botanische Jahrbücher für Systematik*, **103**:145–183.

1983. Monographien der Subtribus Leipoldtiinae. IV. Monographie der Gattung *Vanzijlia* (Mesembryanthemaceae). *Botanische Jahrbücher für Systematik*, **103**:499–538.

Henrici, M. 1935. Germination of Karoo bush seeds. *South African Journal of Science*, **32**:223–234.

1939. Germination of Karoo bush seeds. II. *South African Journal of Science*, **36**:212–219.

Herppich, W. 1989. CAM-Ausprägung in *Plectranthus marrubioides* Benth. (Fam. Lamiaceae). Einfluß der Faktoren Licht, Blattemperatur, Luftfeuchtigkeit, Bodenwasserverfügbarkeit und Blattwasserzustand. Inaugural Dissertation, Münster.

Hilton-Taylor, C. 1987. Phytogeography and origins of the Karoo flora. In: R. M. Cowling and P. W. Roux (eds.), pp. 70–95.

Hoffman, T. M. 1989. Vegetation studies and the impact of grazing in the semi-arid eastern Cape. Thesis, University of Cape Town.

Ihlenfeldt, H.-D. 1971. Some aspects of the biology of dissemination of the Mesembryanthemaceae. In: H. Herre (ed.): *The genera of the Mesembryanthemaceae*, Tafelberg Publications, Cape Town, pp. 28–34.

1983. Dispersal of Mesembryanthemaceae in arid habitats. *Sonderbände des Naturwissenschaftlichen Vereins in Hamburg*, **7**:381–390.

1985. Lebensformen und Überlebensstrategien bei Sukkulenten. *Berichte der Deutschen Botanischen Gesellschaft*, **98**:409–423.

Ihlenfeldt, H.-D. and H. E. K. Hartmann 1982. Leaf surfaces in Mesembryanthemaceae. In: D. F. Cutler, K. L. Alvin and C. E. Price (eds.): *The Plant Cuticle*, Academic Press, London, pp. 397–423.

Jordan, P. W. and P. S. Nobel 1981. Seedling establishment of *Ferocactus acanthodes* in relation to drought. *Ecology*, **62**:901–906.

Jürgens, N. 1985. Konvergente Evolution von Blatt- und Epidermismerkmalen bei blattsukkulenten Familien. *Berichte der Deutschen Botanischen Gesellschaft*, **98**:425–446.

1986. Untersuchungen zur Ökologie sukkulenter Pflanzen des südlichen Afrika. *Mitteilungen aus dem Institut für Allgemeine Botanik, Hamburg*, **21**:139–365.

1990. Remarks on the biogeography of Mesembryanthemaceae and their possible use in other arid regions. *Mitteilungen aus dem Institut für Allgemeine Botanik, Hamburg*, **23B**:1057–1060.

Kaemmer, F. 1974. Klima und Vegetation auf Tenerife, besonders im Hinblick auf den Nebelniederschlag. *Scripta Geobotanica*, Vol. 7, E. Goltze, Göttingen.

Kaiser, E. and W. Beetz 1919. Die Wassererschließung in der südlichen Namib Südwestafrikas. *Zeitschrift für Praktische Geologie*, **27**:165–178 and **27**:183–198.

Kappen, L., O. L. Lange, E.-D. Schulze, U. Buschbom and M. Evenari 1980. Ecophysiological investigations of lichens of the Negev desert. VII. The influence of the habitat exposure on dew imbibition and photosynthetic productivity. *Flora*, **169**:216–229.

Kaul, R. B. 1977. The role of multiple epidermis in foliar succulence of *Peperomia* (Piperaceae). *Botanical Gazette*, **138**:213–218.

Kausch, W. 1965. Beziehungen zwischen Wurzelwachstum, Transpiration und CO_2-Gaswechsel bei einigen Kakteen. *Planta*, **66**:229–238.

Kays, W. M. 1966. *Convective Heat and Mass Transfer*, MacGraw-Hill Book Comp., New York.

Killian, C. and D. Feher 1939. Recherches sur la microbiologie des sols desertiques. In: P. Lechavalier (ed.): *Encyclopedie biologique*, vol. 21, Paris.

King, L. 1963. *South African Scenery*, Oliver & Boyd, Edinburgh.

Kluge, M. 1976. Crassulacean acid metabolism (CAM): CO_2 and water economy. In: O. L. Lange, L. Kappen and E.-D. Schulze (eds.): *Water and Plant Life*, Springer-Verlag, Berlin, Heidelberg, New York, pp. 313–320.

Kluge, M. and I. P. Ting 1978. *Crassulacean Acid Metabolism: Analysis of an Ecological Adaptation*, Ecological Studies Series Vol. 30, Springer-Verlag, Berlin.

Knutsom, R. M. 1975. Heat production and temperature regulation in eastern skunk cabbage. *Science*, **186**:746–747.

Koide, R. T., R. H. Robichaux, S. R. Morse and C. M. Smith 1989. Plant water status, hydraulic resistance. In: R. W. Pearcy *et al.* (eds.): *Plant Physiological Ecology*, Chapman and Hall, London.

Koller, D. 1972. Environmental control of seed germination. In: T. T. Kozlowski (ed.): *Seed Biology*, vol. 2, Academic Press, New York, pp. 1–101.

Lancaster, J., N. Lancaster and M. K. Seely 1984. Climate of the central Namib desert. *Madoqua Series*, **14**:5–16.

Lange, O. L. 1962a. Über die Beziehungen zwischen Wasser- und Wärmehaushalt von Wüstenpflanzen. *Veröffentlichungen des Geobotanischen Instituts der ETH Zürich, Stiftung Rübel*, **37**:155–168.

1962b. Untersuchungen über Wärmehaushalt und Hitzeresistenz mauretanischer Wüsten- und Savannenpflanzen. *Flora*, **147**:595–651.

Lange, O. L. and M. Zuber 1977. *Frerea indica*, a stem succulent CAM plant with deciduous C_3 leaves. *Oecologia*, **31**:67–72.

Larcher, W. 1983. *Physiological Plant Ecology*. Springer-Verlag, Berlin, Heidelberg, New York, Tokyo.

Leeman, V., D. Earing, R. K. Vincent and S. Ladd 1971. *The NASA Earth Resources Spectral Information System: A Data Compilation*, NASA CR–31650–24–T, Ann Arbor.

Leick, E. 1910. *Untersuchungen über die Blütenwärme der Araceen*, Verlag von Bruncken & Co., Greifswald.

Leistner, O. A. 1979. Southern Africa. In: D. W. Goodall and R. A. Perry (eds.), vol. 1, pp. 109–143.

Le Roux, A. 1984. 'n Fitososoiologiese studie van die Hester Malan-Natuurreservaat. Thesis, University of Pretoria. (Abstr.)

Leser, H. 1976. *Südwestafrika: eine geographische Landeskunde*, Verlag der Südwestafrikanischen Wissenschaftlichen Gesellschaft, Windhoek.

Lewis, D. A. and P. S. Nobel 1977. Thermal energy exchange model and water loss of a barrel cactus, *Ferocactus acanthodes*. *Plant Physiology*, **60**:609–616.

Lösch, R. and L. Kappen 1981. The cold resistance of Macaronesian Sempervivoideae. *Oecologia*, **50**:98–102.

1983. Die Temperaturresistenz makaronesischer Sempervivoideae. *Verhandlungen der Gesellschaft für Ökologie (Göttingen)*, **10**:521–528.

Lüttge, U. 1986. Nocturnal water storage in plants having Crassulacean acid metabolism. *Planta*, **168**:287–289.

1988. Day–night changes of citric-acid levels in Crassulacean acid metabolism: phenomenon and ecophysiological significance. *Plant, Cell and Environment*, **11**:445–451.

Lüttge, U. and P. S. Nobel 1984. Day–night variations in malate concentration, osmotic pressure, and hydrostatic pressure in *Cereus validus*. *Plant Physiology*, **75**:804–807.

Lüttge, U., M. Kluge and G. Bauer 1988. *Botanik. Ein grundlegendes Lehrbuch*, VCH, Weinheim.

MacDougal, D. T., E. R. Long and J. G. Brown 1915. End results of desiccation and respiration in succulent plants. *Physiological Research*, **1**:289–325.

MacMahon, J. A. and F. H. Wagner 1985. The Mojave, Sonoran and Chihuahuan deserts of North America. In: M. Evenari, I. Noy-Meir and D. W. Goodall (eds.), pp. 105–202.
McGinnies, W. G. 1979. General descriptions of desert areas. In: D. W. Goodall and R. A. Perry (eds.), vol. 1, pp. 5–19.
McGinnies, W. G., B. J. Goldman and P. Paylore 1968. *Deserts of the World*, University of Arizona Press, Tucson.
Marloth, R. 1909. Die Schutzmittel der Pflanze gegen übermässige Insolation. *Berichte der Deutschen Botanischen Gesellschaft*, **27**:362–371.
Matos, de Cardosa, G. and J. N. Sousa, de Baptista 1968. *Reserva parcial de Mocamedes. Carta da vegetação e memoria descritiva*, Inst. Inv. Agron. Angola, Nova Lisboa, 57pp.
Medina, E. 1982. Temperature and humidity effects on dark CO_2 fixation by *Kalanchoe pinnata. Zeitschrift für Pflanzenphysiologie*, **107**:251–258.
Meigs, P. 1953. World distribution of arid and semi-arid homoclimates. In: *Reviews of Research on Arid Zone Hydrology*, UNESCO, Paris, Arid Zone Programme 1, pp. 203–209.
1966. *The Geography of Coastal Deserts*, UNESCO, Paris, Arid Zone Research 28.
Mendonça, F. A. 1961. Indices fitocorologicos da vegetação de Angola. *Garcia de Orta*, **9**:479–483.
Merxmüller, H. (ed.) 1966–1972. *Prodomus einer Flora von Südwestafrika*, Cramer, Lehre, 35 parts.
Milne, R. 1989. Diurnal water storage in the stems of *Picea sitchensis* (Bong.) Carr. *Plant, Cell and Environment*, **12**:63–72.
Moisel, A. and E. J. Moll 1981. A Braun–Blanquet survey of the vegetation of the Welwitschia Plain. *Dinteria*, **15**:3–11.
Monson, R. K. 1989. On the evolutionary pathways resulting in C_4 photosynthesis and Crassulacean acid metabolism (CAM). *Advances in Ecological Research*, **19**:57–110.
Monteith, J. L. 1973. *Principles of Environmental Physics*, Edward Arnold (Publishers) Ltd, London.
Mulroy, T. W. and P. W. Rundel 1977. Annual plants: adaptations to desert environments. *BioScience*, **27**:109–114.
Neckel, H. and D. Labs 1984. The solar radiation between 3300 and 12500 Å. *Solar Physics*, **90**:205–258.
Neilson, R. P. 1986. High-resolution climatic analysis and southwest biogeography. *Science*, **232**:27–34.
Nobel, P. S. 1974. Boundary layers of air adjacent to cylinders. *Plant Physiology*, **54**:177–181.
1975. Effective thickness and resistance of the air boundary layer adjacent to spherical plant parts. *Journal of Experimental Botany*, **26**:120–130.
1976. Water relations and photosynthesis of a desert CAM-plant *Agave deserti. Plant Physiology*, **58**:576–582.
1977a. Water relations and photosynthesis of a barrel cactus *Ferocactus acanthodes* in the Colorado desert. *Oecologia*, **27**:117–133.
1977b. Water relations of flowering of *Agave deserti. Botanical Gazette*, **138**:1–6.
1982. Low-temperature tolerance and cold hardening of cacti. *Ecology*, **63**:1650–1656.
1983. *Biophysical Plant Physiology and Ecology*, W. H. Freeman and Co., New York.
1988. *Environmental Biology of Agaves and Cacti*. Cambridge University Press, New York.
1989. Shoot temperatures and thermal tolerance for succulent species of *Haworthia* and *Lithops. Plant, Cell and Environment*, **12**:643–651.
Nobel, P. S. and J. Sanderson 1984. Rectifier-like activities of roots of two desert succulents. *Journal of Experimental Botany*, **35**:727–737.

Noy-Meir, I. 1973. Desert ecosystems: environment and producers. *Annual Review of Ecology and Systematics*, **4**:25–51.

Orshan, G. 1988. *Plant Phenomorphological Studies in Mediterranean Type Ecosystems*, Kluwer, Dordrecht, Boston.

Osmond, C. B., M. M. Ludlow, R. Davis, I. R. Cowan, S. B. Powles and K. Winter 1979. Stomatal responses to humidity in *Opuntia inermis* in relation to control of CO_2 and H_2O exchange patterns. *Oecologia*, **41**:65–76.

Osmond, C. B., O. Björkman and D. J. Anderson 1980. *Physiological processes in plant ecology. Towards a synthesis with Atriplex*. Ecological Studies, vol. 36, Springer-Verlag, Berlin, Heidelberg, New York.

Ostrem, J. A., S. W. Olson, J. M. Schmitt and H. J. Bohnert 1987. Salt stress increases the level of translatable mRNA for phosphoenolpyruvate carboxylase in *Mesembryanthemum crystallinum*. *Plant Physiology*, **84**:1270–1275.

Parkhurst, D. F., P. R. Duncan, D. M. Gates and F. Kreith 1968. Convection heat transfer from broad leaves of plants. *Journal of Heat Transfer*, **90**:71–76.

Parlange, J.-Y., P. E. Waggoner and G. H. Heichel 1971. Boundary layer resistance and temperature distribution on still and flapping leaves. *Plant Physiology*, **48**:437–442.

Pate, J. S. and K. W. Dixon 1981. Plants with fleshy underground storage organs – a Western Australian survey. In: J. S. Pate and A. J. McComb (eds.): *The Biology of Australian Plants*, University of Western Australia Press, pp. 181–215.

Pearcy, R. W. and J. Ehleringer 1984. Comparative ecophysiology of C_3 and C_4 plants. *Plant, Cell and Environment*, **7**:1–13.

Pearman, G. I. 1965. Preliminary studies of the loss of heat from leaves under conditions of free and forced convection. *Australian Journal of Botany*, **13**:153–160.

Pearman, G. I., H. L. Weaver and C. B. Tanner 1972. Boundary layer heat transfer coefficient under field conditions. *Agricultural Meteorology*, **10**:83–92.

Prentice, I. C. and M. J. A. Werger 1985. Clump spacing in a desert dwarf shrub community. *Vegetatio*, **63**:133–139.

Range, P. 1932. Die Flora des Namalandes I. *Feddes Repertorium*, **30**:129–158.

Raschke, K. 1956. Über die physikalischen Beziehungen zwischen Wärmeübergangszahl, Strahlungsaustausch, Temperatur und Transpiration eines Blattes. *Planta*, **48**:200–238.

Rauh, W. 1985. The Peruvian–Chilean deserts. In: M. Evernari, I. Noy-Meir and D. W. Goodall (eds.), vol. 12A, pp. 239–268.

1988. *Tropische Hochgebirgspflanzen. Wuchs- und Lebensformen*, Springer, Berlin, New York.

Raunkiaer, C. 1907. *Platerigets Livsformer og deres Betydning for Geografien*, Nordisk Verl., Copenhagen.

Robinson, E. R. 1977. A plant ecological study of the Namib Desert Park. MSc Thesis, University of Natal, Pietermaritzburg. (Abstr.)

Robinson, E. R. and W. Giess 1974. Report on the plants noted in the course of a trip from Lüderitz Bay to Spencer Bay, January 10–21, 1974. *Dinteria*, **10**:13–17.

Rösch, M. W. 1977. Enkele plantekologiese aspekte van die Hester Malan-Natuurreservaat. Thesis, University of Pretoria. (Abstr.)

Rössner, H. and M. Popp 1986. Ionic patterns in some Crassulaceae from Austrian habitats. *Flora*, **178**:1–10.

Rowley, G. D. 1987. *Caudiciform and Pachycaul Succulents*. Strawberry Press, Mill Valley, California.

Ruess, B. R. and B. M. Eller 1985. The correlation between Crassulacean acid metabolism and water uptake in *Senecio medley-woodii*. *Planta*., **166**:57–66.

Rundel, P. W. 1982. Water uptake by organs other than roots. In: A. Pirson and M. H. Zimmermann (eds.): *Encyclopedia of Plant Physiology*, NS, Springer-Verlag, Berlin, Heidelberg, New York, pp. 111–134.

Rutherford, M. C. and R. H. Westfall 1986. Biomes of southern Africa – An objective categorization. *Memoirs of the Botanical Survey of South Africa*, **54**:1–98.

Rygol, J., K.-H. Büchner, K. Winter and U. Zimmermann 1986. Day/night variations in turgor pressure in individual cells of *Mesembryanthemum crystallinum* L. *Oecologia*, **69**:171–175.

Rygol, J., U. Zimmermann and A. Balling 1989. Water relations of individual leaf cells of *Mesembryanthemum crystallinum* plants grown at low and high salinity. *Journal of Membrane Biology*, **107**:203–212.

Schanderl, H. 1935. Untersuchungen über die Lichtverhältnisse im Inneren von Hartlaub- und Sukkulentenblättern. *Planta*, **24**:454–469.

Schulze, E.-D., H. Ziegler and W. Stichler 1976. Environmental control of Crassulacean acid metabolism in *Welwitschia mirabilis* Hock Fil. in its range of natural distribution in the Namib Desert. *Oecologia*, **24**:323–334.

Schulze, E.-D., B. M. Eller, D. A. Thomas, D. J. von Willert and E. Brinckmann 1980. Leaf temperatures and energy balance of *Welwitschia mirabilis* in its natural habitat. *Oecologia*, **44**:258–262.

Schulze, R. 1970. *Strahlenklima der Erde*, Steinkopff Verlag, Darmstadt.

Schulze, R. E. and O. S. McGee 1978. Climatic indices and classification in relation to the biogeography of southern Africa. In: M. J. A. Werger (ed.), pp. 19–52.

Shmida, A. 1985. Biogeography of the desert flora. In: M. Evenari, I. Noy-Meir and D. W. Goodall (eds.), pp. 23–77.

Shmida, A. and T. L. Burgess 1988. Plant growth-form strategies and vegetation types in arid environments. In: M. J. A. Werger *et al.* (eds.): *Plant Form and Vegetation Structure*, SPB Academic Publishers, The Hague, pp. 211–241.

Shreve, F. and I. L. Wiggins 1964. *Vegetation and Flora of the Sonoran Desert*, 2 vols, Stanford University Press, Stanford, California (edition 1977, reprint of 'Vegetation of the Sonoran desert', Carnegie Institute Washington Publication no. 591.)

Silberberg, B. 1910. Das Pflanzenbuch des Abu-Hanifa Ahmed ibn Da'ud ad-Dinawari. *Zeitschrift für Assyriologie*, **24**:225–265.

1911. Das Pflanzenbuch des Abu-Hanifa Ahmed ibn Da'ud ad-Dinawari. *Zeitschrift für Assyriologie*, **25**:39–88.

Sinclair, R. and D. A. Thomas 1970. Optical properties of leaves of some species in arid South Australia. *Australian Journal of Botany*, **18**:261–273.

Skujins, J. 1984. Microbial ecology of desert soils. In: K. C. Marshall (ed.): *Advances in Microbiology*, vol. 7, Plenum, New York, pp. 49–91.

Slavik, B. 1974. Methods of studying plant water relations. *Ecological Studies*, vol. 9, Academy of Science of Czechoslovakia, Prague.

Sloan, R., J. H. Shaw and D. Williams 1955. Infrared emission spectrum of the atmosphere. *Journal of the Optical Society of America*, **45**:455–460.

Smith, W. K. 1978. Temperatures of desert plants: another perspective on the adaptability of leaf size. *Science*, **201**:614–616.

Specht, R. L. 1972. *The Vegetation of South Australia*, Government Printer, Adelaide, S.A., 2nd edition, 328 pp.

Springstubbe, H. 1990. Untersuchungen zur Physiologie der Stomatabewegung bei der CAM Pflanze *Aptenia cordifolia* (L.f.). Inaugural Dissertation, Münster.

Squires, V. R. and W. S. W. Trollope 1979. Allelopathy in the Karoo shrub, *Chrysocoma tenuifolia*. *South African Journal of Science*, **75**:88–89.

Stalfeldt, M. G. 1956. Morphologie und Anatomie des Blattes als Transpirationsorgan. In: O. Stocker (ed.): *Pflanze und Wasser. Handbuch der Pflanzenphysiologie*, vol. III, Springer, Berlin, pp. 324–341.

Stanhill, G. and M. Fuchs 1977. The relative flux density of photosynthetically active radiation. *Journal of Applied Ecology*, **14**:317–322.

Starnecker, G. 1984. *Ökophysiologische Anpassungen im Gasstoffwechsel bei der Gattung Peperomia* Ruiz & Pavon. Dissertationes Botanicae. J. Cramer, Vaduz.
Steenbergh, W. F. and C. H. Lowe 1977. Ecology of the saguaro. II. Reproduction, germination, establishment, growth, and survival of the young plants. *National Park Service Scientific Monographs Series*, No. 8, US Govt. Printing Office, Washington.
Steudle, E., J. A. C. Smith and U. Lüttge 1980. Water relation parameters of individual mesophyll cells of the Crassulaceaen acid metabolism plant *Kalanchoe daigremontiana*. *Plant Physiology*, **66**:1155–1163.
Stoutjesdijk, P. 1974. The open shade, an interesting microclimate. *Acta Botanica Neerlandica*, **23**:125–130.
Szarek, S. R. and I. P. Ting 1974. Seasonal patterns of acid metabolism and gas exchange in *Opuntia basilaris*. *Plant Physiology*, **54**:76–81.
Tanner, V. and B. M. Eller 1986. Epidermis structure and its significance for the optical properties of leaves of the Mesembryanthemaceae. *Journal of Plant Physiology*, **125**:285–294.
Ting, I. P. and Z. Hanscom 1977. Induction of acid metabolism in *Portulacaria afra*. *Plant Physiology*, **59**:511–514.
Tölken, H. R. 1977. A revision of the genus *Crassula* in Southern Africa. *Contributions from the Bolus Herbarium*, **8** parts 1 & 2.
Treichel, S., E. Brinckmann, B. Scheitler and D. J. von Willert 1984. Occurrence and changes of proline content in plants in the southern Namib Desert in relation to increasing and decreasing drought. *Planta*, **162**:236–242.
Usada, H., M. S. B. Ku and G. E. Edwards 1984. Rates of photosynthesis relative to activity of photosynthetic enzymes, chlorophyll and soluble protein content among ten C_4 species. *Australian Journal of Plant Physiology*, **11**:509–517.
Usada, H., M. S. B. Ku and G. E. Edwards 1985. Influence of light intensity during growth on photosynthesis and activity of several key photosynthetic enzymes in a C_4 plant (*Zea mays*). *Physiologia Plantarum*, **63**:65–70.
Van der Walt, P. T. 1967. A plant ecological survey of the Noorsveld. *Journal of South African Botany*, **33**:215–234.
Van Rooyen, M. W. and N. Grobbelaar 1982. Die saadbevolkings in die grond van die Hester Malan Natuurreservaat in die Namakwalandse Gebroke Veld. *South African Journal of Botany*, **1**:41–50.
Van Rooyen, M. W., N. Grobbelaar and G. K. Theron 1979a. Phenology of the vegetation in the Hester Malan Nature Reserve in the Namaqualand Broken Veld. 2. The therophyte population. *Journal of South African Botany*, **45**:433–452.
Van Rooyen, M. W., G. K. Theron and N. Grobbelaar 1979b. Phenology of the vegetation in the Hester Malan Nature Reserve in the Namaqualand Broken Veld. 1. General observations. *Journal of South African Botany*, **45**:279–293.
Van Wijk, W. R. and D. W. Scholte Ubing 1966. Radiation. In: W. R. Van Wijk (ed.): *Physics of Plant Environment*, North-Holland Publishing Company, Amsterdam, pp. 62–101.
Vasek, F. C. 1980. Creosote bush: long-lived clones in the Mohave desert. *American Journal of Botany*, **67**:246–255.
Vernon, D. M., J. A. Ostrem, J. M. Schmitt and H. J. Bohnert 1988. PEPCase transcript levels in *Mesembryanthemum crystallinum* decline rapidly upon relief from salt stress. *Plant Physiology*, **86**:1002–1004.
Vogel, S. 1969. Convective cooling at low airspeeds and the shapes of broad leaves. *Journal of Experimental Botany*, **21**:91–101.
Volk, O. H. 1966. Die Florengebiete von Südwestafrika. *Journal der Südwestafrikanische Wissenschaftliche Gesellschaft*, **20**:25–58.

Volk, O. H. and E. Geyger 1970. Schaumböden als Ursache der Vegetationslosigkeit in ariden Gebieten. *Zeitschrift für Geomorphologie*, N.S., **14**:79–95.
Volk, O. H. and H. Leippert 1971. Vegetationsverhältnisse im Windhoeker Bergland, Südwestafrika. *Journal der Südwestafrikanische Wissenschaftliche Gesellschaft*, **25**:5–44.
von Willert, D. J. 1979. Vorkommen und Regulation des CAM bei Mittagsblumengewächsen (Mesembryanthemaceae). *Berichte der Deutschen Botanischen Gesellschaft*, **92**:133–144.
1985. *Welwitschia mirabilis* – new aspects in the biology of an old plant. *Advances in Botanical Research*, **11**:157–191.
von Willert, D. J. and E. Brinckmann 1986. Sukkulenten und ihr Überleben in Wüsten. *Naturwissenschaften*, **73**:57–69.
von Willert, D. J. and D. Kramer 1972. Feinstruktur und Crassulaceen-Säurestoffwechsel in Blättern von *Mesembryanthemum crystallinum* während natürlicher und NaCl-induzierter Alterung. *Planta*, **107**:227–237.
von Willert, D. J. and K. von Willert 1979. Light modulation of the activity of the PEP-carboxylase in CAM plants in the Mesembryanthemaceae. *Zeitschrift für Pflanzenphysiologie*, **95**:42–49.
von Willert, D. J., G. O. Kirst, S. Treichel and K. von Willert 1976. The effect of leaf age and salt stress on malate accumulation and PEP-carboxylase activity in *Mesembryanthemum crystallinum*. *Plant Science Letters*, **7**:341–346.
von Willert, D. J., E. Curdts and K. von Willert 1977. Veränderungen der PEP-Carboxylase während einer durch NaCl geförderten Ausbildung eines CAM bei *Mesembryanthemum crystallinum*. *Biochemie und Physiologie der Pflanzen*, **171**:101–107.
von Willert, D. J., E. Brinckmann, B. Scheitler, D. A. Thomas and S. Treichel 1979a. The activity and malate inhibition/stimulation of phosphoenolpyruvate-carboxylase in Crassulacean-acid-metabolism plants in their natural environment. *Planta*, **147**:31–36.
von Willert, D. J., E. Brinckmann and E.-D. Schulze 1979b. Ecophysiological investigations of plants in the coastal desert of southern Africa. Ion content and CAM. In: R. J. Jefferies and A. J. Davy (eds.): *Ecological Processes in Coastal Environments*. Blackwell Scientific Publications, Oxford, pp. 321–331.
von Willert, D. J., E. Brinckmann, B. Scheitler and R. Baasch 1982. CO_2 gas exchange and transpiration of *Welwitschia mirabilis* Hook. f. in the Central Namib desert. *Oecologia*, **55**:21–29.
von Willert, D. J., E. Brinckmann, B. Scheitler and B. M. Eller 1983. Responses of CAM to increasing and decreasing water stress in plants in the southern Namib desert. In: R. Marcelle *et al.* (eds.): *Effects of Stress on Photosynthesis*, Nijhoff, Junk, The Hague, Boston, London, pp. 155–163.
von Willert, D. J., E. Brinckmann, B. Scheitler and B. M. Eller 1985. Availability of water controls CAM in succulents of the Richtersveld (Namib desert, South Africa). *Planta*, **164**:44–55.
Walter, H. 1936. Die ökologischen Verhältnisse in der Namib-Nebelwüste (Südwestafrika) unter Auswertung der Aufzeichnungen des Dr. G. Boss (Swakopmund). *Jahrbücher für Wissenschaftliche Botanik*, **84**:58–222.
1960. *Standortslehre, analytisch-ökologische Geobotanik*, 2nd edition, Ulmer-Verlag, Stuttgart.
1973. *Die Vegetation der Erde in öko-physiologischer Betrachtung*, G. Fischer Verlag, Stuttgart.
1985. The Namib desert. In: M. Evenari, I. Noy-Meir and D. W. Goodall (eds.) pp. 245–282.

Walter, H. and S.-W. Breckle 1984. *Ökologie der Erde*, vol. 2: *Spezielle Ökologie der Tropischen und Subtropischen Zonen*, G. Fischer Verlag, Stuttgart.
Ward, D. A. and J. A. Bunce 1986. Novel evidence for a lack of water vapour saturation within the intercellular airspaces of turgid leaves of mesophytic species. *Journal of Experimental Botany*, **37**:504–516.
Waring, R. H. and S. W. Running 1978. Sapwood water storage: its contribution to transpiration and effect upon water conductance through the stems of old-growth Douglas fir. *Plant, Cell and Environment*, **1**:131–140.
Watson, I. and R. R. Lemon 1985. Geomorphology of a coastal desert: The Namib, South West Africa/Namibia. *Journal of Coastal Research*, **1**:329–342.
Weisser, P. 1967. Zur Kenntnis der Erdkakteen in Chile. *Berichte der Deutschen Botanischen Gesellschaft*, **80**:331–338.
Wellington, J. H. 1955. *Southern Africa. A Geographical Study*. Cambridge University Press, Cambridge.
Went, F. W. 1979. Germination and seedling behaviour. In: D. W. Goodall and R. A. Perry (eds.), pp. 477–489.
Werger, M. J. A. (ed.) 1978a. *Biogeography and Ecology of Southern Africa*. Monographiae Biologicae, vol. 31, Junk, The Hague.
1978b. Biogeographical division of southern Africa. In: M. J. A. Werger (ed.), pp. 145–170.
1978c. The Karoo–Namib region. In: M. J. A. Werger (ed.), pp. 231–299.
1983a. Vegetation geographical patterns as a key to the past, with emphasis on the dry vegetation types of South Africa. *Bothalia*, **14**:405–410.
1983b. Wurzel/Sproß-Verhältnis als Merkmal der Pflanzenstrategie. In: L. Kutschera *et al.* (eds.): *Wurzelökologie und ihre Nutzanwendung*, Bundanst, Gumpenheim, Irdning, pp. 323–334.
1983c. Plant form and plant function – an introductory historical outline. *Journal of the South West China Teachers University*, (Beibei, Sichuan), No. 4, 141–165.
1985. The Karoo and southern Kalahari. In: M. Evenari, I. Noy-Meir and D. W. Goodall (eds.), vol. 12B, pp. 283–359.
Werger, M. J. A. and B. J. Coetzee 1977. A phytosociological and phytogeographical study of Augrabies Falls National Park, South Africa. *Koedoe*, **20**:11–51.
Werger, M. J. A. and B. J. Coetzee 1978. The Sudano–Zambesian region. In: M. J. A. Werger (ed.), pp. 301–462.
Werger, M. J. A. and R. P. Ellis 1981. Photosynthetic pathways in the arid region of South Africa. *Flora*, **171**:64–75.
West, N. E. (ed.) 1983. *Temperate Deserts and Semi-deserts*, Ecosystems of the world, vol. 5, Elsevier, Amsterdam, Oxford, New York, Tokyo.
Whittaker, R. H. 1972. Evolution and measurement of species diversity. *Taxon*, **21**:213–251.
1977. Evolution of species diversity in land communities. *Evolutionary Biology*, **10**:1–67.
Winter, K. 1973. CO_2-Gaswechsel von an hohe Salinität adaptiertem *Mesembryanthemum crystallinum* bei Rückführung in glykisches Anzuchtmedium. *Berichte der Deutschen Botanischen Gesellschaft*, **86**:467–476.
Winter, K. and U. Lüttge 1979. C_3-Photosynthese und Crassulaceen-Säurestoffwechsel bei *Mesembryanthemum crystallinum* L. *Berichte der Deutschen Botanischen Gesellschaft*, **92**:117–132.
Winter, K. and D. J. von Willert 1972. NaCl-induzierter Crassulaceen-Säurestoffwechsel bei *Mesembryanthemum crystallinum*. *Zeitschrift für Pflanzenphysiologie*, **67**:166–170.
Winter, K., U. Lüttge, E. Winter and J. H. Troughton 1978. Seasonal shift from C_3 photosynthesis to Crassulacean acid metabolism in *Mesembryanthemum crystallinum* growing in its natural environment. *Oecologia*, **34**:225–237.

Wong, C. L. and W. R. Blevin 1967. Infrared reflectance of plant leaves. *Australian Journal of Biological Science*, **20**:501–508.

Woodhouse, R. M., J. G. Williams and P. S. Nobel 1980. Leaf orientation, radiation interception, and nocturnal acidity increase by the CAM plant *Agave deserti* (Agavaceae). *American Journal of Botany*, **67**:1179–1185.

Woolley, J. T. 1971. Reflectance and transmittance of light by leaves. *Plant Physiology*, **47**:656–662.

Yates, D. J. 1981. Effect of the angle of incidence of light on the net photosynthesis rates of *Sorghum album* leaves. *Australian Journal of Plant Physiology*, **8**:335–346.

Index

absorptance 103–4, 123, 126–8, 141, 148, 151
absorptivity 40, 102ff
  of succulents 104–9, 118, 121, 128–9, 151
Acacia (Mimosaceae) 35, 94
  erioloba 93
  karroo 177, 184
Acanthaceae 165, 173
Acanthopsis disperma (Acanthaceae) 9, 247, 250, 252–3
Achmed effect 194
Adansonia digitata (Bombacaceae) 317
adaptation 99, 102, 107, 142, 151, 155, 161, 188, 191, 244, 264, 275, 297, 312–13, 318
adenosine triphosphate (ATP) 226f, 229, 232
Adromischus (Crassulaceae) 11, 165
  maculatus 165, 180
  triebneri (=alstonii) 165
Aeonium tabuliforme (Crassulaceae) 13
Agavaceae 330
Agave (Agavaceae) 25, 52, 119, 179
  deserti 119, 174, 191, 307, 311–12, 325, 330
agaves 13, 119, 134, 189, 192, 215–16
Aitoniaceae 165, 173
Aizoaceae 165, 173
albedo 39
Aloe (Liliaceae) 16, 20, 25, 62, 65, 141, 210, 216, 218, 289, 293
  arborescens 219, 320
  dichotoma 13, 15, 20, 22, 119–122
  marlothii 17
  pearsonii 9, 11–12, 21, 108, 126, 165, 180, 216, 218, 220, 247–8, 250–2, 286, 290, 293, 296
  pillansii 13, 21, 165, 293
  ramosissima 9, 13, 21, 107, 150, 165, 180, 184–6, 247, 250, 286, 290, 293
aloes 13, 119, 292
Amaranthaceae 165, 173
Amoebophyllum (Mesembryanthemaceae) 317
Anacampseros (Portulacaceae) 124
  alstonii 190
  comptonii 7, 308
  papyracea 124–5
Apocynaceae 64, 165, 173
Aptenia cordifolia (Mesembryanthemaceae) 327
Araceae 324
Argyroderma (Mesembryanthemaceae) 63, 126–7, 148, 215, 315
  delaetii 148–9
  patens 220
  pearsonii 321
Aridaria (Mesembryanthemaceae) *19*
Asclepiadaceae 20, 26, 64, 72, 165, 173, 315
ash content 246–9
aspartate amino transferace (AAT) 228
Asteraceae 10, 23, 64, 72, 93, 164–5, 169, 172–3, 286
Atriplex (Chenodiaceae) 326
Augea capensis (Zygophyllaceae) 16, 182–4, 268–70, 273, 275, 277, 313–14

biomass 224, 300–1, 304, 310, 312
  structural (non-utilizable) 253, 301, 303, 305–6
  utilizable 224, 253, 279, 297, 300–8, 310–15
boundary layer 42, 46, 102, 108, 136, 139, 147, 160ff
  thickness 136, 139, 161, 163, 179
Brachystelma (Asclepiadaceae) 7
Brassicaceae 93
Bromeliaceae 20, 155, 201
Brownanthus (Mesembryanthemaceae) 179
  schlichtianus 54, 97, 150, 169, 175, 179, 184–6, 200, 219–24, 247–8, 250, 287, 296, 315, 317
Bryophyllum daigremontianum (Crassulaceae) 320

Bulbine (Liliaceae) 153–4, 180, 187
cf. *mesembryanthoides* Fig. 4.31 opp. 155
sedifolia (Liliaceae) 61
bundle sheath 231–3, 236, 243

$\delta^{13}C$ value 242
C_3-photosynthesis 184, 230ff, 238–9, 243, 265–6, 297, 307, 313, 316–17
C_3–C_4 intermediates 237–41
C_3 to CAM intermediates 174, 237–41, 288
plants 180, 230f, 235, 242, 259, 263
succulents 179, 181–2, 184, 187–8, 202, 238, 240–1, 249, 264–6, 269–71, 273–5, 277, 292, 294, 296–7, 313
C_4-photosynthesis 184, 230–3, 236, 243, 279, 313
plants 231–3, 235, 242ff, 263
succulents 245
Cactaceae 64, 315
cacti 13, 16, 20, 25, 65, 118–21, 124, 134, 177, 189, 192, 215–16, 278, 315
calcium 173–5, 289, 292–3
CALVIN-cycle 225, 230–1, 234
CAM (Crassulacean acid metabolism) 184, 230, 234–5, 265–6, 289, 297, 307, 313, 316–17
cycling 280, 292, 297, 307
idling 280, 297
plants 224, 234–45
role of 185, 179–200, 275–80
succulents 179–80, 182, 184–8, 195–200, 206–7, 209, 212–14, 216, 218, 221, 235, 239, 244, 248, 255, 257–60, 265, 274, 278–81, 286–7, 292, 296–7, 313
capacitance 194–6
Caralluma (= Frerea) indica (Asclepiadaceae) 317
carbon
balance 152, 224, 252, 254, 300–12
budget 89, 199, 224, 252–97, 306
fluxes 224–97
gain 224, 226, 245, 266, 268, 301
investment 245–52, 302, 312
turnover 302ff
Carnegiea (Cactaceae) 62
gigantea 17, 59
Carpobrotus (Mesembryanthemaceae) 60
Catananche lutea (Asteraceae) 64
caudiciform 7, 23
Cephalocereus hoppenstedtii (Cactaceae) 120–1
Cephalophyllum (Mesembryanthemaceae) 65
Ceraria fruticulosa (Portulacaceae) 9, 54, 165, 180, 247, 250, 252–3, 286, 317
namaquensis 7–9, 165, 180, 247, 250, 286
pygmaea 7, 9, 165, 247, 250, 286
Cereus (Cactaceae) 20
validus 174, 324
Ceropegia (Asclepiadaceae) 62
Chasmatophyllum (Mesembryanthemaceae) 16
Cheiridopsis (Mesembryanthemaceae) 215, 315
cigarettifera 153
robusta 9, 54, 66, 107, 150, 169, 175, 180, 215–16, 218, 220, 253, 288, 296
Chenopodiaceae 68, 165, 173
chlorenchyma 3, 23, 151, 154, 215, 218–19, 222, 245, 265–6, 292
chloride 168, 173–5, 220–1
chloroplast 226ff, 230–1
Chrysocoma tenuifolia (Asteraceae) 53, 327
Cissus quadrangularis (Vitidaceae) 16
citric acid 228–9, 234, 237, 281, 283–92, 294
clear sky condition 39–43, 45–6, 113–15, 117, 126, 128–32, 134, 146–8
Cleretum (Mesembryanthemaceae) 62
papulosum subsp. papulosum 63
climate
desert 28–49
gradient 80, 82, 95
microclimate 17, 89–90, 207
Namib 75–89
CO_2 budget 225–30, 236, 263–4, 277, 279, 282
concentration 226, 232–3
exchange 224, 236, 245, 252–3, 255, 257–9, 262, 266, 270, 272, 280, 282, 285, 302
diurnal course 234–5, 257–9, 261–2, 267, 269, 271–2, 276, 278, 281
fixation 226, 230–2, 242, 274
mode of 230–45, 265–6, 313, 316–17
nocturnal 176, 206, 234, 243, 255, 258, 274, 278, 280–3
uptake 234, 237, 255, 261–2
morning peak 234–6, 261, 267
Codon royenii (Hydrophyllaceae) 180, 184, 247, 250, 262–3, 273
Coleus (Lamiaceae) 264
Commiphora (Burseraceae) 317
condensation 44, 46
Conophytum (Mesembryanthemaceae) 62–3, 125–6, 153, 215, 315
aequale 9, 247, 249–50, 252–3, 288
calculus 220
minutum 9, 180
convection 17, 42, 101–2, 131, 136–42, 144, 151, 155
convergence 23
Cotula barbata (Asteraceae) 59
Cotyledon orbiculata (Crassulaceae) 7, 9, 11, 16, 61, 109–10, 123, 150, 165, 184, 187–8, 247, 286, 311

Crassula (Crassulaceae) 62, 65, 165, 168, Fig. 4.60 opp. 203, 203, 278, 289, 293, 308, 328
alstonii 16
atropurpurea 180
brevifolia 177, 180, 188, 203, 207, 286
clavata 11, 180, Fig. 4.59, opp. 203, 291, 293
columnaris 16
deceptor 9, 16, 168, 180, 247, 250, 253, 286, 291, 293, 296
elegans 9, 180, 252–3, 286, 291, 293
erolusa 180, Fig. 4.56, opp. 202, 286, 293
expansa 65, 180
fusca 168, 286
grisea 54, 175, 180, 293
macowaniana 247, 250, 286
nudicaulis 205
pseudohemisphaerica 13, 180
rupestris 180
sericea 180, 286
tetragona 204
Crassulaceae 10, 19, 65, 72, 165, 169, 173, 285–6, 289, 326
crassulas 203, 205–6, 292
Crataegus monogyna (Rosaceae) 132
cross-section
leaf 3, 8, 11, 12, 19, 21, 107, 118, 151–2, 203
root 8
stem 8, 14, 15, 22, 23
Cucumis meeusei (Cucurbitaceae) 180, 266, 268, 292, 294
Cucurbitaceae 165, 173
cuticle 17, 20, 21, 151, 161
Cyphostemma (Vitaceae) 210
cytosol 226–7

decarboxylation 228–9, 231–2
of malic acid 227–9, 236
degree of sclerophylly 246, 248–9
degree of succulence 3ff, 246, 248–50, 296
dehydration–rehydration cycle 216, 218
Delosperma (Mesembryanthemaceae) 110–12, 134
cooperi 110–12, 266, 268
lehmannii 112
nubigenum 25, 134
pergamentaceum 9, 55, 57–8, 180, 247–50
*Δ*W, *see* water vapour
desert 32–3, 178
climate 28–49
coastal 30
cold 29
definition 29–30
extreme 29–33, 39, 245
hot 40
semi-desert 29, 32–3
desiccation 50, 167, 189, 196, 198, 208, 211, 213, 215–17, 254–5, 264–6
dew 25, 30, 37, 44, 47, 50, 68, 101, 130, 132, 142, 145–6, 185, 189, 198, 201ff
dew point temperature 45, 47, 202
Didelta carnosa (Asteraceae) 61, 150, 165, 184, 247, 250
dispersal 63–5, 301
agents 64
requirements 63–4
systems 64–5
dissemination 63–5
dormancy 52, 55, 307, 315
Dorotheanthus (Mesembryanthemaceae) 62–3, 179, 301, 306
Drosanthemum (Mesembryanthemaceae) 17, 19, 150, 168, 175–7, 179, 192–3, 195, 207, 220, 287
drought 25, 240–1, 254, 266, 280, 287, 297, 309
atmospheric drought (*see also* wind, bergwind) 254–64
long-term 4, 6, 11, 55, 97, 167, 170, 177, 184, 192–3, 215, 217, 254, 264–75, 278, 284, 290–2, 294, 318
short-term 6, 254–64
stress 187, 189, 192, 198, 203, 219, 221–2, 245, 264, 267, 285, 292, 303

emittance of thermal radiation 102, 116, 128–9, 144
Encelia californica (Asteraceae) 124
farinosa 110, 124, 181, 321
endemism 91ff
endogenous CO_2 228, 230, 236, 268–70, 283
diurnal course 269, 271
energy
balance 116, 133, 139, 141–55
expenditure 246, 248, 252
fluxes 40, 42, 100–55
investment 246–9
Enneapogon desvauxii (Poaceae) 64
epicuticular wax 17, 18, 102, 107–10, 121, 123–4, 142, 161
epidermis 11, 14, 17–20, 23, 102, 106, 112, 123–4, 136, 151–3, 155, 161, 179, 181, 202, 233, 246, 279
Eriospermum paradoxum (Liliaceae) 61
Eucalyptus (Myrtaceae) 320
Euclea pseudebenus (Ebenaceae) 180, 247, 250
Euphorbia (Euphorbiaceae) 140, 190
candelabrum 13, 17
conspicua (cooperi) 13
dregeana 165, 168, 247, 259
gariepina 165
gregaria 94
gummifera 54, 165, 247–8, 252–3

ingens 13
mauritanica 61
peltigera 165, 247
Euphorbiaceae 20, 72, 164–5, 173, 315
evaporation 44, 47, 48, 132, 148–9
evaporative demand 17, 31, 46, 279
external CO_2 229–2, 243, 283

Faucaria (Mesembryanthemaceae) 16
Fenestraria (Mesembryanthemaceae) 64, 153–4, 187, 323
Ferocactus (Cactaceae) 6
acanthodes 177, 191, 312, 323–5
Fingerhuthia africana (Poaceae) 184, 263
fluxes 99ff
carbon 224–97
energy 40, 42, 100–55
heat 41–2, 100–2, 135–7, 139
water 155–224, 252
Fockea crispa (Asclepiadaceae) 7
fog 30, 32–3, 35, 37, 58, 68, 77ff, 86ff, 146, 201ff
Forsskaolea candida (Urticaceae) 9, 184, 247, 250, 253, 261–3, 273
Frithia (Mesembryanthemaceae) 315
pulchra 135

Galenia dregeana (Aizoaceae) 19, 177, 180, 269, 271, 273, 275, 277
Gasteria pillansii (Liliaceae) 3, 11
Gazania lichtensteinii (Asteraceae) 145
geomorphology 73–5
Geraniaceae 72, 165, 173
Gibbaeum pubescens (Mesembryanthemaceae) 110
Glottiphyllum (Mesembryanthemaceae) 315
grass 68, 92ff, 201, 263
growth 251–2, 264, 279, 281, 297, 301, 304, 306, 310, 315
growth form 5, 13, 49, 246, 251, 298–300
definition 300

Haageocereus pacalaensis (Cactaceae) 16
Haworthia (Liliaceae) 20, 134, 153, 187, 325
attenuata 21
cymbiformis 153
maughanii 154
tesselata 13, 16, 180
heat 115, 127
conduction 42, 135–6
latent 132–41
sensible 41, 102ff, 132–41
transfer 41, 100–2, 135–7, 139
volumetric specific heat of succulents 135, 142
Hellmann quotient 24, 26, 31
Hermbstaedtia glauca (Amaranthaceae) 9, 247, 250
Herrea (= Conicosia) (Mesembryanthemaceae) 64
blanda (= C. elongata) 61
Hertrichocereus beneckei (Cactaceae) 124
humidity 30, 44–7, 281
absolute 46, 47
diurnal course 47
relative 44–7, 78–80, 85, 130, 158–9, 178, 198
hydathodes Fig. 4.56 opp. 202, Figs. 4.59 and 4.60 opp. 203, 203–9
Hydrophyllaceae 165, 173
Hypertelis salsoloides (Aizoaceae) 61

idioblast 17, 19, 20, 23, 110–11, 153, 174, 176, 179, 181
inorganic ions 167–9, 174, 219, 222–4
irrigation 192–4, 196, 213, 221–2, 258, 260, 274, 276–8
isocitric acid 234, 283–94

Jacobsenia (Mesembryanthemaceae) *19*
Jensenobotrya lossowiana (Mesembryanthemaceae) 322
Justicia orchioides (Acanthaceae) 177, 184

Kalanchoe (Crassulaceae) 177
blossfeldiana 240
daigremontiana 174, 264, 328
pinnata 282, 325
pumila 109, 123–4, 321
tomentosa 110
Karoo Domain 91ff
Kirchhoff's radiation law 105, 116, 128
Knersvlakte 74–5, 79, 87, 93–4, 149, 169, 182, 190

Lapidaria margaretae (Mesembryanthemaceae) 16
Larrea divaricata (Zygophyllaceae) 69
leaf
anatomy 3, 8, 12–20, 160, 179, 231, 233
conductance 178–9, 181–6, 200, 286, 270, 282
diurnal course of conductance 181, 186, 200, 269, 271
orientation 119–22, 127, 137, 155
Ledebouria ovatifolia (Liliaceae) 144, 151, 321
Leipoldtia constricta (Mesembryanthemaceae) 180, 255
Leipoldtiinae 322–3
lichens, 36–7, 50, 93, 201
life cycle 51–65, 172, 239, 300–12
life form 13, 164, 203, 292, 298–300, 312–13
definition 300
life strategy 49, 208, 246, 297, 298–318
definition 300, 312ff

Liliaceae 10, 72, 153, 164–5, 173, 286, 289
Lithops (Mesembryanthemaceae) 2, 6, 17, 20, 129–30, 133–5, 153–4, 187, 215, 315, 320–1, 325
  karasmontana Fig. 4.30 opp. 154
  lesliei 18, 134, 154, 187
  turbiniformis 133

malate dehydrogenase (MDH) 227ff, 232, 234
Malephora crocea (Mesembryanthemaceae) 220
  lutea 57
  uitenhagensis 16
malic acid (malate) 185–6, 196–200, 206, 208–9, 219, 227–9, 231, 234–7, 240–1, 279, 284
  decarboxylation 227–9, 263
  nocturnal accumulation 238–41, 254–5, 260, 279–81, 288, 292, 295–6
  oscillation 235, 260, 280, 283, 286–92, 294
malic enzyme 228, 231, 236
Massonia depressa (Liliaceae) 61
Maytenus cf. *tenuifolia* (Celastraceae) 247, 250
Mesembryanthemaceae 10, 13, 19, 23, 62, 64–5, 72, 92–3, 124, 151, 153, 164–9, 172–5, 208, 219, 246, 248, 254, 278, 286–8, 315, 319, 321, 323, 328–9
Mesembryanthemoideae 19
Mesembryanthemum (Mesembryanthemaceae) 55, 176, 179
  aitonis 57
  barklyi 9, 150, 184, 247, 249–50, 287, 313
  crystallinum 63, 174, 177, 220, 238–40, 279, 288, 297, 301, 304–5, 313, 326–30
  nodiflorum 58, 67
  pellitum 9, 55, 175, 184–6, 247, 249–50, 252–3, 306
mesophyll 231–2, 236, 243, 279
Mimosaceae 48, 165, 173
mitochondria 226ff, 231, 236
Mitrophyllum (Mesembryanthemaceae) 13, 19, 125
  clivorum 9, 247, 249–50
  mitratum 287–8
Mojave 30, 32–3, 59
Monanthes (Crassulaceae) 19
Monechma mollissimum (Acanthaceae) 177, 184, 188, 247, 250, 263–4, 273
Monilaria (Mesembryanthemaceae) 179
  globosa 280–1, 317
Montiniaceae 165, 173

Namaland Domain 91ff
Namaqualand 9, 25, 60–1, 68, 75
Namib 30, 32–3, 35, 37, 41, 43, 49, 68, 70–98, 113–14, 130–1, 137–8, 143, 146, 201, 203, 242, 254
  central 35, 42, 59, 73, 77, 79–80, 138, 143, 179–80, 184, 208
  climate 75–89
  geomorphology 73–5
  northern 31, 36, 73
  plant geography 89–98
  southern 7, 11, 30–1, 55, 73, 76, 83–9, 106, 146, 150, 181–3, 189–90, 197, 199–201, 203, 205, 208–9, 219, 269, 273, 280, 286–7, 306, 308
  vegetation 89–98
Namib Domain 90ff
Nananthus albipunctus (Mesembryanthemaceae) 153
Neochilenia napina (Cactaceae) 189
neoglucogenesis 228ff, 236
Nicotiana glauca (Solanaceae) 67, 247, 250
Numees 46–7, 54–6, 75, 78, 84–90, 95, 97, 108, 119, 147, 150, 164, 178, 186, 191, 196, 199–200, 211, 247–51, 253, 255–6, 259, 263, 265, 277, 284–6, 288–9, 292, 296

Ohm's law 156
ontogeny 23, 239–41
Oophytum (Mesembryanthemaceae) 315
Ophthalmophyllum (Mesembryanthemaceae) 62, 154
Opophytum aquosum (Mesembryanthemaceae) 9, 55, 57, 175, 184, 247–50, 252–3, 306
optical properties (spectra) 105–15, 117, 123
  leaf 106–11, 121
  plant 102–13, 127, 155
  soil 112, 148
  succulents 105–11, 118, 121, 127
Opuntia (Cactaceae) 16
  basilaris 176, 192–3, 292, 295, 328
  bigelovii 65
  inermis 282, 326
Opuntioideae 65
Orchidaceae 26
organic acids
  content 176, 186, 219
  overnight accumulation 176, 185, 196–200, 206, 208–9, 238, 241, 266, 280, 283–92, 294–5
Osteospermum microcarpum (Asteraceae) 9, 165, 168, 247, 253
Othonna cylindrica (Asteraceae) 317
  herrei 20, 22, 165, 317
  opima 11, 16, 107, 121, 150, 165, 180, 184, 187–8, 202, 215, 217, 220, 247, 249–53, 292, 294, 311

Oxalidaceae 19
Oxalis (Oxalidaceae) 19
  succulenta 7
oxaloacetic acid (OAA) 227–9, 231ff
Ozoroa dispar (Anacardiaceae) 9, 107, 150, 188, 247, 249–52, 259, 311

Pachypodium brevicaule (Apocynaceae) 7
  namaquanum 9, 121, 180, 247, 250, 265–8, 292, 294, 317–18
Papilionaceae 48
parenchyma 3, 14, 153
partial pressure
  of CO_2 225–6, 232, 236, 268
  of O_2 225–6
Pelargonium (Geraniaceae) 180, 190, 308
  crithmifolium 61
  incrassatum 61
Pelliciera rhizophoreae (Pellicieraceae) 311
Peperomia (Piperaceae) 3–4, 23, 296, 323, 327
  asperula 154
  columella 154
phenological rhythm 60–3
Phoenix dactylifera (Arecaceae) 145
phosphoenolpyruvate (PEP) 227–9, 232–3, 236–7, 243
phosphoenolpyruvate carboxykinase (PEPCK) 229
phosphoenolpyruvate carboxylase (PEPC) 227, 230ff, 238ff, 243, 278–9, 282, 288–9
phosphoglycolic acid (PGA) 225, 230
photorespiration 225–6, 228, 230, 232–3, 236–7, 244–5, 279
photosynthesis 102, 115, 127, 144, 181, 184, 226, 228, 230, 232–3, 235, 237, 244, 269
pigments 104, 106, 108, 112, 115, 151–4, Fig. 4.30 opp. 154
Piperaceae 23
plant communities 95ff
plants
  annual 19, 34, 50, 52, 59–60, 62–3, 68, 166, 170, 174–5, 185, 224, 245, 248, 252, 279, 297, 301, 304–6, 313–15
  deciduous perennial 308–12, 314–18
  drought deciduous 60, 181, 224, 253, 264, 297, 316–17
  ephemeral 34, 52, 59–60, 174, 238, 301, 304–6, 314
  evergreen 60, 181, 215, 224, 245, 278–9, 297, 314
  geophytic 24, 50, 60, 307–8, 314–15
  hapaxanthous 52, 306–7, 311
  homoiohydric 50
  perennial 19, 34, 52, 59, 62, 69, 224, 252, 297, 303, 313–18
  poikilohydric 50, 156, 201
  seasonal deciduous 60, 224, 248, 253, 264–5, 278, 280–1, 297, 308–13, 316–17
Plectranthus marrubioides (Lamiaceae) 174, 178, 195, 210, 240–1, 266, 268, 280–3, 292, 295, 323
Pleiospilos (Mesembryanthemaceae) 151, 153
  bolusii 153
pollination 62–3
Portulacaceae 10, 72, 165, 173, 285–6
Portulacaria (Portulacaceae) 241
  afra 240, 322, 328
potassium 167, 171, 173–5, 220–3, 279
Prenia sladeniana (Mesembryanthemaceae) 3, 9, 146–7, 150, 200, 211–17, 220, 222, 247, 249–52, 303, 311, 315
pressure–volume curve 172, 174, 178, 195, 210
Primula glutinosa (Primulaceae) 26
productivity 226, 244–5
proline 167, 222
Psammophora nissenii (Mesembryanthemaceae) 141
Psilocaulon (Mesembryanthemaceae) 148
  subnodosum 9, 150, 175, 177, 180, 184, 207–9, 247, 249–50, 255, 257–8, 260–1, 287, 289
pyruvate 228, 231ff, 236
pyruvate-P_i-dikinase 232

radiation 37–40, 87, 228, 230
  blackbody 101, 116–17, 127–8
  budget 40, 46, 116–32
  diffuse 37–8, 101ff, 113–15
  direct 17, 37–8, 101ff, 113–15
  diurnal course 88, 90, 115, 122, 130–2, 138, 146–9, 186, 261, 269, 276
  global 38–40, 101ff, 113–15, 117–23, 129, 131–2, 143–4
  loss 101, 130–5, 143, 145, 148, 150, 185
  photosynthetically active (PAR) 89–90, 104, 115, 123–4, 151–2, 155, 244, 261, 269, 276
  solar 37–41, 43, 46, 113ff, 128, 141, 145, 148, 154
  thermal 40, 46, 101ff, 113–17, 127–32, 143
rainfall 30–7, 49–50, 67, 69, 80, 164, 176, 191, 264, 306
  abundant 3, 4, 11, 167–8, 170, 172, 184, 192–4, 198, 254–5, 259
  annual 10, 24, 30–1, 35, 77, 201, 215, 218
  distribution 30–1, 55, 83–4
  effective 31, 34, 87, 306
  erratic 31–3, 77ff
  off-seasonal 31, 55, 57

rainfall (*cont.*)
run-off 34, 48, 69
run-on 34, 69
seasonal 31
summer 31–3, 68, 75ff, 244–5, 313
winter 31–3, 68, 75ff, 89ff, 92, 166, 217, 306, 310, 313
Ranunculus glacialis (Ranunculaceae) 26–7
Raoult's law 158
refixation of respiratory CO_2 227–8, 230, 236–7, 245, 255, 266, 279–80, 285, 289
reflectance 103–4, 123
reflection 37–8, 101, 106
reflectivity 102ff
of succulents 105–6
reproduction 60–3
vegetative 65
resistance
boundary layer 160ff, 179
stomatal 160ff
to water flow 157, 160ff
respiration 102, 225ff, 230, 234, 237, 245, 282, 311
response to rainfall (or irrigation) 4–5, 50, 192–3, 196, 213, 216, 221–2, 258, 260–1, 274, 276–8, 286
Rhigozum trichotomum (Bignoniaceae) 69, 94
Rhus populifolia (Anacardiaceae) 177, 184
ribulose-1, 5-bisphosphate (RuBP) 225, 230ff, 243
Richtersveld 46, 53–4, 56, 74–6, 78–9, 82–90, 95, 97, 108, 150, 164–77, 179–80, 182, 186, 190–2, 196–7, 199–201, 203, 206–7, 211, 215, 217, 219, 222, 247–51, 253–6, 263–6, 275, 277, 284–9, 296–7, 306, 310–11
root/shoot ratio 251–3
roots
rain roots 192
rooting horizon 43, 48, 87, 189, 191, 202
root systems 189–90, 192, 197
Rubiaceae 26
rubisco 225–6, 228, 230ff, 242ff, 279
Ruschia (Mesembryanthemaceae) 9, 11, 14, 148–9, 167, 169–70, 177, 234, 278, 296
schneiderana 190
sedoides 153, Fig. 4.30 opp. 154
subaphylla 180
utilis 13
Ruschianthemum gigas (Mesembryanthemaceae) 13
Ruschioideae 19

Salicornia (Chenopodiaceae) 2
europaea 314
Salsola (Chenopodiaceae) 177
kali 67
Sarcostemma viminale (Asclepiadaceae) 16
Saxifraga (Saxifragaceae) 26
Sedum acre (Crassulaceae) 26
album 26
seed germination 31, 34, 52–8, 239, 301, 304, 306
delay of 55, 57–8
seedbank 50, 52–8
seedling establishment 34, 58–9, 129, 301, 306–7
mortality 56, 59
Selenicereus (Cactaceae) 16
Sempervivoideae 324
Sempervivum (Crassulaceae) 25–6, 134
Senecio corymbiferus (Asteraceae) 9, 153, 165, 180, 207, 247, 250, 252–3, 286, 311
herreianus 153
longiflorus 165, 175, 247, 286, 296
medley-woodii 11, 110, 124, 212–13, 326
sodium 167, 171, 173–5, 220–1
sodium chloride 238–40
soil 41, 47–9, 76, 97, 154, 163
inverse texture effect 67
moisture 34, 57–8, 159, 189–99, 254, 266, 283–4
optical properties 112, 148
salt content 168–9, 175, 208
temperature 42–3, 133–5, 148–9
Solanaceae 165, 173
Solanum incanum (Solanaceae) 263–4, 273
namaquanum 184
Sonora 31–3, 41, 48, 59, 68, 145, 176, 179, 292
Sorghum album (Poaceae) 331
Spartina (Poaceae) 244
species diversity 97–8
spectral distribution 38, 104
Spergula (Caryophyllaceae) 26
Sphalmanthus (Mesembryanthemaceae) 62, 168
decurvatus 19
scintillans 255
trichotomus 9, 175, 283–5, 287, 296
Spinacea oleracea (Chenopodiaceae) 319
Stapelia (Asclepiadaceae) 62, 65
similis 188
Stefan-Boltzmann law 40, 101, 116, 128
stem
anatomy 13–17, 20–3
periderm 20, 23, 265
Stenocereus (Machaerocereus) eruca (Cactaceae) 16
Stoeberia beetzii (Mesembryanthemaceae) 5, 9, 13–14, 169, 180, 184, 188, 191–2, 196–7, 207, 253, 288, 296
stomata 8, 17, 20–1, 23, 160–1, 179, 181, 183, 185, 187, 202–5, 226, 230, 233, 236–7, 255, 260, 266, 268–70, 279, 286

density 180
Stomatium (Mesembryanthemaceae) 16
succulence 1ff, 155, 209, 224, 245–54, 265–6, 273, 275, 279, 292–7, 302, 313
carbon investment 245–52
evolution 23, 24, 151
mesophyll 4, 296–7
significance 60, 151–5, 209–24, 252–97, 302
succulence quotient 246–50, 252, 273, 297
succulents 1–27, 61, 68, 156, 161, 164, 171–2, 177, 179–81, 188ff, 201, 219, 222, 230, 241, 145–246, 264, 275, 289, 298–318
all-cell succulents 2ff, 23, 25, 166, 172, 210–11, 216, 218–20, 222, 245–6, 251, 286, 287
C_3 succulents 179, 181–2, 184, 187–8, 202, 238, 240–1, 249, 264–6, 269–71, 273–5, 277, 292, 294, 296–7, 313
C_4 succulents 245
CAM succulents 179–80, 182, 184–8, 195–200, 206–7, 209, 212–14, 216, 218, 221, 235, 239, 244, 248, 255, 257–60, 265, 274, 278–281, 286–7, 292, 296–7, 313
CO_2 exchange 234–5, 257–9, 261–2, 267, 269–72, 276, 278, 281
definition 2–6
distribution 24–27, 92–94
leaf 6ff, 19, 93, 118, 145, 151, 170, 181, 185, 216, 246–7, 266–7, 283, 316
leaf water potential 176–8, 192–4, 208, 273–4
life strategies 298–318
osmotic potential 164–78
partially succulent 3ff, 10, 11, 25, 166, 168, 172, 210, 215–16, 218, 220, 222, 245–6, 251, 286–7
root 6ff
stem 6ff, 20, 24, 68, 93, 181, 185, 222–3, 247–8, 266–7, 308, 316
transpiration 181–2, 184–6, 188, 202, 270–1, 276, 308
turgor potential 172–9
surface expansion 4

temperature 40–4
air 41–2, 47, 80, 82ff, 129, 133–4, 146, 149–50, 178, 185, 254, 256
diurnal course 42, 47, 88, 90, 146–7, 149, 186, 256–7, 259, 271, 276
effect on CO_2 fixation 281–2
frost 24–6, 67–8, 133–5, 145
leaf 11, 129, 132–3, 135, 142, 144–7, 149–50, 185, 226, 228
maximum 40, 78–9, 84ff
minimum 40, 78–9, 84ff
overtemperature 150
range 24, 32–3, 40
soil 42–3, 133–5, 148–9
tolerance 24–6
undertemperature 150, 201
Tetragonia reduplicata (Tetragoniaceae) 177
Tillandsia recurvata (Bromeliaceae) 20
Trachyandra falcata (Liliaceae) 308
transmissivity 102ff
of succulents 105–6
transmittance 104, 123
transpiration 101, 132–3, 143–5, 160–3, 178–88, 202, 230, 264, 270, 308, 311
cuticular 161, 202
diurnal course 181–2, 186, 271, 276
driving forces 160
of succulents 181–2, 184–6, 188, 202, 270–1, 276, 308
reversed 160, 162, 202, 206
stomatal 161ff
Trianthema triquetra (Aizoaceae) 9, 177, 181–4, 247, 250, 252–3, 269–70, 273, 275, 277
tricarbonic acid (TCA)-cycle 227–8, 234
Trichodesma africanum (Boraginaceae) 9, 180, 247, 250, 252–253
Trichodiadema sp. aff. *barbatum* (Mesembryanthemaceae) 112
Tylecodon (Crassulaceae) 19, 210, 252–3, 278, 285
paniculatus 9, 20, 61, 105–7, 150, 180, 184, 253, 265–8, 285, 308–10, 313, 316–18
pearsonii 9, 180, 253
reticulatus 9, 180, 253
wallichii 9, 61, 180

Urticaceae 165, 173

van't Hoff law 158, 218
Vanzijlia annulata (Mesembryanthemaceae) 58, 323
vegetation 65–70, 89–98
floristic composition 65, 67, 95
pattern 67–9, 92ff
types 32–3
Vitaceae 64

Water 30–7
availability 30, 48–9, 60, 63, 69, 89, 93, 167, 181, 187, 189, 194, 198, 212–13, 238, 244, 251, 253–5, 260, 264, 283–5, 305, 310
balance 209–24, 300–12
budget 155, 178, 192, 198, 202, 205, 208, 210, 224, 263, 277
content 2ff, 9–10, 135, 155, 188, 192–4, 212, 218, 249–50, 292, 295–6

Water (*cont.*)
fluxes 155–224, 252
gain 155, 197, 199, 202, 206–9, 260
loss 17, 155, 178–88, 208, 230, 237, 260, 264, 308, 311
non-utilizable 301, 303–12
relative water content 195–6
status 155, 164–78, 197ff, 260
storage 3ff, 11, 19, 194, 197, 210, 218–19, 222, 246, 260
translocation 156, 163–4, 167, 177, 213–24, 253, 292, 297, 314
uptake 37, 50, 155, 163, 188–97, 246, 260, 279
utilizable 5ff, 7–11, 17, 63, 156, 187, 192, 210–15, 224, 245, 253, 255, 260, 264, 266, 270, 297, 301, 303–12, 314–15
water potential 156–78, 254, 295
atmospheric 158ff
gradient 156–7, 189, 194, 197
leaf 176–8, 192–4, 208, 272–4
osmotic 157ff, 164–72, 177–8, 192, 194, 196, 198, 208, 211–12, 218–19
soil 159, 163, 189, 191, 198–200, 282
turgor 157ff, 172–9
water storage tissue 3ff, 10–12, 19, 23, 192, 194, 197, 210, 215, 218–19, 222, 246, 251, 260, 292
irreversible 216, 245
reversible 210, 216, 245
shrinkage 10–12, 20, 188, 212–14, 219, 308–9
water use efficiency (WUE) 274
C_3 plants 230, 263, 277
C_4 plants 233, 263
CAM plants 237
instantaneous 274–5, 277
water vapour 38, 44
diurnal course of ΔW 182, 271, 276
diurnal course of VPD 47, 186, 256–9, 261–2
partial pressure 44–5, 159
partial pressure deficit (VDP) 44–7, 87, 146, 186–7, 191, 254, 256, 260–1, 281–3, 308
partial pressure difference (ΔW) 162, 178, 181–4, 225, 237, 283
saturation pressure 44–5, 159
wavelength range 38, 103–4, 110–13, 126–7
wax bloom 107–10, 121, 123–4, 141–2, 151, 181
Welwitschia (Welwitschiaceae) 144–5
mirabilis 43, 59, 93, 131, 140, 143–4, 146, 152, 180, 184, 222, 242, 321, 327, 329
Welwitschia Plain 42, 82, 114, 131–2, 138, 141, 143, 146
Western Cape Domain 92ff
wind 41, 101, 136–41, 144, 146–7, 179
bergwind 77, 85ff, 145, 184, 187, 191, 198, 206–7, 254–5, 257–9, 261–3, 286, 308
foehn 77
window plants 153–5, 187

xeromorphism 1, 2, 17, 110–11, 152–3, 166, 174, 248, 302
xerophytes 1, 2, 252, 302

Zea mays (Poaceae) 232, 328
Zygophyllaceae 72, 165, 172–3
Zygophyllum longicapsulare (Zygophyllaceae) 180, 247, 250
meyeri 61, 317
prismatocarpum 177, 184, 247, 250, 273–74, 276–7
simplex 245, *306*
stapfii 180